Mathematics and Visualization

Series Editors

Torsten Möller
Bernd Hamann
Robert D. Russell

Editors

Mathematical Foundations of Scientific Visualization, Computer Graphics, and Massive Data Exploration

With 183 Figures, 134 in Color and 15 Tables

Torsten Möller
School of Computing Science
Simon Fraser University
8888 University Drive
Burnaby BC, V5A 1S6
Canada
torsten@cs.sfu.ca

Robert D. Russell
Department of Mathematics
Simon Fraser University
8888 University Drive
Burnaby BC, V5A 1S6
Canada
rdr@cs.sfu.ca

Bernd Hamann
Department of Computer Science
University of California, Davis
1 Shields Avenue
Davis, CA 95616-8562
USA
hamann@cs.ucdavis.edu

ISBN: 978-3-540-25076-0 e-ISBN: 978-3-540-49926-8
DOI: 10.1007/978-3-540-49926-8

Mathematics and Visualization ISSN 1612-3786

Library of Congress Control Number: 2008944010

Mathematics Subject Classification (2000): 35-XX, 65Dxx, 41-XX, 51-XX, 54-XX, 65-XX, 76-XX

Cover design: deblik, Berlin

Printed on acid-free paper

springer.com

Preface

The goal of visualization is the accurate, interactive, and intuitive presentation of data. Complex numerical simulations, high-resolution imaging devices and increasingly common environment-embedded sensors are the primary generators of massive data sets. Being able to derive scientific insight from data increasingly depends on having mathematical and perceptual models to provide the necessary foundation for effective data analysis and comprehension. The peer-reviewed state-of-the-art research papers included in this book focus on continuous data models, such as is common in medical imaging or computational modeling.

From the viewpoint of a visualization scientist, we typically collaborate with an application scientist or engineer who needs to visually explore or study an object which is given by a set of sample points, which originally may or may not have been connected by a mesh. At some point, one generally employs low-order piecewise polynomial approximations of an object, using one or several dependent functions.

In order to have an understanding of a higher-dimensional geometrical "object" or function, efficient algorithms supporting real-time analysis and manipulation (rotation, zooming) are needed. Often, the data represents 3D or even time-varying 3D phenomena (such as medical data), and the access to different layers (slices) and structures (the underlying topology) comprising such data is needed. It has become evident over recent years that, due to the ever-increasing complexity inherent in today's data sets, it is necessary to develop feature extraction algorithms that facilitate sensible mappings of physical data values to visual attributes, enhancing the understanding of structures and structure relationships. It is crucially important that visualization algorithms support precise, error-controlled quantitative visual analysis, especially in applications like medical data analysis for diagnosis and surgical planning.

Over the last 20 years the profound impact of scientific computing on nearly every area of science and engineering has become more and more evident. Visualization, being a very young scientific field which has evolved as a branch of computer graphics, has in turn become an important driver for the development of exciting new directions in mathematics and computer science. Many common approaches used in contemporary visualization algorithms and software are still quite "ad-hoc," and considerable work remains to be done to establish the much-needed mathematical foundation for the growing field of scientific visualization.

Most current visualization algorithms break down for very large data sets. While standard approaches use multiresolution data structures, approximations, and visualization paradigms, peta-size data sets cannot be handled with the presently used approaches and software. New algorithms based on sophisticated mathematical modeling techniques must be devised that permit the extraction of high-level topological structures that can be visualized and understood.

We organized a workshop at the Banff International Research Station (BIRS), at the Banff Centre, Canada, from May 22 to May 27, 2004. The workshop focused specifically on *mathematical issues* as they relate to the challenges posed by the need to more effectively perform data processing and analysis on very large and highly complex data sets for visual exploration. The primary objective of the workshop was to bring together the leading researchers focusing on mathematical and foundational research in visualization. Scientists presented their recent research results and also shared their views concerning the most pressing research challenges facing this field in the near future. The workshop was organized in the following five topical areas:

- Topology and discrete methods
- Signal and geometry processing
- Partial differential equations
- Data approximation techniques
- Massive data applications

While a large portion of the workshop consisted of presentations by participants from of state-of-the-art research in the various fields, a significant amount of time was reserved for open-ended brainstorming sessions. In three such sessions, the participants were split into four groups which discussed these focus areas in detail. The group leaders were asked to obtain answers to a number of questions that were distributed among the participants beforehand. The group leaders summarized these sessions and the results. The questions distributed before the workshop were:

- What are the scientifically challenging problems to be tackled in your topic area?
- What are the driving applications in this field?
- Which journals and conferences exist today that are appropriate venues for publishing mathematically oriented methods in this field?
- Which good on-line resources exist today supporting research in this subfield, e.g., data sets, commercial and free software libraries, publication databases, benchmarking sites, etc.?
- Which scientific domains and subfields are needed to solve successfully and elegantly the identified problems?

The brainstorming sessions were welcomed by the participants. As far as we know, this format of discussing specialized topics in a question-driven fashion has not previously been used in visualization workshops. Participants commented positively on the format, and it seems to us that sharing ideas and perspectives in this way is a highly effective means for defining relevant new directions in visualization.

This book contains papers authored by participants at the workshop. We hope that they are inspiring and convey some of the excitement we all experienced during the sunny days at the Banff workshop. We would like to thank the following colleagues for helping with the organization of the workshop or serving as group discussion leaders: Herbert Edelsbrunner, Hans Hagen, Chris Johnson, Ken Joy, Raghu Machiraju, Tamara Munzner, Greg Nielsen, Jack Snoeyink, Gabriel Taubin, and Ross Whitaker.

Torsten Möller
Bernd Hamann
Robert D. Russell

Contents

Maximizing Adaptivity in Hierarchical Topological Models Using Cancellation Trees

Peer-Timo Bremer[1], Valerio Pascucci[2], and Bernd Hamann[3]

[1] Department of Computer Science, University of Illinois, Urbana-Champaign
ptbremer@acm.org
[2] Center for Applied Scientific Computing, Lawrence Livermore National Laboratory
pascucci@llnl.gov
[3] Institute for Data Analysis and Visualization, Department of Computer Science, University of California, Davis
hamann@cs.ucdavis.edu

Summary. We present a highly adaptive hierarchical representation of the topology of functions defined over two-manifold domains. Guided by the theory of Morse–Smale complexes, we encode dependencies between cancellations of critical points using two independent structures: a traditional mesh hierarchy to store connectivity information and a new structure called cancellation trees to encode the configuration of critical points. Cancellation trees provide a powerful method to increase adaptivity while using a simple, easy-to-implement data structure. The resulting hierarchy is significantly more flexible than the one previously reported (IEEE Trans. Vis. Comput. Graph. 10(4):385–396, 2004). In particular, the resulting hierarchy is guaranteed to be of logarithmic height.

1 Introduction

Topology-based methods used for visualization and analysis of scientific data are becoming increasingly popular. Their main advantage lies in the capability to provide a concise description of the overall structure of a scientific data set. Subtle features can easily be missed when using "traditional" visualization methods like volume rendering or isocontouring, unless "correct" transfer functions and isovalues are chosen. On the other hand, the presence of a large number of small features creates a "noisy visualization," in which larger features can be overlooked. By visualizing topology directly, one can guarantee that no feature is missed. Furthermore, one can use sound mathematical principles to simplify a topological structure. The topology of functions is also often used for feature detection and segmentation (e.g., in surface segmentation based on curvature).

However, for topology-based data analysis one needs flexible, hierarchical models able to adaptively remove noise or features not relevant for a particular

T. Möller et al. (eds.), *Mathematical Foundations of Scientific Visualization, Computer Graphics, and Massive Data Exploration*, Mathematics and Visualization,
DOI: 10.1007/978-3-540-49926-8,

segmentation. In practice, the simplification/refinement should be fast (preferably interactive) and highly adaptive in order to be useful in a large variety of situations. Requiring interactivity inadvertently leads to the use of hierarchical encodings rather than simplification schemes. Hierarchical models often reduce the adaptivity of a representation to gain the ability to perform incremental changes for varying queries.

We address the need for adaptive topology-based data exploration by improving significantly the topological hierarchy proposed in [4]. Creating two largely independent hierarchies, we show how one can remove many of the dependencies in the original hierarchy, making the structure simpler, more compact, and more adaptive than the original one.

1.1 Related Work

The topological structure of a scalar field can be described partially by its contour tree [5, 17, 18], which describes the relations between the connected components of its level sets. This structure provides a user with a compact representation of the topology [1] and can be used to accelerate the computation of isosurfaces [24]. However, the contour tree provides little information about the embedding of the level sets and therefore remains somewhat abstract. Morse theory [15, 16], on the other hand, provides methods to analyze the complete topology of a function over a manifold as well as its embedding. Early approaches for the bivariate case are provided in [6, 14, 19]. More recently, the Morse–Smale complex was introduced by Edelsbrunner et al. [8, 9] as a description of the topology of scalar-valued functions over two- and three-dimensional manifolds. Applications of this theory vary from implicit geometry modeling [21] to shape description [13]. Related concepts are also used in flow visualization. Helman and Hesselink [12] showed how to find and classify critical points in flow fields and propose a structure similar to the Morse–Smale complex for vector fields. Later, methods to analyze and simplify this complex were proposed by de Leeuw and van Liere [7] and Tricoche et al. [22, 23].

The first multiresolution encoding of a Morse–Smale complex we are aware of was proposed by Pfaltz [20], which has been improved and extended by Edelsbrunner et al. [9] and Bremer et al. [3, 4]. More recent hierarchical structures are based on the concept of *persistence* [10], which relates the difference in function value of critical point pairs to the importance of a topological feature. Given a Morse–Smale complex, we:

1. Provide an improved hierarchical encoding of the Morse–Smale complex
2. Prove that the resulting hierarchy is of logarithmic height
3. Demonstrate our methods for various data sets

We first review necessary concepts from Morse theory and the construction of a Morse–Smale complex (Sect. 2). In Sect. 3, we describe cancellation trees and the resulting hierarchy in Sect. 4. We conclude with results and possibilities for future research (Sect. 6).

2 Morse–Smale Complex

We base our algorithms on intuitions derived from the study of smooth functions. We review key aspects from Morse theory [15, 16] for smooth functions and discuss how these can be used in the piecewise linear case.

2.1 Morse Theory

Given a smooth function $f : \mathbb{M} \to \mathbb{R}$, a point $a \in \mathbb{M}$ is called *critical* when its *gradient* $\nabla f(a) = (\delta f/\delta x, \delta f/\delta y)$ vanishes; it is called *regular* otherwise. For two-manifolds, (nondegenerate) critical points are maxima (f decreases in all directions), minima (f increases in all directions), or saddles (f switches between decreasing and increasing four times around the point). Using a local coordinate frame at a, we compute the *Hessian* H of f, which is the matrix of second partial derivatives. If H is nonsingular we can construct a local coordinate system such that f has the form $f(x_1, x_2) = f(a) \pm x_1^2 \pm x_2^2$ in a neighborhood of a. The number of minus signs is the *index* of a and distinguishes the different types of critical points: minima have index 0, saddles have index 1, and maxima have index 2.

At any regular point, the gradient (vector) is nonzero, and when we follow the gradient we trace out an *integral line*, which starts at a critical point and ends at a critical point, while technically not containing either of them. Since f is smooth, two integral lines are either disjoint or the same. The *descending manifold* $D(a)$ of a critical point a is the set of points that flow toward a. More formally, it is the union of a and all integral lines that end at a. The collection of descending manifolds is a complex in the sense that the boundary of a cell is the union of lower-dimensional cells. Symmetrically, we define the *ascending manifold* $A(a)$ of a as the union of a and all integral lines that start at a. If no integral line starts and ends at a saddle, see [9], we can overlay these two complexes and obtain what we call the *Morse–Smale complex* of f. Its vertices are the vertices of the two overlayed complexes, which are the minima, maxima, and saddles of f. Its cells are four-sided regions bounded by parts of integral lines between saddles and extrema. An example is shown in Fig. 1.

Using the insight gained from smooth Morse theory when applied to piecewise linear functions, we follow the concepts described in [3].

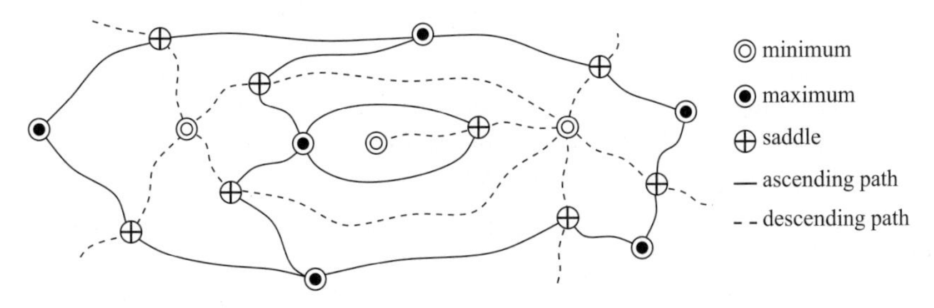

Fig. 1. Morse–Smale complex

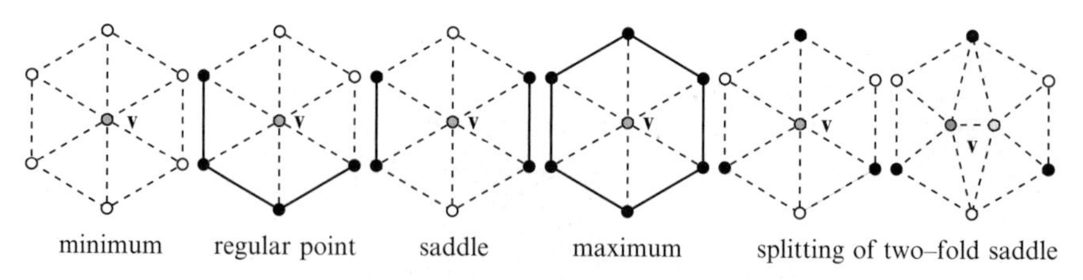

Fig. 2. Classification of a vertex v based on relative height of its edge-connected neighbors, where light vertices/edges mark higher neighbors and solid vertices/edges lower neighbors

We follow the concepts described in [3] to apply the concepts of smooth Morse theory to piecewise linear functions. Critical points are identified and classified based on their local neighborhood, see [2, 9]. If all vertices that are edge-connected to a point u have function values below that of u, we call it a maximum; if all are above u, then we call it a minimum, etc., see Fig. 2. In general, there can exist saddles with high multiplicity that we split into simple ones, as shown on the far right in Fig. 2.

2.2 Persistence

As a numerical measure of the importance of critical points we define pairs of critical points and use the absolute difference between their height/function values. The underlying intuition is the following: We imagine sweeping the two-manifold $\mathbb{M}$ in the direction of increasing height (w.r.t. the scalar field value.) The topology of the part of $\mathbb{M}$ below the sweep line changes whenever we add a critical vertex, and it remains unchanged whenever we add a regular vertex. Each change either *creates* a component, *destroys* a component, or changes its genus. We pair a vertex v that creates a component with the vertex u that destroys the component. The *persistence* of u and of v is the "delay" between the two events: $p = f(v) - f(u)$, see [10].

2.3 Construction

In practice, we construct the Morse–Smale complex by successively computing its edges, starting from the saddles, see [3]. Starting from each saddle, we compute two lines of steepest ascent and two lines of steepest descent connecting the saddle to two maxima and two minima. We call these lines *ascending* or *descending paths*. Two paths in the same direction (ascending or descending) can merge; two paths with different direction must remain separate. Once two paths have been merged they never split. Following these rules, we are guaranteed to produce a nondegenerate Morse–Smale complex. A more detailed analysis can be found in [3]. Having computed all paths, we partition the surface into four-sided regions forming the cells of the Morse–Smale complex. Specifically, we grow each quadrangle from a triangle incident to a saddle without ever crossing a path.

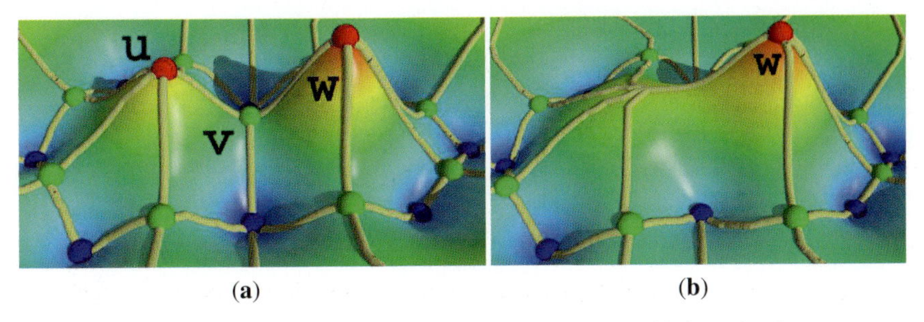

Fig. 3. Graph of a function before (**a**) and after (**b**) cancellation of pair u, v

2.4 Simplification

To simplify an Morse–Smale complex locally we use a *cancellation* that eliminates two critical points. The inverse operation to refine the complex is called an *anti-cancellation*. Only two adjacent critical points in an Morse–Smale complex can be canceled. The possible configurations are a minimum and a saddle or a saddle and a maximum. Since the two cases are symmetric we limit our discussion to the second case, which is illustrated in Fig. 3.

Only if v is a simple saddle adjacent to two distinct maxima u, w with $f(w) > f(v)$ the pair u, v can be canceled. In particular, a cancellation or anticancellation must always maintain a *valid* Morse–Smale complex. An Morse–Smale complex is called valid, if all cells have four (not necessarily distinct) corners and every path between a saddle and maximum/minimum is ascending/descending. Alternatively, an adaptively refined Morse–Smale complex is valid if it can be created from the highest resolution one using a sequence of cancellations.

3 Cancellation Forest

The information an Morse–Smale complex provides can be separated into the critical points and their connectivity. The critical points information includes position, type, and function value and we refer to this as *critical point configuration* . The connectivity encodes which paths (edges) define a Morse cell and the neighboring information between cells. As with most mesh encoding schemes the critical point configuration provides most (but not all) information about the Morse–Smale complex. Especially during simplification, the connectivity of the Morse–Smale complex can often be inferred from the critical point configuration. For example, in Fig. 3 after u and v have been removed all saddles that were connected to u are now connected to w.

When encoding a cancellation the separation between critical point configuration and connectivity is very intuitive. The top row of Fig. 4 shows three consecutive cancellations $C1$, $C2$, and $C3$ of minima. To reverse any of these cancellations one first needs to know how the connectivity of the Morse–Smale complex changes. For example, in Fig. 4d $m4$ must be created on the left of $m3$ (not on its right). This information is provided by the neighborhood relations between Morse cells, see Sect. 4.

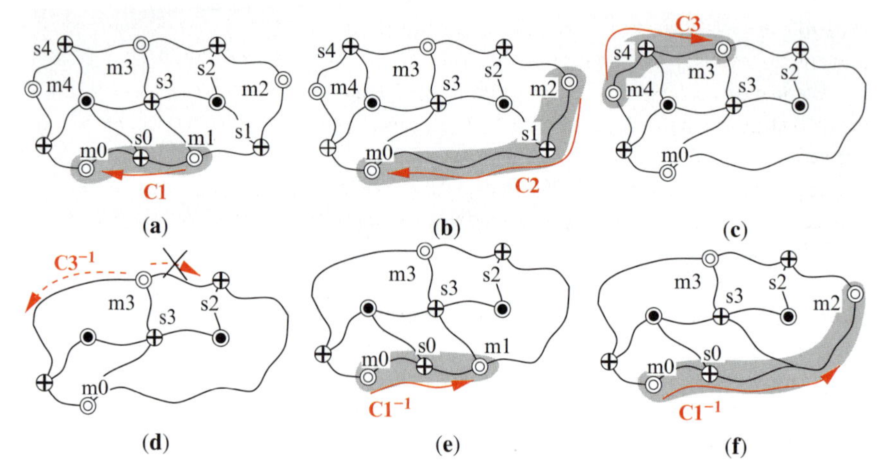

Fig. 4. Morse–Smale complex (**a**) shown after three successive cancellations (**b**), (**c**), and (**d**). The configurations in (**e**) and (**f**) have the same connectivity but a different critical point configuration

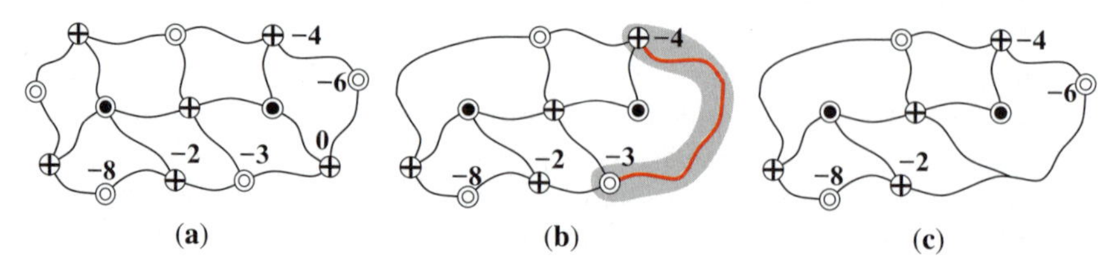

Fig. 5. Morse–Smale complex of Fig. 4 with function values. (**a**) Original complex. (**b**) Invalid critical point configuration (the path marked in *red* cannot be descending.) (**c**) Valid critical point configuration requires anticancellation $C1^{-1}$ to create $m2$ rather than $m1$

One important aspect when encoding (anti)cancellations is whether the operations can be performed out of order. The less ordered dependent the encoding is the more flexible the resulting hierarchy becomes. However, when reversing the order of anticancellations the connectivity alone does not uniquely encode a Morse–Smale complex. For example, starting from Fig. 4d and performing $C1^{-1}$ before $C2^{-1}$ seems to result in the structure of Fig. 4e. Nevertheless, the Morse–Smale complex drawn in (f) has the same connectivity but a different critical point configuration.

The straightforward solution to encoding the critical point configuration is to link it directly to each cancellation. If a cancellation removed the critical point pair u, v then the corresponding anticancellation would introduce u, v. However, this imposes restrictions on the order of cancellations and anticancellations. Figure 5 shows the example of Fig. 4 enhanced by labeling some critical points with function values. In this situation the configuration after reversing $C1$ must be the one shown in Figs. 5c and 4f, respectively. The saddle $s2$ cannot be connected to $m0$ since the resulting path could not be descending from saddle to minimum. However, $C1$ removed $s0$, $m1$ and linking the critical point configuration directly to each cancellation would create an invalid Morse–Smale complex. The algorithm proposed in [4] avoids these complications by imposing additional restrictions on the order of operations, see Sect. 4.

We propose a different strategy that allows us to store connectivity and critical point configuration independently of each other using a simple data structure. The core idea is to view the cancellation shown in Fig. 3 not as removing u and v but as merging the triple u, v, and w into w. After a sequence of cancellations we think of every extremum as the *representative* of itself plus all extrema merged with it. Maxima only merge with maxima and minima only with minima. We keep track of these merges by creating a graph for every extremum. Initially, each extremum is represented by itself as a graph with a single node. During each cancellation an arc is added between the two extrema that were connected to the corresponding saddle in the initial Morse–Smale complex. Notice, that these two extrema are not necessarily the ones involved in the current cancellation, which merges their representatives. Since no extremum can merge with itself these graphs are trees, called *cancellation trees* which form the *cancellation forest*. Figure 6 shows several cancellations and the resulting trees. Figure 17a shows the cancellation trees of a typical terrain data set. Notice, that the cancellation trees provide a very intuitive description of the orientation and general shape of the dominate ridges and valleys in the data.

Even though the data structure used for cancellation trees is simple, it is also very powerful due to two key properties. First, recall that during a cancellation the higher maximum or lower minimum always prevails in the Morse–Smale complex. This fact implies that, for example, the representative of a tree of maxima is always the highest node of the tree. Second, arcs of a cancellation tree correspond to saddles and/or cancellations. In fact, given a cancellation forest created, for example, during an earlier simplification, it is possible to derive a (nearly) complete Morse–Smale

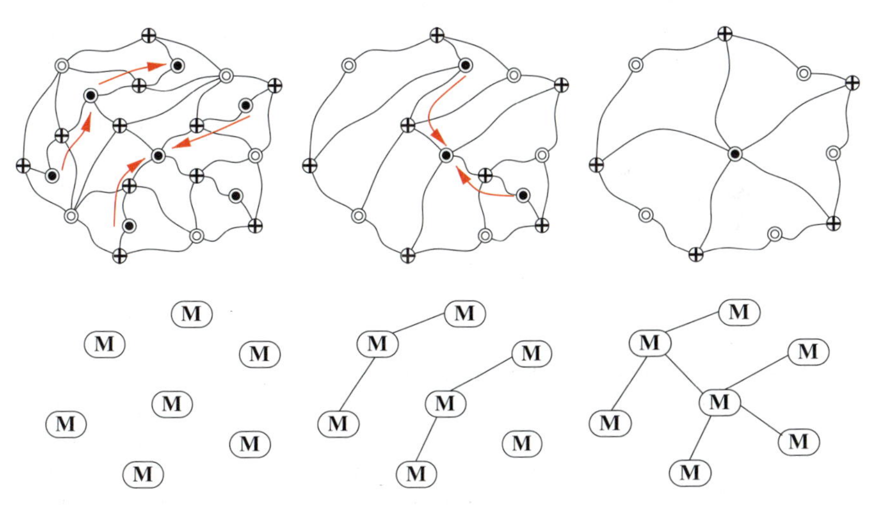

Fig. 6. Example of cancellation trees of maxima resulting from multiple cancellations. Morse–Smale complex with some cancellations indicated in *red* (*top*). Corresponding cancellation trees of all maxima (*bottom*). Note, that arcs are added between extrema incident to the same saddle in the initial complex not the extrema merged by the current cancellation

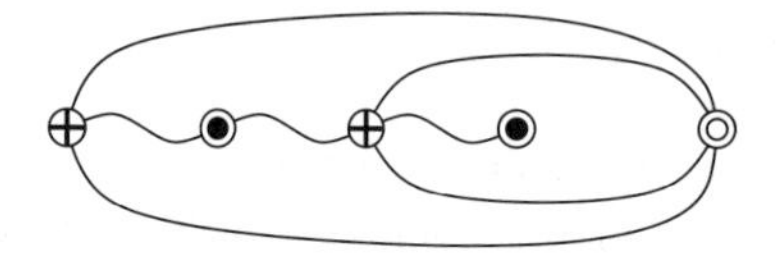

Fig. 7. Strangulation where two Morse cells have the same corners

complex based only on a set of saddles. Assume one is given a highly simplified Morse–Smale complex and the corresponding cancellation forest; Furthermore, assume a refinement of the Morse–Smale complex is described by a set of saddles $S = \{s_0, \ldots, s_n\}$ that must appear in the refined complex, for example all saddles within a view frustrum. First, one removes all arcs corresponding to a saddle in S from the cancellation forest resulting in another forest with more but smaller trees. Subsequently, one can reconstruct the Morse–Smale complex in the following manner: Each saddle s_i was initially connected to two maxima M_0, M_1 and two minima m_0, m_1. All of these extrema are part of a tree, and the saddle is connected to the four representatives of these trees. This defines the adaptive Morse–Smale complex to the level of the embedding of the paths. The saddles are given, the remaining critical points are the representatives of the cancellation trees, and the paths embedding can be derived from concatenating original paths.

Nevertheless, the connectivity between Morse cells is not uniquely defined by the construction described above. This is due to the fact that in an Morse–Smale complex paths are not uniquely defined by their end points, see Fig. 7. As a result, Morse cells are not identified by their corners and the connectivity must still be stored explicitly. Section 4 describes how the connectivity as well as the configuration of saddles can be stored hierarchically.

In general, a cancellation tree can be split anywhere at any time. As a result, the search for the representative of a subtree does not map to a union-find approach traditionally employed in similar situations. Therefore, maintaining the cancellation forest involves a linear search during an anticancellation and is a constant-time operation during a cancellation. While more sophisticated structures are possible our experiments suggest that cancellation trees have an overall low branching factor. This would likely diminishes any advantage of more complicated structures and would make implementation more difficult.

4 Hierarchy

Using cancellation trees to maintain the critical point configuration allows us to create a mesh hierarchy geared completely toward connectivity. The main objective is to construct a hierarchy that supports as many different configurations as possible. Following traditional triangle mesh hierarchies, (anti)cancellations are stored in a dependency graph representing a partial order among operations. All configurations that can be created by observing the partial order should result in a valid Morse–Smale complex.

4.1 Hierarchy Construction

Following the approach discussed in [4], we split each Morse cell into two *Morse triangles* by introducing the diagonal connecting the minimum to the maximum into the complex. As a result, the neighborhood around a saddle then consists of four triangles that form the *diamond* around the saddle, as indicated in gray in Fig. 8a. Each cancellation removes one diamond from the Morse–Smale complex. We create a hierarchy in a bottom-up fashion by successively canceling critical points, see Fig. 9 for an example. Two cancellations are called *independent* if it is irrelevant in what order they are performed and *dependent* otherwise. The *extended dependency graph* contains a node for every cancellation and an arc between dependent cancellations. The *dependency graph* is derived from the extended one using path compression. The *height* of the dependency graph is defined as the maximal distance from a root to a leaf. In practice, one is interested in constructing a shallow graph with few edges since this implies the possibility of a large number of different configurations.

Clearly, the definition of dependencies between cancellations determines the shape of the dependency graph. In [4], the *region of interference* of the cancellation in Fig. 8 is defined as all Morse cells incident to either u, v, or w. Two cancellations

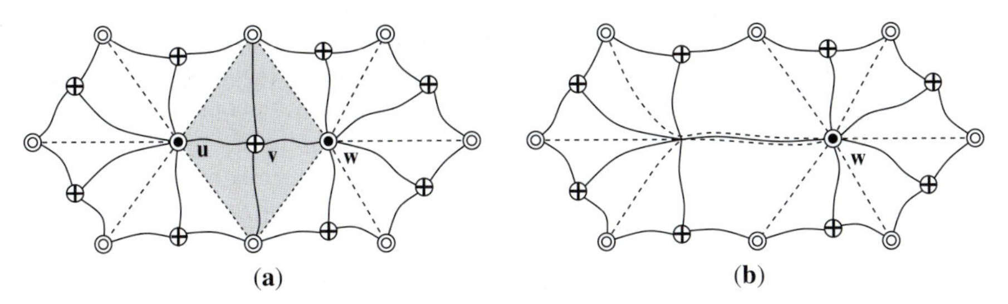

Fig. 8. Morse–Smale complex corresponding to Fig. 3 (**a**) before and (**b**) after cancellation of pair u, v. Diagonals indicating diamonds are shown as *dotted lines*

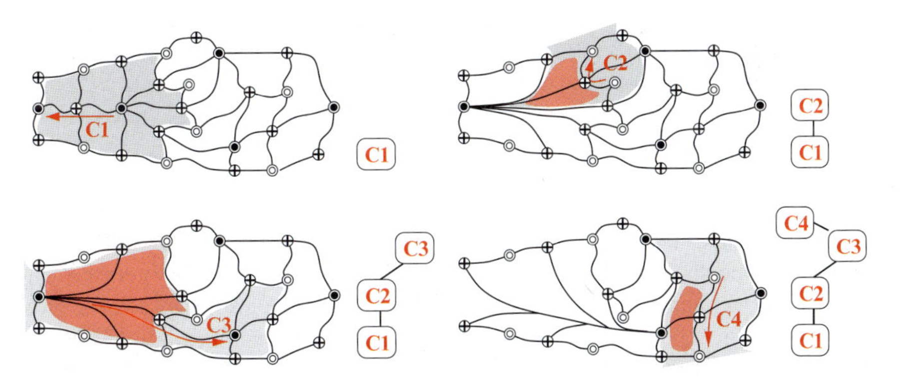

Fig. 9. Hierarchy construction as described in [4]. Cancellations are indicated by *arrows*, the corresponding region of interference is shaded in *gray*, and regions of overlap with previous cancellations are shaded in *red*. The corresponding dependency graphs are shown next to the Morse–Smale complexes. After four cancellations the dependency graph is a line

are defined as dependent if their regions of interference have a (true) intersection. This large region of interference is necessary to avoid the problems discussed in Sect. 3. Given the large region of interference, storing the hierarchy is straightforward. Each cancellation replaces Morse cells around three critical points by Morse cells around the remaining one. The boundary of the region does not change and the dependencies ensure that a (anti)cancellation is only performed if the Morse–Smale complex is locally identical to the one encountered during construction. This can be viewed as a special case of the concepts described for general multiresolution structures described, for example, by De Floriani et al. [11]. An example of several cancellations and the resulting dependency graphs using the old hierarchy is shown in Fig. 9. Due to the large regions of interference the final dependency graph (lower right corner) is a line allowing no adaptations beyond the ones encountered during construction.

Using cancellation trees one can ignore the configuration of minima and maxima, requiring us to encode only the connectivity and saddle configuration. Since each cancellation removes the diamond around a saddle it is natural to link the saddle information directly to a diamond. Therefore, if we can store the diamond information (the connectivity) hierarchically, cancellation trees provide the remaining information.

To store the connectivity information we use the concepts from [11] but now with a significantly smaller region of interference. Each cancellation removes one diamond replacing eight triangles around a vertex by four. An anticancellation reintroduces a diamond replacing four triangles by eight, introducing two vertices. Some possible configurations are shown in Fig. 10. The cancellation of a diamond changes a reduced Morse–Smale complex only for the neighboring (edge-connected) diamonds. Therefore, the region of interference of a cancellation is defined as the corresponding diamond plus its edge-connected neighbors. The smaller regions of interference produce a smaller set of dependencies. In fact, the number of ancestors and the number of children of each node in the dependency graph is bounded (assuming path compression). Each diamond has at most four edge-connected neighbors and therefore, a node cannot have more than four children. Canceling a diamond merges its four neighbors into two. As a result, a node can have no more than two ancestors. Figure 11 shows the example of Fig. 9 using cancellation trees.

We create a hierarchy by removing diamonds from the highest-resolution Morse–Smale complex in “batches” of independent cancellations. However, this strategy can result in cancellations of high persistence to be dependent on cancellations with much lower persistence, which is undesirable for most applications. Therefore, we limit the batches such that the largest persistence in a batch is not larger than twice the maximal persistence of the previous batch. The resulting hierarchy performs significantly better than the unrestricted one in terms of the error cause for a given number of critical points and shows practically no difference in flexibility. However, theoretically, the restricted algorithm can create a hierarchy of linear height. Without this restriction, it is guaranteed that each batch contains about one quarter of the remaining diamonds in the complex and therefore the algorithm creates a hierarchy of logarithmic height.

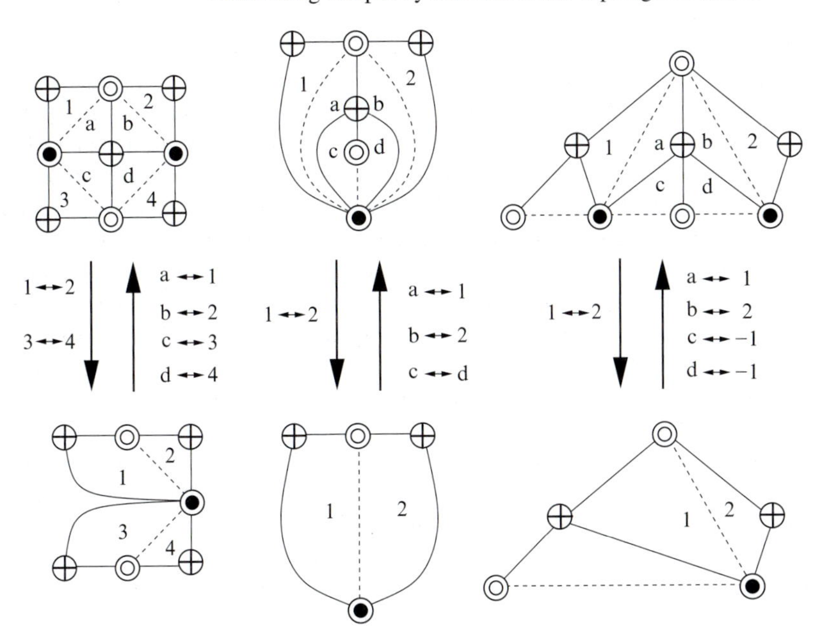

Fig. 10. Three examples for encoding the connectivity during cancellations. The triangulation before (*top*) and after (*bottom*) the cancellation of the diamond a, b, c, d is shown. The *middle row* shows how the change in neighborhood structure for an (anti)cancellation is encoded as a list of triangle pairs (-1 indicating a boundary edge)

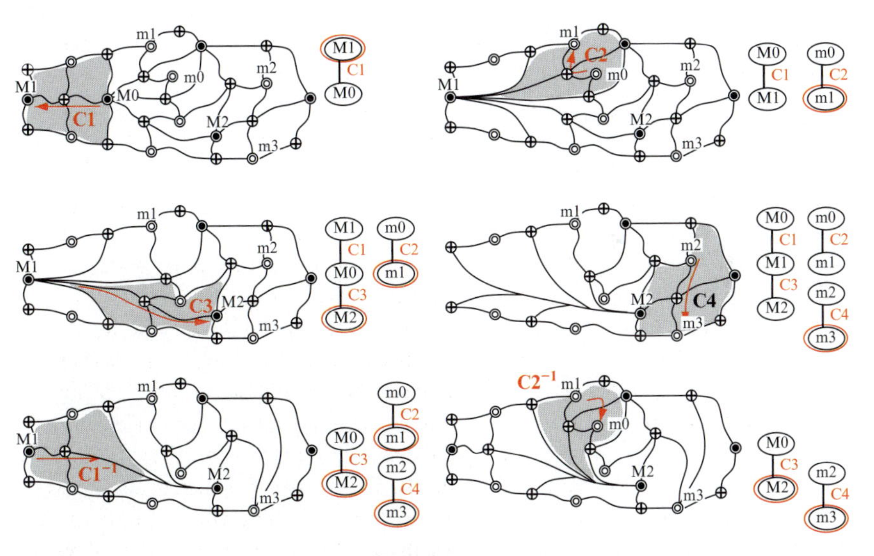

Fig. 11. The *top two rows* show the example of Fig. 9 using cancellation trees to encode the hierarchy. The regions of interference are shaded in *gray*, and the corresponding cancellation trees are drawn on the right side of each figure with the representative marked in *red*. Using the reduced Morse–Smale complex all cancellations are independent. The *bottom row* shows the complex after the anticancellation of $C1$ (*left*) and $C2$ (*right*). Note that $C1^{-1}$ correctly creates $M1$ rather than $M0$ ($M1$ is higher than $M0$)

5 Results

To compare the new hierarchy with the one proposed in [4] we applied both strategies to a 1,201-by-1,201 single-byte integer value terrain data set of the Grand Canyon. Figure 12 shows a rendering (a) and the initial Morse–Smale complex (b) of the Grand Canyon data set with 11,620 critical points. We assess quality via a fly-over comparing the adaptivity of the cell-based hierarchy with the one using cancellation trees. A narrow view-frustum is defined where the topology is refined to the highest resolution. Outside the given view-frustum only dependent topology is used. Figures 13 and 14 show two frames of the fly-over for two distinct stages of the fly-over path.

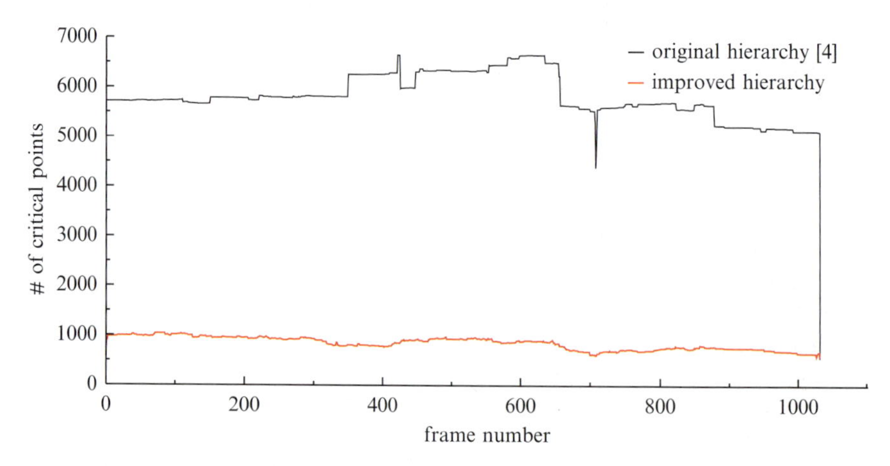

Fig. 12. Number of critical points used during a fly-over (Grand Canyon data set)

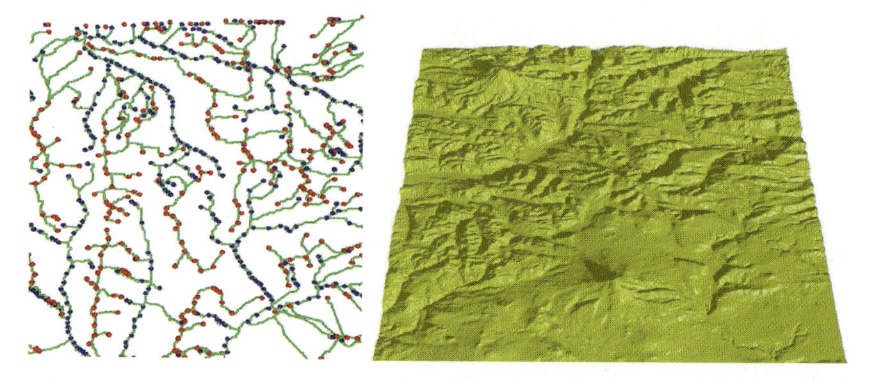

Fig. 13. *Left:* Typical cancellation trees of a terrain. Maxima are shown in *red*, minima in *blue*, and arcs in *green*. Note the overall low branching factor. *Right:* Rendering of original Yakima data set

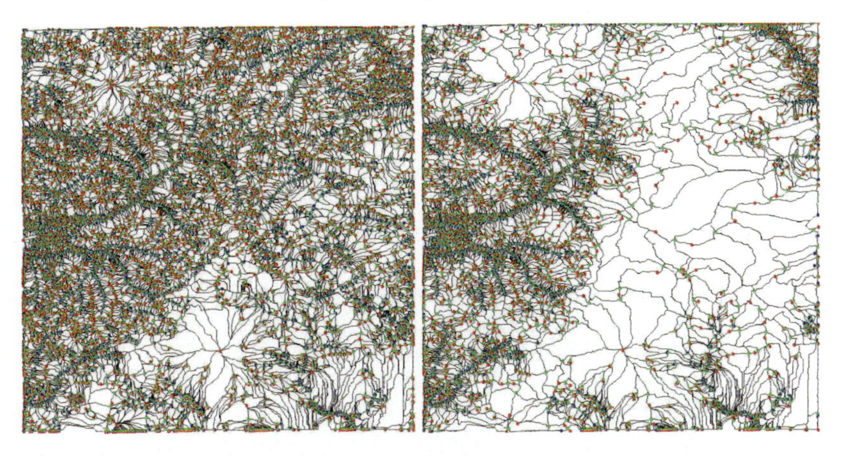

Fig. 14. *Left:* Original Morse–Smale complex of the Yakima data set (17,691 critical points); (*right*) adaptively refined Morse–Smale complex, where only features below function value of 0.14 are preserved (8,063 critical points)

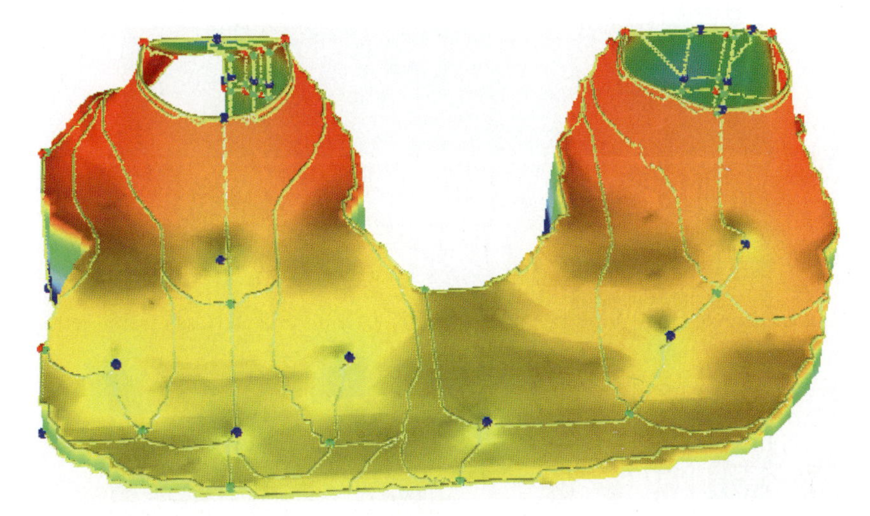

Fig. 15. Pseudo-colored rendering and simplified Morse–Smale complex of oil-pressure data set

Figure 15 shows the number of critical points in the adaptive Morse–Smale complex during the fly-over for both methods used for hierarchy construction. The hierarchy using cancellation trees is clearly superior to the original encoding. One explanation for the large differences in quality is the presence of high-valency extrema in the Morse–Smale complex. Often, data sets (especially terrains) are biased to contain significantly more maxima than minima (or the reverse), which results in some extrema of the Morse–Smale complex having high valency values. Using the original large region of interference, the hierarchy around a high-valency extremum degenerates into a linear sequence. The smaller region of interference proposed in this paper, however, is based on saddles which always have valence four. Therefore, the shape of the hierarchy remains largely unaffected by high valency extrema.

The adaptive refinement and display of topology is useful for many areas. Figure 16 shows the oil pressure of an underground oil reservoir. The figure shows an isosurface of water saturation, pseudo-colored by oil pressure. The linear color

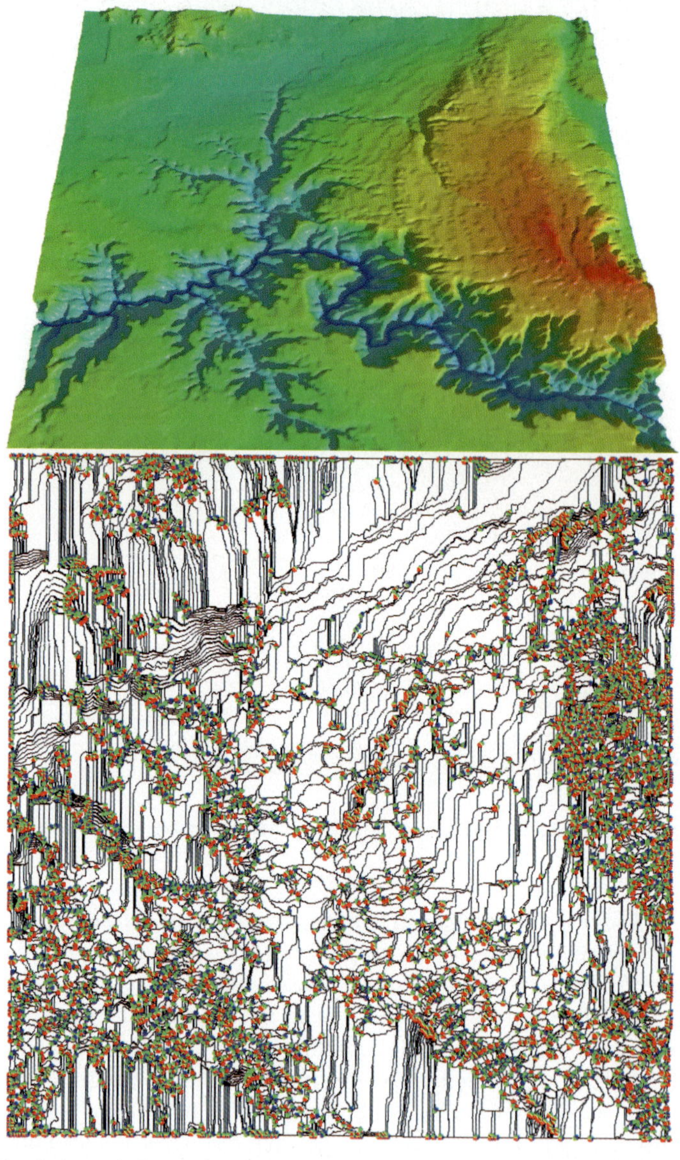

Fig. 16. Rendering of Grand Canyon data set; (*Top*) original Morse–Smale complex of (*Bottom*) using 11,620 critical points (minima shown in *blue*, maxima in *red*, and saddles in *green*)

map used in Fig. 16 provides little structural information. However, the seven oil extraction sites are visible as local minima in the simplified Morse–Smale complex.

Figure 17b shows a rendering of the Yakima terrain data set consisting of $1{,}201 \times 1{,}201$ single-byte integer height values. Figure 18 shows the corresponding

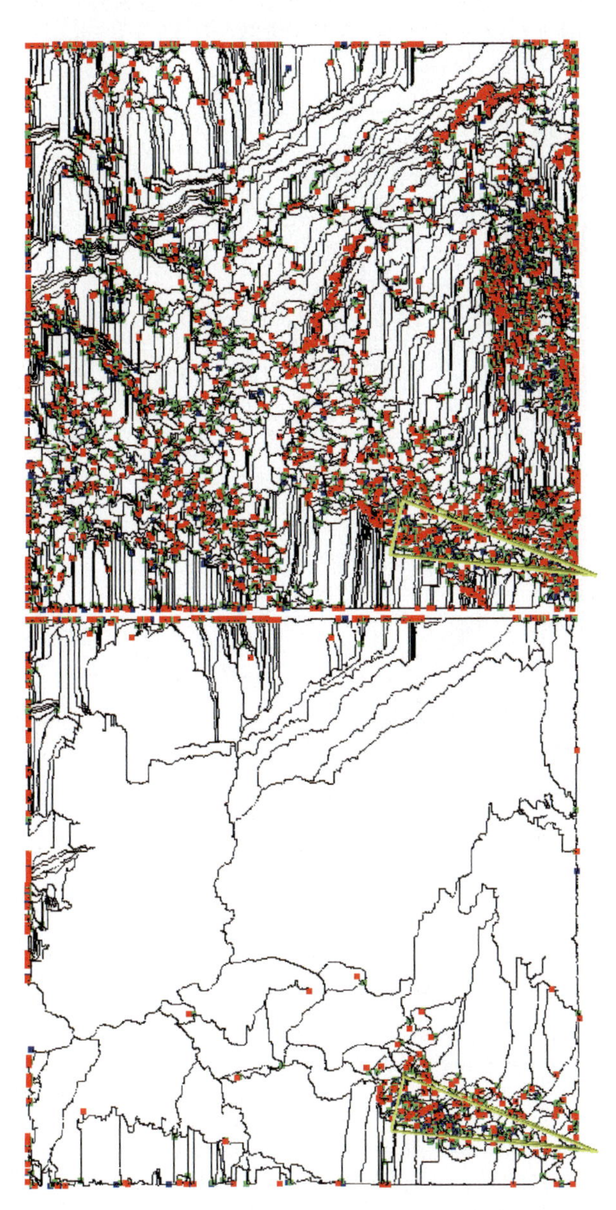

Fig. 17. Global view of a fly-over of Grand Canyon data set. Inside the local view frustum (*yellow*) the finest resolution topology is shown on the outside only dependent topology is used. (*Top*) The results of the hierarchy in [4]; (*Bottom*) refinement using the improved hierarchy introduced in this paper

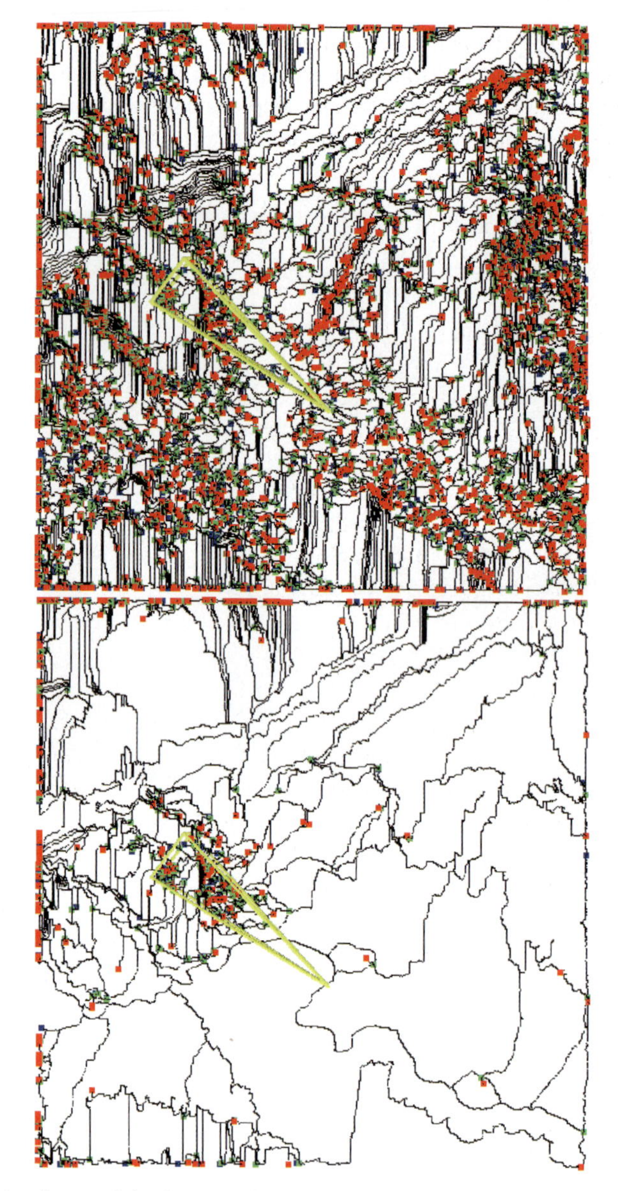

Fig. 18. Another frame of the fly-over of the Grand Canyon data set. (*Top*) Using the original hierarchy; (*Bottom*) using the cancellation forest

Morse–Smale complex with 17,691 critical points and the same complex refined to preserve only features below a function value of 0.14 (with function values scaled to [0, 1]) using 8,063 critical points. The density of the Morse–Smale complex shows how the region around the canyons remains highly refined.

One disadvantage of the new technique is that the hierarchy is so flexible that it becomes impossible to precompute function values corresponding to all possible topological refinements. However, for any topological refinement we can compute a function with the given topology using the concepts of [4]. The general idea of this computation is indicated in Fig. 8. Canceling the maximum u with the saddle v requires us to lower the function within a region around u and to raise the function along the path $u - v - w$.

6 Conclusions and Future Research

We have improved our original results discussed in [4] significantly in several different ways, moving toward the practical application of topology for data visualization and analysis. Using cancellation trees, the hierarchy is smaller, more adaptable, and supports the use of larger, more complicated Morse–Smale complexes. Furthermore, cancellation trees are easy to implement and to maintain during refinement. Currently, we only display the adapted topology, not the corresponding adapted function interactively. We plan to develop new techniques computing high-quality topological approximation on-the-fly.

Acknowledgments

This work was performed under the auspices of the US Department of Energy by University of California Lawrence Livermore National Laboratory under contract No. W-7405-Eng-48. B. Hamann is supported by National Science Foundation under contract ACI 9624034 (CAREER Award), through the Large Scientific and Software Data Set Visualization (LSSDSV) program under contract ACI 9982251, through the National Partnership for Advanced Computational Infrastructure (NPACI), and through a large Information Technology Research (ITR) grant. We thank the members of Data Science thrust from the Center for Applied Scientific Computing (CASC) at Lawrence Livermore National Laboratory, and the Visualization and Computer Graphics Research Group at the University of California, Davis.

References

1. C. L. Bajaj, V. Pascucci, and D. R. Schikore. Visualization of scalar topology for structural enhancement. In D. Ebert, H. Hagen, and H. Rushmeier, editors, *Proc. IEEE Visualization 1998*, pages 51–58, Los Alamitos, California, 1998. IEEE Computer Society Press.
2. T. F. Banchoff. Critical points and curvature for embedded polyhedral surfaces. *American Mathematical Monthly*, 77(5):457–485, May 1970.
3. P.-T. Bremer, H. Edelsbrunner, B. Hamann, and V. Pascucci. A multi-resolution data structure for two-dimensional Morse–Smale functions. In G. Turk, J. J. van Wijk, and R. Moorhead, editors, *Proc. IEEE Visualization 2003*, pages 139–146, Los Alamitos, California, 2003. IEEE Computer Society Press.

4. P.-T. Bremer, H. Edelsbrunner, B. Hamann, and V. Pascucci. A topological hierarchy for functions on triangulated surfaces. *IEEE Transactions on Visualization and Computer Graphics*, 10(4):385–396, 2004.
5. H. Carr, J. Snoeyink, and U. Axen. Computing contour trees in all dimensions. *Computational Geometry: Theory and Applications*, 24(3):75–94, 2003.
6. A. Cayley. On contour and slope lines. *The London, Edinburgh and Dublin Philosophical Magazine and Journal of Science*, XVIII:264–268, 1859.
7. W. de Leeuw and R. van Liere. Collapsing flow topology using area metrics. In *Proc. IEEE Visualization 1999*, pages 349–354. IEEE Computer Society Press, 1999.
8. H. Edelsbrunner, J. Harer, V. Natarajan, and V. Pascucci. Morse–Smale complexes for piecewise linear 3-manifolds. In *Proc. 19th Sympos. Comput. Geom.*, pages 361–370, 2003.
9. H. Edelsbrunner, J. Harer, and A. Zomorodian. Hierarchical Morse–Smale complexes for piecewise linear 2-manifolds. *Discrete and Computational Geometry*, 30:87–107, 2003.
10. H. Edelsbrunner, D. Letscher, and A. Zomorodian. Topological persistence and simplification. *Discrete and Computational Geometry*, 28:511–533, 2002.
11. L. De Floriani, E. Puppo, and P. Magillo. A formal approach to multiresolution modeling. In W. Straßer, R. Klein, and R. Rau, editors, *Theory and Practice of Geometric Modeling*. Springer, Berlin, 1996.
12. J. L. Helman and L. Hesselink. Visualizing vector field topology in fluid flows. *IEEE Computer Graphics and Applications*, 11(3):36–46, 1991.
13. M. Hilaga, Y. Shinagawa, T. Kohmura, and T. L. Kunii. Topology matching for fully automatic similarity estimation of 3D shapes. In E. Fiume, editor, *Proceedings of ACM SIGGRPAH 2001*, pages 203–212, New York, NY, USA, 2001. ACM.
14. J. C. Maxwell. On hills and dales. *The London, Edinburgh and Dublin Philosophical Magazine and Journal of Science*, XL:421–427, 1870.
15. J. Milnor. *Morse Theory*. Princeton University Press, New Jersey, 1963.
16. M. Morse. Relations between the critical points of a real functions of n independent variables. *Transactions of the American Mathematical Society*, 27:345–396, July 1925.
17. S. P. Morse. A mathematical model of the analysis on contour-line data. *Journal of the Association for Computing Machinery*, 15(2):205–220, 1968.
18. V. Pascucci and K. Cole-McLaughlin. Efficient computation of the topology of level sets. In M. Gross, K. I. Joy, and R. J. Moorhead, editors, *Proc. IEEE Visualization 2002*, pages 187–194, Los Alamitos, California, 2002. IEEE Computer Society Press.
19. J. Pfaltz. Surface networks. *Geographical Analysis*, 8:77–93, 1976.
20. J. Pfaltz. A graph grammar that describes the set of two-dimensional surface networks. *Graph-Grammars and Their Application to Computer Science and Biology. Lecture Notes in Computer Science, vol. 73*. Springer, Berlin, 1979.
21. B. T. Stander and J. C. Hart. Guaranteeing the topology of implicit surface polygonization for interactive modeling. In *Proc. of ACM SIGGRPAH 1997*, volume 31, pages 279–286, New York, USA, Aug. 1997. ACM Press / ACM SIGGRAPH.
22. X. Tricoche, G. Scheuermann, and H. Hagen. A topology simplification method for 2D vector fields. In *Proc. IEEE Visualization 2000*, pages 359–366, Los Alamitos, California, 2000. IEEE Computer Society Press.
23. X. Tricoche, G. Scheuermann, and H. Hagen. Continuous topology simplification of planar vector fields. In *Proc. IEEE Visualization 2001*, pages 159–166, Piscataway, NJ, Oct. 2001. IEEE Computer Society Press.
24. M. J. van Kreveld, R. van Oostrum, C. L. Bajaj, V. Pascucci, and D. Schikore. Contour trees and small seed sets for isosurface traversal. In *Symposium on Computational Geometry*, pages 212–220, 1997.

The Toporrery: Computation and Presentation of Multiresolution Topology

Valerio Pascucci[1], Kree Cole-McLaughlin[2], and Giorgio Scorzelli[1]

[1] Scientific Computing and Imaging Institute, School of Computing, University of Utah, UT, USA

[2] Dept. of Mathematics, UCLA, CA, USA, School of Engineering, University of Roma Tre, Roma, Italy

Summary. The Contour Tree of a scalar field is the graph obtained by contracting all the connected components of the level sets of the field into points. This is a powerful abstraction for representing the structure of the field with explicit description of the topological changes of its level sets. It has proven effective as a data-structure for fast extraction of isosurfaces and its application has been advocated as a user interface component guiding interactive data exploration sessions. In practice, this use has been limited to trivial examples due to the problem of presenting a graph that may be overwhelming in size and in which a planar embedding may have self-intersections. We propose a new metaphor for visualizing the Contour Tree borrowed from the classical design of a mechanical orrery – see Fig. 1a – reproducing a hierarchy of orbits of the planets around the sun or moons around a planet. In the toporrery – see Fig. 1b – the hierarchy of stars, planets and moons is replaced with a hierarchy of maxima, minima and saddles that can be interactively filtered, both uniformly and adaptively, by importance with respect to a given metric.

The implementation of the system is based on (1) a hierarchical graph model allowing coarse-to-fine traversal for selective refinements and (2) a new algorithm for constructing a multiresolution Contour Tree with guaranteed topological correctness independently of the simplification metric. We have tested the approach using topological persistence as the main metric for constructing the tree hierarchy, and using geometric position as a secondary metric for adaptive refinements. The result is presented in linked views of the abstract toporrery and the geometric embedding of the input data.

1 Introduction

A Morse function over a domain $\mathcal{D}$, is a smooth mapping, $f : \mathcal{D} \rightarrow \mathbb{R}$, such that all its critical points (maxima, minima and saddles) are distinct. Complex natural phenomena, both sampled and simulated, are often modeled as Morse functions.[1]

[1] Technically the definition of Morse function is often weakened to allow multiple critical points or other degeneracies present in real data.

T. Möller et al. (eds.), *Mathematical Foundations of Scientific Visualization, Computer Graphics, and Massive Data Exploration*, Mathematics and Visualization,
DOI: 10.1007/978-3-540-49926-8,

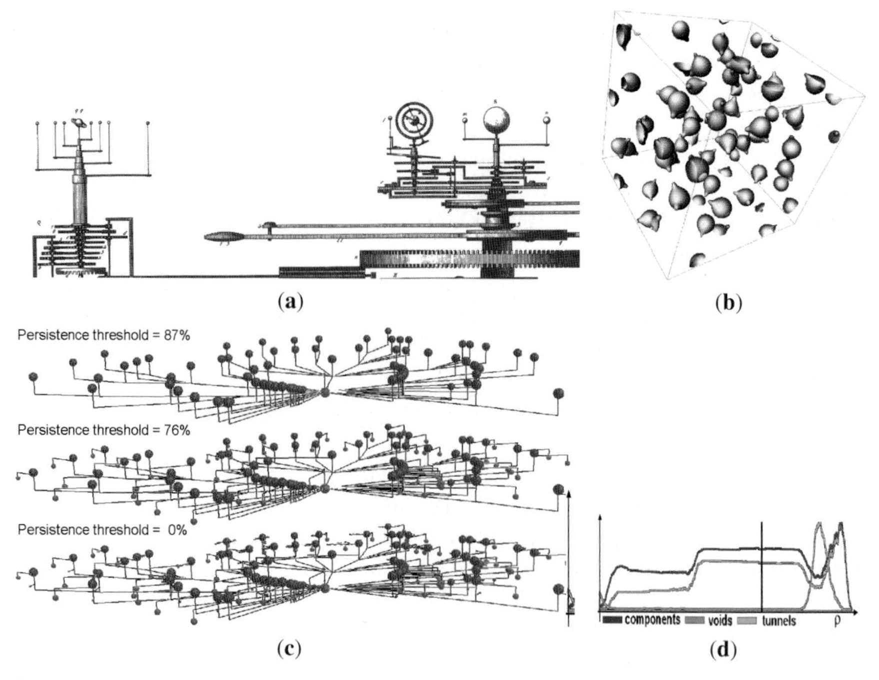

Fig. 1. (**a**) Orrery reproducing the hierarchical relationship between the orbits of the sun, the planets and their moons. Original design (1812) by A. Janvier reprinted recently by E. Tufte [24]. (**b**) Toporrery representing the hierarchical relationship between the critical points in a scalar field. The particular field is an electron density distribution (ρ) computed with an ab initio simulation for water molecules at high pressure. The three levels of pruning of the topology by persistence highlight: (*top*) the water molecules, (*middle*) the dipole hydrogen–oxygen structure for each molecule, and (*bottom*) the detailed topological features including possible numerical noise (useful for debugging the simulation). (**c**) A level set of the scalar field ρ selected on the basis of the topological features and highlighting the coarse structure of the water molecules. Note that at this high pressure even for water molecules the topology characterizes molecular structures better than the distance between the atoms. (**d**) Spectral diagram [3] providing a global summary of the topological information

MRI scans generate Morse functions that are used in medical imaging to reconstruct human tissues. Electron density distributions computed by high-resolution molecular simulations are Morse functions whose topologies express bonds among the atoms in molecular structures. The structure of geometric models used in computer graphics and CAD applications can be effectively represented in terms of the topology of a Morse function [12].

The Reeb graph [16] is a simple structure that summarizes the topology of a Morse function. For functions with simply connected domains this graph is also simply connected and is called the Contour Tree. The Reeb graph has been used to analyze the evolution of teeth contact interfaces in the chewing process [18], and to compute indices of topological similarity for databases of geometric models [12].

Topological information has been used to guide the construction of transfer function for volume rendering of scientific data [21, 26]. A more extensive discussion of the use of the Reeb graph and its variations in geometric modeling and visualization can be found in [11].

The first algorithm for constructing Reeb graphs of Morse functions with two-dimensional domains is due to [19]. Given a triangulated surface, this scheme takes as input the set of all distinct level lines and therefore has worst case time complexity $O(n^2)$, where n is the number of vertices in the triangulation. An $O(n \log n)$ algorithm for computing Contour Trees in any dimension was introduced in [5]. This scheme has been extended in three dimensions to include the genus of all isosurfaces [15]. The first multiresolution representation of the Reeb graph was introduced in [12]. Their method hierarchically samples the range space of f while concurrently refining the Reeb graph. They obtain a multiresolution model that is suitable for fast comparison of graphs. However, this hierarchy does not represent the topology of f at multiple levels of detail. A formal framework for ranking topological features by persistence has been introduced in [10] and applied to two-dimensional Morse functions in [2]. Topological simplification is used in [20] to design transfer function that highlight only the major features in the data. Topological simplification is also widely used in vector field visualization to highlight the most important structures present in the data [22, 23].

Early techniques for the simplification of the Contour Tree are reported in [17], where a simple greedy approach is used to prune the arcs of the tree corresponding to local features of small area. This work has been recently extended by [6], with a new simplification algorithm that applies to several classes of data approximation metrics. In the extended abstract of this paper [8] where we first introduce a simplification algorithm with proper cancellation of single pairs of critical points (see Theorem 3.28 in [14]). The information is then collected in a multiresolution representation amenable for coarse-to-fine adaptive traversal of the Contour Tree.

Integration of the Contour Tree in user interfaces to help selecting isosurfaces was first suggested [62] but only fully developed 6 years later in [3]. The latter work is particularly interesting for the use of a new concept of "Path Seeds" that explicitly links the arcs in the Contour Tree to distinct connected components (contours) of the level sets. This introduces the powerful new paradigm of selecting contours instead of entire isosurfaces. Our scheme introduces a user interface based on a multiresolution representation of the Contour Tree (not just a simplification) and a metaphor for presenting the tree in an abstract 3D embedding similar to a mechanical orrery. In particular, we adapt a simple radial graph drawing algorithm [9] and combine it with an embedding typically used for large tree hierarchies [13].

Contributions

Our results are summarized in Figs. 1 and 2 with presentations of the topology of two scalar fields defined on 3D domains (electron density distribution of water at high pressure) and 2D domains (armadillo). (1) We provide a multiresolution representation for the Contour Tree with algorithms for uniform and adaptive refinement on the basis of precomputed metrics. (2) We provide a simple scheme for laying

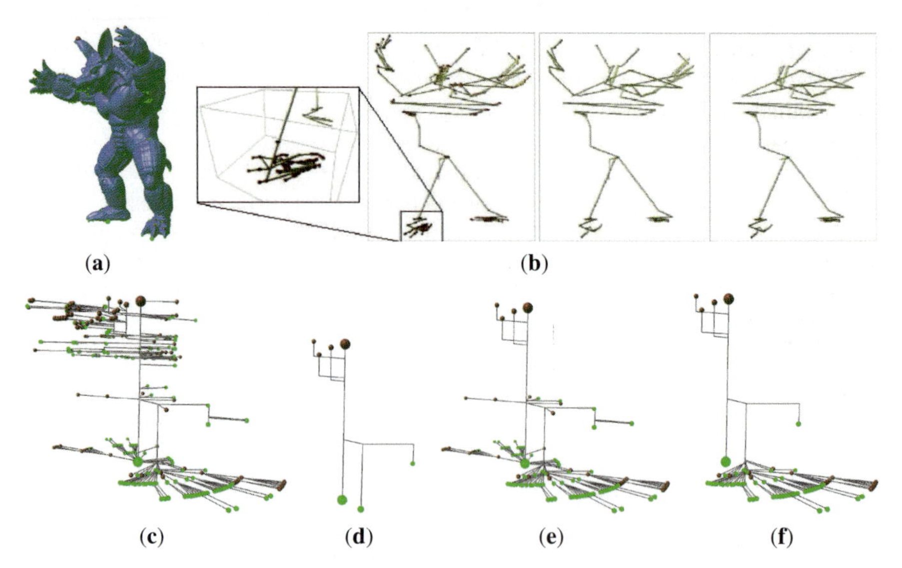

Fig. 2. (**a**) A polygonal armadillo model with 172,974 vertices and 345,944 polygons. The Morse function f is the height in the vertical direction. The maxima (*red*) and minima (*green*) of f are marked with small spheres. (**b**) The Contour Tree of f presented with the critical points in their original position. Several version of the tree with adaptive (one foot) or uniform refinement. (**c**) Full resolution toporrery of f. (**d**) Simplification down to 58% of persistence showing a skeletal structure with arms, legs, ears and tail. Adaptive refinement with full resolution for lower half of the body (**e**) or only the left foot (**f**)

out the tree in an way that highlights the hierarchical relationship among the critical points and can be integrated in an interactive graphical user interface. (3) We provide an algorithm for computing a multiresolution Contour Tree directly from join and split trees. We discuss for what class of function the topological hierarchy we compute corresponds to actual topological simplifications that can be constructed for the function f. (4) The results are demonstrated on datasets of different nature, such as terrain, surface models and volumetric scientific data.

2 Multiresolution Contour Trees

Contour Tree of a Morse Function

Let $\mathcal{D}$ be a triangulated domain and $f : \mathcal{D} \to \mathbb{R}$ be a function obtained by linear interpolation of the value of f at the vertices of $\mathcal{D}$. Morse theory provides a formal framework for understanding the topology of $\mathcal{D}$ by analyzing the function f. The fundamental tool in Morse theory is the characterization of each point of $\mathcal{D}$ as being either regular or critical.

We assume that $\mathcal{D}$ is a simplicial complex. Therefore, every k-cell c of $\mathcal{D}$ is the convex hull of $k + 1$ vertices of $\mathcal{D}$. Moreover, a cell c' is a called *face* of c if its vertices are a subset of those of c. If $c \in \mathcal{D}$ then all its faces must be in $\mathcal{D}$. For a vertex $v \in \mathcal{D}$, its *link* Lk_v is the set of cells that do not contain v but that are faces of some cell containing v. Furthermore, the *lower link* of v, Lk_v^-, is the set of all cells in Lk_v that only have vertices with function value smaller than $f(v)$. The *upper link* Lk_v^+ is the set of cells in Lk_v that have only vertices with function value greater than $f(v)$.

Definition 1. *Let $\mathcal{D}$ be a triangulated manifold with boundary and $f : \mathcal{D} \rightarrow \mathbb{R}$ be a piecewise-linear function. A vertex $v \in \mathcal{D}$ is called* regular *if both Lk_v^- and Lk_v^+ have exactly one connected component. Otherwise v is called a* critical point *and $f(v)$ is called a* critical value.

We can now define a Morse function. Since the definition only refers to critical points it applies equally well in the smooth and discrete settings.

Definition 2. *f is a* Morse function *iff all its critical values are distinct.*

On piecewise-linear functions this condition can be enforced by symbolically perturbing the critical values. If the vertices $v_i, v_j \in \mathcal{D}$ are critical points such that $f(v_i) = f(v_j)$, then we define $f(v_i) < f(v_j)$ if and only if $i < j$. In practice we apply the symbolic perturbations to the function value at all the vertices $v \in \mathcal{D}$. This allows us to sort the vertices by their function value and to simply define $f(v_i) \equiv i$.

Definition 3. *A* level set *of f is the preimage of a real value ω, $L_f(\omega) = f^{-1}(\omega)$. Given a level set, $L_f(\omega)$, we call a connected component of $L_f(\omega)$ a* contour.

Morse theory describes how the topology of $L_f(\omega)$ changes as the field value, ω, changes. One of the main results states that if a and b are such that the range $[a, b]$ contains no critical values, then $L_f(\omega)$ is homeomorphic to $L_f(\nu)$ for all $\omega, \nu \in [a, b]$. On the other hand if the range $[a, b]$ contains a single critical value, ω_0, then for $\omega \in [a, \omega_0)$ and $\nu \in (\omega_0, b]$ the difference in the topology of $L_f(\omega)$ and $L_f(\nu)$ can be completely described as follows: (1) If ω_0 is a local minimum, a new contour is created in $L_f(\nu)$ that did not exist in $L_f(\omega)$. (2) If ω_0 is a local maximum, a contour of $L_f(\omega)$ is destroyed. (3) If ω_0 is a saddle point, either two contours of $L_f(\omega)$ merge into a single contour of $L_f(\nu)$ or one contour of $L_f(\omega)$ divides into two contours of $L_f(\nu)$. For volumetric or higher dimensional domains, a saddle point can also induce a topological change in a single contour of L_f (no split nor merge).

The Contour Tree encodes the changes in the number of contours of the level set.

Definition 4. *Consider the graph obtained by contracting each contour of every level set of f to a point. For general Morse functions this graph is called the* Reeb graph *and can have any number of cycles, depending on the topology of $\mathcal{D}$ [7]. For simply connected $\mathcal{D}$ the* Reeb graph *is also simply connected and is called* Contour Tree.

From the definition it can be seen that the nodes of the Contour Tree correspond critical points of f and are therefore associated with the relative critical value. Furthermore, nodes that correspond to extrema are leaf nodes, and nodes that correspond to saddle points must have degree three (or higher in degenerate cases). Figure 3a shows a simple terrain as an example of Morse function, where the elevation of each

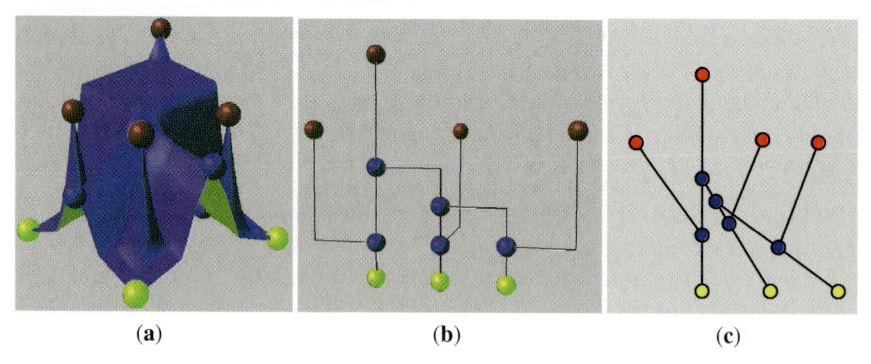

(a) (b) (c)

Fig. 3. (**a**) A simple terrain model. The Morse function f is the vertical elevation, which critical points are highlighted with spheres of different colors: red for maxima, *yellow* for minima and *blue* for saddles. (**b**) Contour Tree of f embedded in 3D as a toporrery. The z coordinate of each node is equal to the corresponding critical value of f. (**c**) 2D layout of the same tree with y coordinate equal to the corresponding critical value of f. Note that the 2D layout is not planar and cannot be drawn without self intersection because of the constraint on the y coordinate of its nodes

point is the value of f. Figure 3b show the corresponding Contour Tree. Figure 3c shows the planar layout proposed in [62] where the y coordinate of each node is constrained to be equal to the corresponding critical value of f. Note that with this constraint the graph cannot be drawn in the plane without self-intersections – see Fig. 3c.

Hierarchical Graph Representation

We define a multiresolution representation of the Contour Tree that allows linear time access to simplified representations of the topology. Typically finite graphs are represented as a list of nodes and a list of arcs, where each arc is defined as a node pair. In this section we discuss an alternative representation called a branch decomposition. A *branch* is a monotone path in the graph traversing a sequence of nodes with nondecreasing (or nonincreasing) value of f. The first and last nodes in the sequence are called the endpoints of the branch. All other nodes are said to be interior to the branch. Note that a branch can be thought of equally as a sequence of nodes or a sequence of arcs. A set of branches is called a *branch decomposition* of a graph if every arc in the graph appears in exactly one branch of the set. The standard representation of a graph satisfies this definition, where every branch is a single arc. We call this the trivial branch decomposition.

Definition 5. *A branch decomposition of a tree is a* hierarchical tree *if: (1) there is exactly one branch connecting two leaves (called* root branch*), (2) every other branch connects a leaf to a node that is interior to another branch.*

We wish to construct a branch decomposition representing the Contour Tree of a scalar field $f : \mathcal{D} \to \mathbb{R}$, such that the endpoints of each branch (except the root)

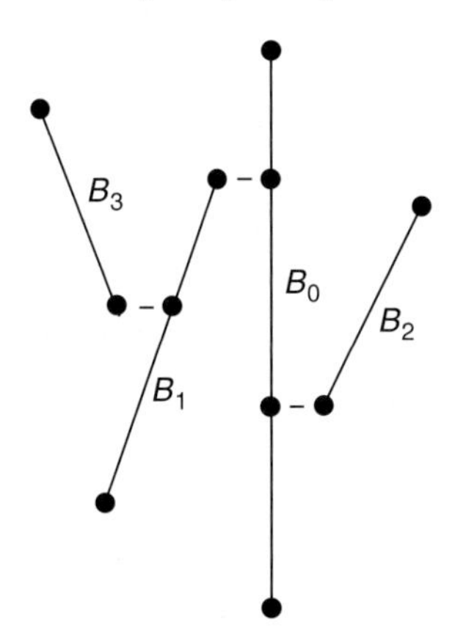

(**a**) A Contour Tree decomposed into branches. The root branch B_0 of the tree is the only one connecting two extrema. These are the only critical points of the field that cannot be canceled. B_1 is a minimum paired with a join saddle, which *cannot* be canceled before B_3. The branches B_2 and B_3 are maxima paired with split saddles. They can each be canceled independently

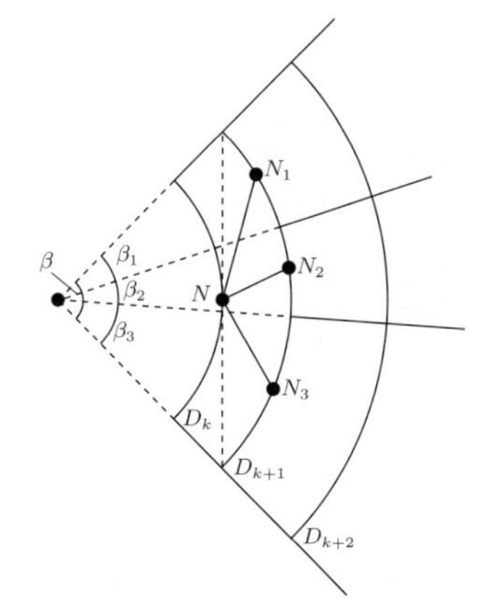

(**b**) Computation of the radial layout for the orrery interface of the Contour Tree. The diagram in figure shows the arrangement used to compute the angular wedges β_1, β_2, and β_3 for the nodes N_1, N_2, and N_3 that are children of N. This scheme is applied recursively to the branches of a hierarchical Contour Tree so that the most important branches (higher persistence) are located at the center of the scene

Fig. 4. Branch decomposition and angular arrangement used to build the hierarchical Contour Tree and its layout for presentation in a graphics user interface

represent an extremum paired with a saddle point of the scalar field. See Fig. 4a. The tree can be simplified by removing a branch that does not disconnect the tree. This corresponds to the cancellation of two critical points in the scalar field. This simplification process defines a hierarchy of cancellations where a branch $B1$ is said to be the parent of branch $B3$ if one endpoint of $B3$ is interior to $B1$. The root branch has no parent and cannot be simplified. Removal of a parent before one of its children disconnects the tree. In the next section we will discuss the construction of a branch decomposition based on the persistence of critical point pairs.

Once the decomposition is constructed and the parent–child relations are defined, we can build any approximation of the original tree by incrementally connecting child branches to their parent. In particular, we associate values to each branch for several metrics (such as persistence or geometric location) and artificially enforce a nesting condition that requires, for all the metrics, the value of the parent to be greater than or equal to the value of its children. Given a tolerance threshold for several metrics at the same time, we start from the root branch and iteratively select children with metrics above the required thresholds.

Tree Layout and Presentation

We define an embedding of the Contour Tree, which can be used as a user interface tool. The vertical coordinate-axis is fixed to represent the value of the scalar field. In doing so we lose one degree of freedom, which makes it impossible, in general, to build a planar embedding without self-intersections. Figure 3 is an example of a simple scalar field with a Contour Tree that cannot be embedded in the plane without self-intersections. Thus, in this section, we describe a three-dimensional embedding of the Contour Tree that uses the z-coordinate to represent the field value, and such that the projection of the tree onto the plane $z = 0$ has no self-intersections. We also provide a progressive construction of this embedding using the multiresolution representation given above.

Our visualization scheme can use any algorithm for the layout of rooted trees [9]. We chose a radial layout algorithm that positions the root node of the tree at the origin and positions its descendants in concentric circles.

The main idea of the layout algorithm is to define a sequence of consecutive disks, $D_1 \subset D_2 \subset D_3 \subset \cdots$, with radii $r_1 < r_2 < r_3 < \cdots$. We then compute an angular wedge at each node such that the subtree rooted at that node is contained entirely within the angular wedge. The root node is positioned at the origin and the nodes of depth k are arranged on the boundary of the disk D_k. We require the ratio of consecutive radii to be a constant, $\rho = \frac{r_{k+1}}{r_k} > 1$. This guarantees the branches will be spread out nicely. If instead we fix the difference between consecutive radii, then the ratio $\frac{r_{k+1}}{r_k} \to 1$ as k approaches ∞, and the maximal size of the angular wedges goes to 0. Thus the subtrees of nodes far away from the origin will appear to be arranged along a straight line.

Figure 4b demonstrates the algorithm for computing the angular wedge of a node N, which is on the boundary of the disk D_k. Let β be the angular wedge that has been computed for N. First, we can guarantee no self-intersections by ensuring that all arcs drawn from N to one of its children lie to the right of the tangent to the disk D_k at N. Otherwise, an arc could cross into the interior of the disk D_k, and may intersect an edge of the tree that has already been drawn. To ensure this is not the case we must restrict $\beta \leq 2cos^{-1}(\frac{r_k}{r_{k+1}}) = 2cos^{-1}(\frac{1}{\rho})$. In the figure we show the limiting case where $\beta = 2cos^{-1}(\frac{1}{\rho})$. In our implementation we use $\rho = \sqrt{2}$, thus we restrict $\beta < 2cos^{-1}(\frac{1}{\sqrt{2}}) = \frac{\pi}{2}$. However, one can see that it is only necessary to enforce this condition for the nodes on the boundary of the disk D_1, since we have chosen $\frac{r_{k+1}}{r_k}$ to be constant.

In Fig. 4b the children of the node N are the nodes N_1, N_2, and N_3. To compute the angular wedges β_i we partition the angle β proportionally to the sizes of the subtrees rooted at each node. If we let n_i be the number of leaves of the subtree rooted at N_i and n the number of leaves of the subtree rooted at N, then we have the following relations: $n = n_1 + n_2 + n_3$, $\beta = \beta_1 + \beta_2 + \beta_3$, and $n_1 : n_2 : n_3 = \beta_1 : \beta_2 : \beta_3$.

Therefore, we have that $\beta_i = \frac{n_i}{n}\beta \leq \beta$. Since $\beta < \frac{\pi}{2}$ then $\beta_i < \frac{\pi}{2}$ and we can guarantee that the subtree rooted at N_i is free of self-intersections.

To compute the embedding of a hierarchical tree we use the parent–child relationship between branches to construct a rooted tree whose nodes are the branches of the hierarchical tree. Applying the layout algorithm above to this tree produces a planar embedding, which we use for the (x, y)-coordinates of nodes in the hierarchical tree. For each branch we assign these (x, y)-coordinates to all its interior nodes and its unpaired endpoint(s). As stated above the z-coordinate of each node is assigned the function value of the corresponding critical point in the scalar field. The branches are then visualized as "L" shapes, where the base of the "L" connects the branch to its parent along a horizontal line at the height of the paired endpoint.

3 Hierarchical Morse Functions

In this section we develop the tools used to compute a hierarchical representation of the Contour Tree for a given Morse function. While we do not simplify the input Morse function we establish criteria to determine in which cases the simplified Contour Tree corresponds to a Morse function that can be constructed from the input data by cancellations of critical points.

In summary our algorithm is *robust* in the sense that for any input field it constructs a valid branch decomposition of the Contour Tree. In fact, it produces a valid branch decomposition even if the input data has a Reeb graph with loops.

Unfortunately, the simplification is guaranteed, in general, to produce simplifications that have a topological equivalent only if all the saddles in the data merge or split contours. In 3D, for example, there may be pairs of critical points that only change the genus of the contours and that may need to be canceled in pairs. To resolve this, we plan to extend our representation to allow nodes of degree two as in [15].

For simplicity of presentation we assume in the following that the domain $\mathcal{D}$ is a simply connected, compact surface.

Simplification

We develop a multiresolution framework for distinguishing fine resolution topological features from persistent, coarse resolution structures. There are two operations that are known to construct Morse functions from Morse function: cancellations and handle slides. A cancellation transforms a Morse function into a topologically "simpler" function. Handle slides are more subtle transformations the details of which are not relevant here.

Let $m \in \mathcal{D}$ be an extremum and $v \in \mathcal{D}$ be a saddle point of the Morse function f, such that there is a gradient curve connecting them. We consider the problem of defining a new Morse function f' such that m and v are regular points of f' and all other critical points of f are critical points of f'. When such an f' can be found we say that the pair of critical points (m, v) can be canceled.

A method for computing a sequence of paired critical points, called persistence pairing, was described in [10]. It is based on the definition of the persistent homology groups. The persistence of a pair, (m, v), is defined to be $|f(v) - f(m)|$. One thinks of the lower valued critical point as creating a topological feature and the greater valued one as destroying it. A hierarchy is constructed on these features by sorting them according to their persistence. This hierarchy defines an ideal sequence of simplifications. However, it is known that the critical point pairs cannot, in general, be canceled in this order. The authors introduce the notion of topological obstructions to explain why a cancellation in the sequence cannot be performed.

The algorithm we present constructs a similar hierarchy, but one that defines an order of pairs such that the next pair can always be canceled. Conceptually, we produce a sequence that guarantees that for any given pair of critical points all obstructions are canceled before canceling that pair. In this section, we prove that it is possible to construct such a sequence for any Morse function over $\mathcal{D}$, where $\mathcal{D}$ is a simply connected, closed surface.

Consider a saddle point $v \in \mathcal{D}$ with $f(v) = \omega$. Let C be the contour of the level set $L_f(\omega)$ that contains v. C is the union of two simple closed curves, called *petals*, which intersect at v and do not intersect at any other point in $\mathcal{D}$. A petal of v partitions $\mathcal{D}$ into disjoint regions. The region that contains no other petals of v is said to be enclosed by the petal.

Lemma 1 *Let $f : \mathcal{D} \to \mathbb{R}$ be a Morse function, if f has more than two critical points then it must have at least one saddle point.*

Proof. Since $\mathcal{D}$ is compact, f must have one global maximum and one global minimum. If there is another critical point, it is either a saddle (which proves the theorem) or another extremum. In the latter case the Contour Tree of f has at least three leaf nodes. Since the Contour Tree is connected there must be a node with degree three, which corresponds to a saddle point of f.

Lemma 2 *Let $f : \mathcal{D} \to \mathbb{R}$ be a Morse function, if f has more than two critical points then there exists a saddle point, $v \in \mathcal{D}$ with a petal that encloses exactly one critical point of f.*

Proof. By Lemma 1, f must have a saddle point v_0. Choose a petal of v_0 and the region M_0 enclosed by it. We assume, without loss of generality, that the descending gradient curves starting from to the boundary of $\mathcal{D}_0$ point toward its interior. Thus, there must be a local minimum, m_0, in the interior of M_0. Let f_0 be the restriction of f to $\mathcal{D}_0$. By a symbolic perturbation of the function values on boundary of $\mathcal{D}_0$ we can make v_0 a maximum of f_0 and make all the other points on the boundary of $\mathcal{D}_0$ regular points. If m_0 is the only critical point in the interior of $\mathcal{D}_0$ the theorem is proved. So assume that there are $n_0 > 1$ critical points of f in the interior of $\mathcal{D}_0$. This implies that f_0 has $n_0 + 1 > 2$ critical points, so there must be a saddle point $v_1 \in \mathcal{D}_0$. But v_0 is the only critical point of f_0 on the boundary of $\mathcal{D}_0$ so v_1 must be in the interior of $\mathcal{D}_0$ and therefore it is a saddle point of the entire function f.

Now apply the above construction to v_1 and recursively create a sequence of saddle points $v_0, v_1, \ldots \in \mathcal{D}$. Thus the corresponding sequence of regions $\mathcal{D}_i$, enclosed by the petals of the v_i, satisfy the inclusion relations $\mathcal{D}_0 \supset \mathcal{D}_1 \supset \cdots$. Finally, this implies that the numbers n_i of critical points of f in the interiors of $\mathcal{D}_i$ form a decreasing sequence, $n_0 > n_1 > \cdots$. Since there are only a finite number of critical points and $n_i > 0$ for all i, there must be some number k such that $n_k = 1$. Therefore $v = v_k$ is the required saddle point.

Lemma 3 *Let $f : \mathcal{D} \to \mathbb{R}$ be a Morse function, v be a saddle point as in Lemma 2, and m be the unique critical point enclosed by a petal of v. Then there exists a function f' that cancels the pair (m, v). Moreover, the size of the region where the sign of the gradient of f' differs from that of f can be made arbitrarily small.*

Proof. First, we distinguish between the topological condition on f' and the geometric one. The topological condition states that f' cancels the pair (m, v). The proof of this fact is a well known theorem of Morse theory and can be found in Sect. 3.4 of [14].

On the other hand the geometric condition states that we can make the region where we must change the sign of gradient flow as small as we like. This condition is slightly stronger than what is typically found in the Morse theory literature. We will demonstrate this is possible by using a triangulation of $\mathcal{D}$. Consider the region of $\mathcal{D}$ shown in Fig. 5, such a region exists by Lemma 2. Without loss of generality we assume that m is a local minimum. Let c be the steepest descending edge path from v to m. If $f(v) = \omega$ then we subdivide the mesh along the portion of the curve $L_f(\omega + 2\delta)$ shown in Fig. 5, for δ small enough. We also subdivide the mesh along the curve N_ϵ, which is defined such that the arcs with endpoints in N_ϵ and c each have length less than ϵ.

The endpoints of the portion of $L_f(\omega + 2\delta)$ drawn in the figure can be connected by following the steepest decent paths that flow into v to form a simple closed curve. Call the region bounded by this curve, $\mathcal{D}_\delta$. Similarly, we can define a simple closed curve by connecting the endpoints of N_ϵ. The region enclosed by this curve will be called $\mathcal{D}_\epsilon \subset \mathcal{D}_\delta$. We now explicitly construct f' by redefining the function values of all the vertices in the region $\mathcal{D}_\delta$ and define $f'(x) = f(x)$ for all $x \notin \mathcal{D}_\delta$. Figure 6 shows how the ranges of the vertices in $\mathcal{D}_\delta - \mathcal{D}_\epsilon$ and $\mathcal{D}_\epsilon$ are scaled. These transformations are reported here:

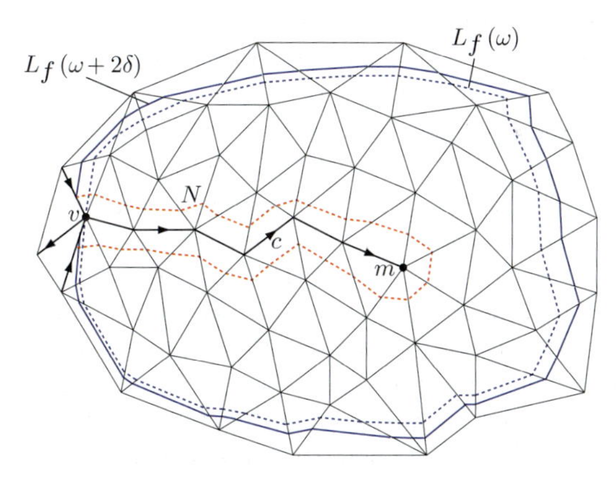

Fig. 5. Two critical points, a saddle point, v, and a minimum, m, are shown, where $f(v) = \omega$. *Bold lines with arrows* indicate the steepest descending edge paths. The steepest descent path, c, and an ϵ-neighborhood, N, of c are also shown. The region of $\mathcal{D}$ enclosed by N is the only place where the sign of the gradient must be inverted. Furthermore, we show parts of the level sets $L_f(\omega)$ and $L_f(\omega + 2\delta)$. The area inside the curve $L_f(\omega + 2\delta)$ is the only region where the function value of f must be modified

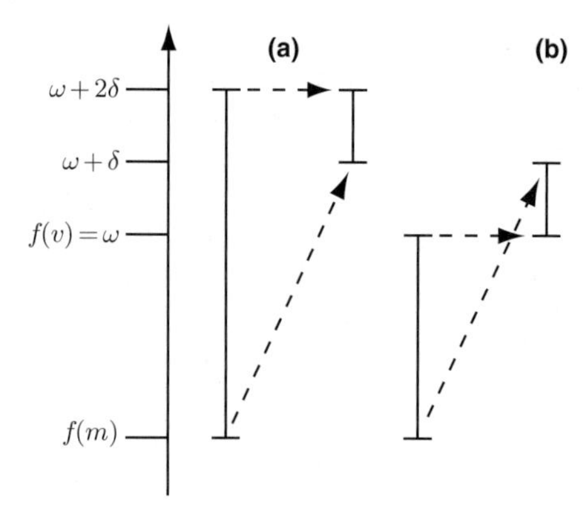

Fig. 6. A pictorial representation of the scaling factors used to construct f'. (**a**) show how the range of the vertices in the region $\mathcal{D}_\delta - \mathcal{D}_\epsilon$ are scaled. (**b**) shows how the range of the vertices in the region S_ϵ are scaled

$$f'(x) = \begin{cases} \dfrac{\delta(f(x) - f(m))}{(\omega + 2\delta) - f(m)} + (\omega + \delta) & x \in \mathcal{D}_\delta - \mathcal{D}_\epsilon \\ \dfrac{\delta(f(x) - \omega)}{f(m) - \omega} + \omega & x \in \mathcal{D}_\epsilon \end{cases}$$

The equation for $x \in \mathcal{D}_\delta - \mathcal{D}_\epsilon$ corresponds to Fig. 6a, which scales the range $[f(m), \omega + 2\delta]$ to the range $[\omega + \delta, \omega + 2\delta]$. On the other hand the range of the vertices $x \in \mathcal{D}_\epsilon$, which is $[f(m), \omega]$, is inverted and scaled to $[\omega, \omega+\delta]$, see Fig. 6b. It is easy to see that using these equations the sign of the gradient is only changed for points in the region $\mathcal{D}_\epsilon$. Since we can make $\mathcal{D}_\epsilon$ as small as we like, the theorem is proved. Furthermore, the construction demonstrates that the only region where the function value has to be modified is $\mathcal{D}_\delta$. □

4 Multiresolution Contour Trees

We present an algorithm for computing a representation of the Contour Tree that allows linear time access to simplified trees, either by uniform or adaptive simplification. Algorithms for computing the Contour Tree can be found in [5] and [15]. In both cases the algorithms first make two passes through the data to compute a join tree and a split tree. The degree three nodes of the join tree represent the saddle points where contours are merged, and the those of the split tree represent saddle points where contours are divided. These trees are then merged to construct the Contour Tree. We use the same approach to construct a hierarchical representation, however, we must store all our trees as branch decompositions and modify the algorithm that merges the join and split trees.

In addition to the basic hierarchical data structure discussed in the previous sections we take into consideration the function value of the vertices associated with each node. Thus we can sort the nodes in a branch by increasing function value. We call the first node the starting node of the branch and the last node the ending node. The length of a branch is defined to be the absolute value of the difference in function value of the endpoints. This value is returned by the function Length(B). Leaf nodes can now be classified as either minima or maxima, by checking if the node is a starting node or ending node respectively. Furthermore, we can now characterize saddle points as either join saddles or split saddles. An interior node is a join saddle if it is the ending point of some branch, but a split saddle if it is the starting point of some branch. In this characterization a join saddle corresponds to a saddle point of f where two contours merge, and a split saddle to a saddle where one contour divides.

This data structure allows us to make certain queries that we can use to determine if a branch can be simplified. Our algorithm checks the criteria for simplification in a procedure called CanSimplify(G, B), which returns true if the branch B in the graph G represents a valid cancellation. The first criterion for this to be true is that the branch must have no children. If a branch has any children then we say that the child branch is obstructing the parent branch. This condition is necessary but not sufficient for determining if a branch is able to be simplified.

Given a pair of critical points that can be canceled we always think of the first point as creating a topological feature that the second as one destroying it. For example, a minimum creates a new contour. Thus a minimum must be paired with a saddle that destroys that contour, which occurs at a join saddle. Similarly we can see that a maximum must be paired with a split saddle. So the other criterion that must be checked by CanSimplify(G, B) is that the endpoints of B are either a minimum and a join saddle or a split saddle and a maximum.

Once a tree is constructed we can perform several queries on it. First, we include the function GetTree(B) that returns the tree that contains the branch B. For an arbitrary branch decomposition it is possible to have degree 2 nodes. We can check if a node, N, has degree two with the function IsRegular(T, N). If a node is a starting point we can perform the query UpBranch(T, N), which returns the branch that starts at the node. Likewise, we can call DownBranch(T, N) on ending points to access the branch that ends at the node. If CanSimplify(B) returns true for a branch B, then exactly one of it endpoints represents a saddle point. In this case we can access the unique saddle point of the branch by calling GetSaddle(B). Finally, a branch is defined to be a leaf branch if it has no interior nodes and one of its endpoints is a leaf node. The function IsLeafBranch(T, E) returns true if E is a leaf branch.

Join and Split Trees. Any of the standard algorithms for computing the join and split trees can be implemented, but the resulting trees must be stored as trivial branch decompositions. In these algorithms every node in each tree represents a critical point. Thus there will be some degree two nodes in each tree, which correspond to saddle points from the other tree.

For completeness we briefly describe the algorithm for constructing the join and split trees that given in [5]. However, this algorithm has been improved upon in [15]. First, the vertices of $\mathcal{D}$ are sorted by function value. The idea is then to keep track of a Union-Find data structure as one sweeps through the vertices in increasing and decreasing order. During the increasing sweep we build the join tree and during the decreasing sweep we build the split tree. We present an algorithm for computing the join tree and describe the differences in the split tree algorithm.

In the pseudo-code given in Table 1 (left) we make use of a simple Union-Find data structure that includes the functions NewUF(), NewSet(UF, i), Find(UF, i), and Union(UF, i, j). These functions respectively create the data structure, add a new class, return the class containing a given index, and merge two classes. Finally, we require the boolean functions IsMin(v) and IsCritical(v) that return true if v is a local minimum of f and a critical point of f respectively (v is the only argument since $f(v)$ is implicitly determined by its order in a sorted array.)

The algorithm for constructing the split tree, ST, is almost identical, the only differences are: on line 3 the *for* statement goes from $n-1$ to 0, on line 8 the test IsMin(v_i) is replace by the test IsMax(v_i), and on line 9 the edges with $j > i$ are considered. When the ST is constructed in this manner the start of each branch has greater function value than the end. So it is also necessary to reverse the direction of all the branches in the ST in order to make all the saddle points in ST split saddles. This can be done in a subroutine and either included in the algorithm or as a postprocessing step before constructing the contour tree.

Computing the Multiresolution *CT*

In previous Contour Tree algorithms the Contour Tree, CT, is constructed form the JT and ST by "peeling off" leaves of the JT and ST and adding them to the CT. This approach uses a queue to store the leaves during processing, which can be removed in any order. Our algorithm uses the same approach, however, we must impose a strong condition on the order in which the leaves are "peeled off." We enforce this condition by using a priority queue such that we always remove the next shortest leaf branch that represents a valid simplification. Once a branch is removed

Table 1. Pseudocode for the computation of JoinTree and PopValid

Algorithm 1. JoinTree
Input: Sorted array of n vertices ($\{v_i\}$) and a triangulated surface ($\mathcal{D}$).
Output: Join tree (JT).

1. JT = NewGraph()
2. UF = NewUF()
3. ***for*** $i = 0$ ***to*** $n-1$ ***do:***
4.   ***if*** IsCritical(v_i) ***then*** AddNode(JT, i)
5.   ***if*** IsMin(v_i) ***then*** NewSet(UF, i)
6. i' = Find(UF, i)
7.   ***for each edge*** $v_i v_j$ ***with*** $j < i$ ***do:***
8. j' = Find(UF, j)
9.   ***if*** $j' \neq i'$ ***then*** AddBranch(JT, j', i')
10.   Union(UF, i', j')
11. ***return*** JT

Algorithm 2. PopValid
Input: Priority Queue (PQ) of branches.
Output: Branch (B) that is a valid simplification.

1. B = Pop(PQ)
2. $isValid = false$
3. ***while not*** $isValid$ ***do:***
4.   ***if not*** CanSimplify(B) ***then:***
5. B = Pop(PQ)
6.   ***else:***
7.   ***if*** Length(B) $\neq$ Priority(B) ***then:***
8.   Push(PQ, B)
9. B = Pop(PQ)
10.   ***else:***
11. $isValid = true$
12. ***return*** B

from the queue the adjacent branches in the JT and ST are merged, which is why JT and ST must be stored as branch decompositions. These merges can change the length of branches that are already in the queue. Thus one of the major difficulties in the algorithm is maintaining a valid priority queue. We do this on the fly by simply checking if the top branch of the queue is valid. The condition that must be checked is complex enough to warrant a subroutine, so we present the routine PopValid(PQ).

There are two possibilities for why the branch at the top of the priority queue is not valid. First of all, the current length of the branch might not be the same as its length when it was entered into the queue. It is possible that the branch was merged with another one, so it could be longer, thus it might have a lower priority than some other branch in the queue. In this case we simply return the branch to the queue with its new priority. Additionally, it might be the case that for the top branch, B, CanSimplify(B) returns false. This can come about by removal and merging of branches as well. In this case the branch has become invalidated in a more essential way and we simply remove it from the queue altogether.

The priority queue is a standard data-structure that uses the operations: Pop(PQ) and Push(PQ, B), that retrieve the top element of the queue, and push a branch onto the queue respectively. It also supports the test IsEmpty(PQ) that returns true if there are no elements in the queue. In our case the priority is the length of a branch, B, and Pop(B) is guaranteed to return the branch with the lowest priority. The priority of a branch, B, when it was entered into the queue can be queried using the function Priority(B). The complete pseudocode for PopValid is reported Table 1 (right).

The procedure PopValid(PQ) ensures that we can pull the first branch that represents a valid cancellation from the queue. In this way we can ensure that each branch represents a topological simplification of f. It was proved in the previous section that we can always find a simplification, thus PopValid(PQ) will always return a value as long as the priority is not empty.

For readability we introduce another subroutine of the Contour Tree algorithm that does the work of "peeling off" a leaf branch. In this routine we make use of the function MergeBranches(B_1, B_2) that merges the branches B_1 and B_2 into the single branch B_1. The pseudocode of PeelOffBranch is reported in Table 2.

Using the subroutines PopValid(PQ) and PeelOffBranch(B, JT, ST) the code for our main algorithm, BuildContourTree(JT, ST), is relatively simple and is reported in Table 3.

It is clear from the discussion in this and the previous section that this algorithm produces a multiresolution Contour Tree, such that each branch represents a valid topological simplification. We can now define an order on the branches that allows one to extract a Contour Tree after any number of simplifications in linear time. First, we define the persistence of a branch to be the greater of its length and the persistence of each of its children. This definition differs from the definition of persistence given in [10] because it takes into consideration the topological obstructions. Thus a pair of critical points is never assigned a persistence value that is less than any of its obstructions.

Table 2. Pseudocode for the computation of PeelOffBranch

Algorithm 3. PeelOffBranch
Input: Branch (B), Join Tree (JT) and Split Tree (ST).
Output: A branch representing a valid simplification or *null*.

```
1.  XT = WhichTree(B)
2.  N = GetSaddle(B)
3.  RemoveBranch(XT, B)
4.  if IsRegular(JT, N) then:
5.      B1 = DownBranch(JT, N)
6.      B2 = UpBranch(JT, N)
7.      MergeBranches(B1, B2)
8.      if XT == ST and CanSimplify(B1) then:
9.          return B1
10. if IsRegular(ST, N) then:
11.     B1 = UpBranch(ST, N)
12.     B2 = DownBranch(ST, N)
13.     MergeBranches(B1, B2)
14.     if XT == JT and CanSimplify(B1) then:
15.         return B1
16. return null
```

Table 3. Pseudocode for the computation of BuildContourTree

Algorithm 4. BuildContourTree
Input: Join Tree (JT) and Split Tree (ST).
Output: Contour Tree (CT).

```
1.  CT = NewGraph
2.  PQ = NewPQ
3.  for each B ∈ JT do:
4.      if IsLeafBranch(B) and CanSimplify(B) then:
5.          Push(PQ, B)
6.  for each B ∈ ST do:
7.      if IsLeafBranch(B) and CanSimplify(B) then:
8.          Push(PQ, B)
9.  while not IsEmpty(PQ) do:
10.     B_top = PopValid(PQ)
11.     AddBranch(CT, B_top)
12.     B_next = PeelOffBranch(B_top, JT, ST)
13.     if not B_next ≠ null then Push(PQ, B_next)
14. return CT
```

The same analysis as in [11] can be used to show that the complexity of BuildContourTree is $O(n \log n)$ where n is the number of nodes in JT and ST. Using a simple FIFO queue instead of a priority queue would yield a complexity of $O(n)$ but with the risk of building an unbalanced tree. In practice this does not seem to be a problem and a linear queue may be advisable for a faster implementation with lighter data-structures.

5 Output Sensitive Visualization

In this section we demonstrate the use of the multiresolution Contour Tree by showing how to produce output sensitive visualizations. We visualize the Contour Tree by embedding the nodes of the tree at the location of the corresponding critical points in 3D space. Arcs are drawn as straight line segments connecting the endpoints. Thus a branch is drawn as a chain of connected arc segments. Since the branches are sorted by their persistence we always draw the branches with greater persistence first (see Figs. 7 and 8).

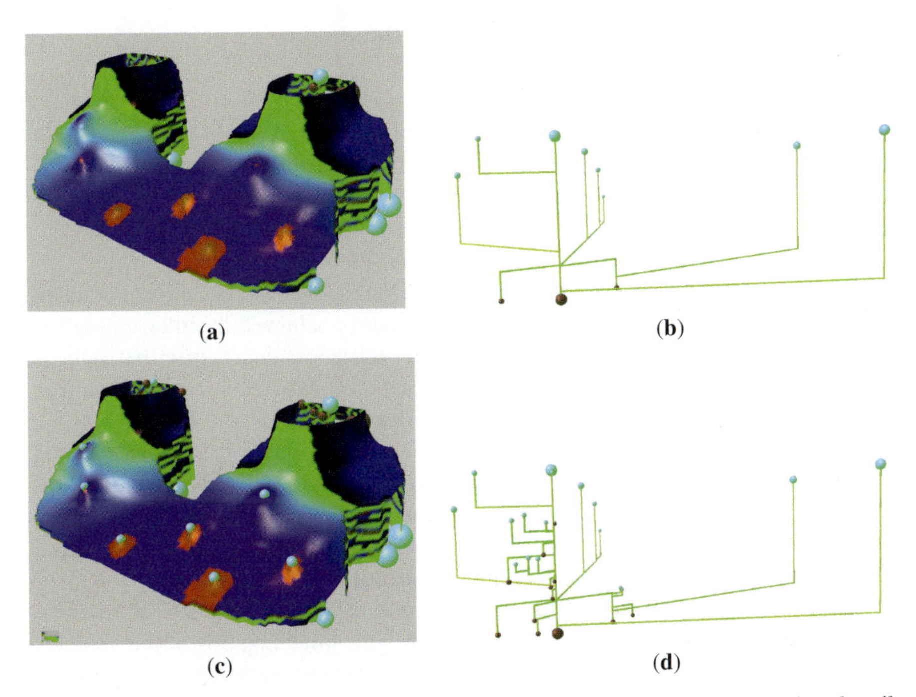

Fig. 7. Isosurface of water saturation with a function (pseudocolored) representing the oil pressure in an oil reservoir. (**a–b**) Surface with critical points and the Contour Tree for a persistence threshold of 65% of persistence. Only the shape of the surface dictates the topology of the function. (**c–d**) Persistence filter reduced to 35% introducing the seven critical points corresponding to the oil wells drilling in the reservoir

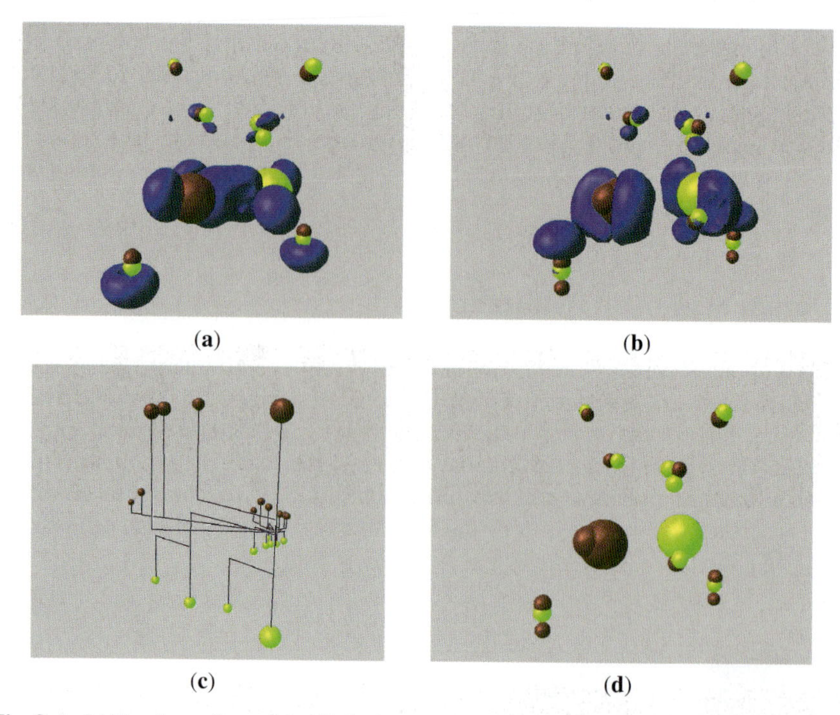

Fig. 8. (**a**, **b**) Two isosurface of the Hipip dataset together with the topology filtered by 53% of persistence. (**c**) Contour Tree of the simplified topology. (**d**) simplified topology in the original embedding

The branch decomposition representation of the CT allows for uniform or adaptive refinement of the tree. Uniform simplification is achieved by interrupting the drawing process when the first branch with persistence less than a specified value is reached. Adaptive simplification is almost as easy, before each branch is visualized it is tested to see if it satisfies the adaptive criterion (see Figs. 9, 10, 11, and 5).

We demonstrated the adaptive visualization of the Contour Tree using a spatial criterion. The criterion we use in our examples is a simple test if the bounding box of a branch intersects a given bounding box that describes a region of interest. The bounding box of a branch, B, is defined to be the bounding box of the region in space that contains all the critical points that must be canceled in order to simplify B.

The bounding boxes of each branch in the multiresolution Contour Tree are computed by a simple recursive algorithm that merges the bounding box of the branch's endpoints with the bounding boxes of each of its child branches. After the bounding boxes have been computed the user is allowed to select a region of interest by manipulating a bounding box in the embedding space. We can now adaptively extract a CT that has a greater level of detail in the region of interest by only visualizing those branches whose bounding boxes intersect the user defined box.

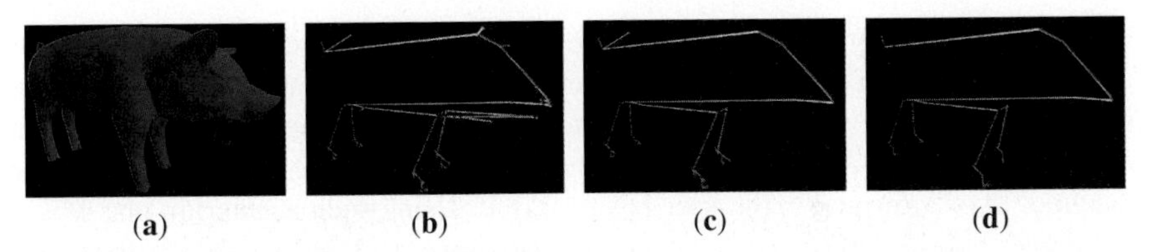

Fig. 9. (**a**) A pig model. (**b**) The Contour Tree of the pig, where the function is the displacement along the vertical axis. (**c**) Simplified tree with persistence value 0.02. (**d**) Simplified tree with persistence value 0.05

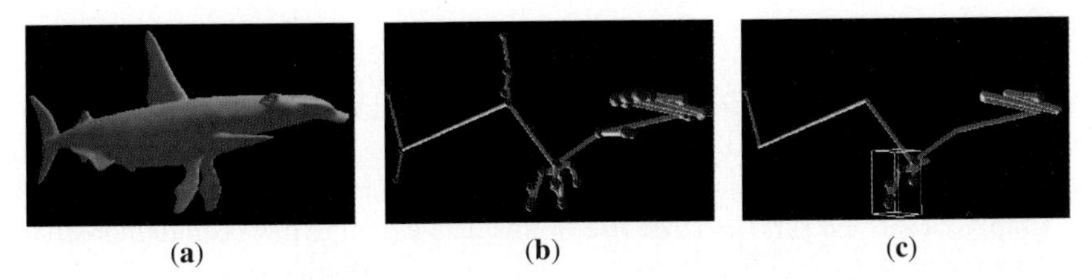

Fig. 10. (**a**) A shark model. (**b**) The Contour Tree of the shark, where the function is the displacement along the horizontal axis. (**c**) Adaptively refined tree showing detail around the shark's fin

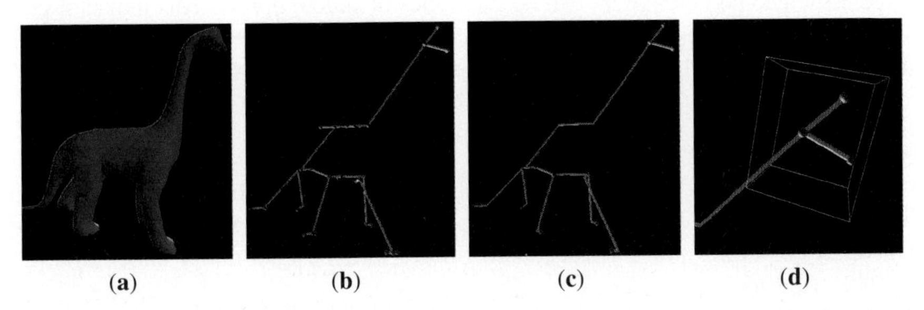

Fig. 11. (**a**) A dinosaur model. (**b**) The Contour Tree of the dinosaur, where the function is the displacement along the vertical axis. (**c**) Simplified tree with persistence value 0.02. (**d**) Adaptively refined tree showing detail around the dinosaur's head

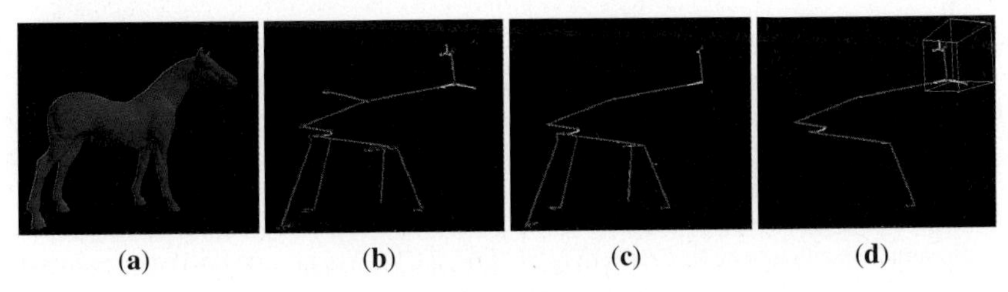

Fig. 12. (**a**) A horse model. (**b**) The Contour Tree of the horse, where the function is the displacement along the vertical axis. (**c**) Simplified tree with persistence value 0.02. (**d**) Adaptively refined tree showing detail around the horse's head

Although this test is very simple it demonstrates a wide range of possibilities for user interaction with the CT. Adaptive refinement becomes increasingly important when one begins to visualize large Contour Trees. In this case a view-dependent criterion would be useful in speeding up the rendering of the tree, as well as reducing the amount of information the user has to take it. Using such a criterion one would not render branches if they are outside the viewing area, too far away, or too small to see.

6 Conclusions

We have provided a data structure for representing multiresolution Contour Trees. We have presented a robust algorithm for constructing these trees for general domains. Guaranteed topological correctness is proved for piecewise-linear functions on simply-connected surfaces. The use of this data structure has been demonstrated by showing how to adaptively extract output sensitive Contour Trees.

We plan to extend the work on proving topological correctness for general volumetric domains. These datasets pose a problem because it is more difficult to determine cancellations for critical points that do not create junctions or bifurcations. In three-dimensional fields there can be two types of saddle points. This makes it possible for the two types of saddle points to form a pair than can be canceled, however, it is difficult to test if a pair of this type can be canceled.

Acknowledgments

This work was performed under the auspices of the US Department of Energy by University of California, Lawrence Livermore National Laboratory under Contract W-7405-Eng-48.

References

1. C. L. Bajaj, V. Pascucci, and D. R. Schikore. The contour spectrum. In Roni Yagel and Hans Hagen, editors, *IEEE Visualization 97*, pages 167–175. IEEE, November 1997.
2. P.-T. Bremer, H. Edelsbrunner, B. Hamann, and V. Pascucci. A multi-resolution data structure for two-dimensional Morse functions. In *Proceeding of IEEE Conference on Visualization*, pages 139–146, October 2003.
3. H. Carr and J. Snoeyink. Path seeds and flexible isosurfaces – using topology for exploratory visualization. In *Proceeding of IEEE TCVG Symposium on Visualization (VisSym '03)*, pages 49–58, Grenoble, Fr, May 2003.
4. H. Carr, J. Snoeyink, and U. Axen. Computing contour trees in all dimensions. In *Proceedings of the eleventh annual ACM-SIAM symposium on Discrete algorithms*, pages 918–926, January 2000.

5. H. Carr, J. Snoeyink, and U. Axen. Computing contour trees in all dimensions. *Computational Geometry Theory and Applications*, 24(2):75–94, February 2003. To appear (extended abstract appeared at SODA 2000).
6. H. Carr, J. Snoeyink, and M. van de Panne. Simplifying flexible isosurfaces using local geometric measures. In *IEEE Visualization*, pages 497–504, October 2004.
7. K. Cole-McLaughlin, H. Edelsbrunner, J. Harer, V. Natarajan, and V. Pascucci. Loops in reeb graphs of 2-manifolds. In *ACM Symposium on Computational Geometry*, pages 344–350, July 2003.
8. K. Cole-McLaughlin and V. Pascucci. Multiresolution representation of topology. In *Proceedings of the 4th IASTED International Conference on Visualization, Imaging, And Image Processing (VIIP 2004)*, pages 282–289, Marbella, Sapin, September 2004.
9. G. di Battista, P. Eades, R. Tamassia, and I. G. Tollis. *Graph Drawing: Algorithms for the Visualization of Graphs*. Prentice-Hall, Upper Saddle River, NJ, 1999.
10. H. Edelsbrunner, D. Letscher, and A. Zomorodian. Topological persistence and simplification. In *Proceeding of The 41st Annual Symposium on Foundations of Computer Science*. IEEE, November 2000.
11. A. T. Fomenko and T. L. Kunii, editors. *Topological Modeling for Visualization*. Springer, Tokyo, 1997.
12. M. Hilaga, Y. Shinagawa, T. Komura, and T. L. Kunii. Topology matching for full automatic similarity estimation of 3d shapes. In *Proceedings of ACM SIGGRAPH 2001*, pages 203–212, August 2001.
13. E. Kleiberg, H. van de Wetering, and J. J. van Wijk. Botanical visualization of huge hierarchies. In *Proceedings IEEE Symposium on Information Visualization (InfoVis'2001)*, pages 87–94, 2001.
14. Y. Matsumoto. *An Introduction to Morse Theory*. AMS, 1997.
15. V. Pascucci and K. Cole-McLaughlin. Efficient computation of the topology of the level sets. In *IEEE Visualization*, pages 187–194, October 2002.
16. G. Reeb. Sur les points singuliers d'une forme de pfaff completement integrable ou d'une fonction numerique. *Comptes Rendus Acad. Sciences Paris*, 222:847–849, 1946.
17. P. J. Scott. *An Application of Surface Networks in Surface Texture*, chapter 11: Efficient contour tree and minimum seed set construction, pages 157–166. John Wiley & Sons, May 2004.
18. Y. Shinagawa, T. L. Kunii, H. Sato, and M. Ibusuki. Modeling the contact of two complex objects: With an application to characterizing dental articulations. *Computers and Graphics*, 19:21–28, 1995.
19. Y. Shinagawa and T. L. Kunii. Constructing a Reeb graph automatically from cross sections. *IEEE Computer Graphics and Applications*, 11:44–51, November 1991.
20. S. Takahashi, G. M. Nielson, Y. Takeshima, and I. Fujishiro. Topological volume skeletonization using adaptive tetrahedralization. In *Proceeding of Geometric Modeling and Processing*, pages 227–236, 2004.
21. S. Takahashi, Y. Takeshima, and I. Fujishiro. Topological volume skeletonization and its application to transfer function design. *Graphical Models*, 66(1):24–49, 2004.
22. A. Telea and J. J. van Wijk. Simplified representation of vector fields. In IEEE Computer Society Press, editor, *Proceedings of the IEEE conference on Visualization '99*, pages 35–42, 1999.
23. X. Tricoche, G. Scheuermann, and H. Hagen. A topology simplification method for 2d vector fields. In *Proceedings of the IEEE conference on Visualization*, pages 359–366. IEEE Computer Society Press, 2000.
24. E. R. Tufte. *Envisioning Information*. Graphics Press LLC, Cheshire, CT, 1990.

25. M. van Kreveld, R. van Oostrum, C. Bajaj, V. Pascucci, and D. Schikore. Contour trees and small seed sets for isosurface traversal. In *Proceedings of the 13th International Annual Symposium on Computational Geometry (SCG-97)*, pages 212–220, June 1997. Extended version. Techincal report UCRL-JC-132016 Lawrence Livermore National Laboratory.
26. G. H. Weber and G. Scheuermann. *Automating Transfer Function Design Based on Topology Analysis*, chapter IV:5, pages 293–308. Mathematics and Visualization. Springer, Berlin, 2004.

Isocontour Based Visualization of Time-Varying Scalar Fields

Ajith Mascarenhas and Jack Snoeyink

Department of Computer Science, University of North Carolina at Chapel Hill, Chapel Hill, NC, USA
ajith@cs.unc.edu, snoeyink@cs.unc.edu

Summary. Time-varying scalar fields are produced by measurements or simulation of physical processes over time, and must be interpreted with the assistance of computational tools. A useful tool in interpreting the data is graphical visualization, often through *level sets*, or *isocontours* of a continuous function derived from the data. In this paper we survey isocontour based visualization techniques for time-varying scalar fields. We focus on techniques that aid selection of meaningful isocontours, and algorithms to extract chosen isocontours.

1 Introduction

Physical processes that are measured over time, or that are modeled and simulated on a computer, can produce large amounts of time-varying data that must be interpreted with the assistance of computational tools. Such data arises in a wide variety of studies including computational fluid dynamics [15], oceanography [6], medical imaging [60], and climate modeling [46]. The data typically consists of finitely many points in space–time and measured or computed values for each point. Often the values are scalar, with perhaps several scalar values for each sample point. E.g., Pressure, temperature, density. A study of motion or velocity of some kind will result in vector-valued data. In this paper, we focus on scalar-valued data, often called *scalar fields*. Irrespective of how the points are sampled, we can connect them into a mesh and interpolate the values to obtain a continuous function over the entire domain. Piecewise-linear interpolation is common for large amounts of data, because of its relative ease; multilinear interpolation is also used for regular grids.

The goal is to understand the data, usually by exploring it for important features. A medical researcher might be interested in tumors, while a climatologist might be interested in regions of high pressure. Because humans possess a highly developed visual system, transforming the data into images and movies that can be displayed, and providing the scientist with tools to control them can be a powerful *visualization* method [47]. Popular techniques employed to create such visualizations are *direct volume rendering*, *slicing*, and *isocontouring*. Direct volume rendering employs two classes of algorithms to display all the data: image-space projection and

T. Möller et al. (eds.), *Mathematical Foundations of Scientific Visualization, Computer Graphics, and Massive Data Exploration*, Mathematics and Visualization,
DOI: 10.1007/978-3-540-49926-8,

volume-space projection. In image-space projection algorithms we cast rays into the three-dimensional volumetric data, map the data at each volume element to a user determined color and opacity value, accumulate these values in front-to-back order and display them on screen [38, 39, 41]. In volume-space projection algorithms, we traverse the volume, compute the color and opacity contribution for each volume element and project it onto the image screen [32, 63]. Direct volume rendering displays all the data simultaneously and requires recomputing the image for each new viewing direction. On the other hand, techniques for computing slices and isocontours (also called level sets) of the data have been used to reduce three-dimensional volumetric data to a two-dimensional form suitable for interactive display. In slicing, we restrict the data to a suitable plane, and display this restriction on a computer screen. By mapping values to colors, we can view the variation on the plane, and by varying the plane of slicing we can explore the variation of data in space. In isocontour based visualization, we fix a scalar value s, compute the points in space with that value and display the results [43]. By varying s we can explore the variation in the data.

Because time-varying data is four-dimensional, we restrict it along two dimensions; one choice is to create time-slices to study temporal behavior, and isocontouring within the time-slice to study variation over space. Although there can be other choices for dimension reduction, most of the research on time-varying visualization use this choice. This is perhaps because visualization of volumetric data preceded that of time-varying data, and a natural step to apply existing algorithms is to consider each time-step as a static volume.

In this paper we survey theory and algorithms for isocontour-based visualization for time-varying scalar fields. We are guided by two goals. The first, given slicing and isovalue parameters extract isocontours for display. The second, and in our opinion more interesting goal, is find "interesting" parameters to aid the scientist in the visualization.

Because visualization is an interactive process, algorithms for isocontour extraction focus on efficiency. Moreover time-varying scalar fields are usually very large and may not fit into physical memory. Therefore reducing storage overhead and improving I/O efficiency are also important.

To help us address the second goal we pose some questions: How do the different components of an isocontour interact as the time and isovalue are continuously varied? At what time and isovalues do new components appear, disappear, merge, split or change genus? What tools exist that can provide this topological information without actually computing each possible isocontour? As we will see in this survey, some of the answers to these questions have resulted in algorithms to compute topological structures such as the Reeb graph that are useful aids in visualization. The Reeb graph encodes isocontour topological features such as number of components, component merge, split and genus change. We will survey algorithms for the Reeb graph of static functions and its extension to time-varying functions. These algorithms are based on the mathematical fields of topology and Morse theory, and work on the real-world piecewise linear approximations of the smooth spaces required by the theory. We see this trend as an important development; there is a rich and well developed body of work in these mathematical fields that visualization research can benefit from.

Outline

This survey is structured into two parts. The first part, consisting of Sect. 2, reviews algorithms for extracting isocontours from time-varying data. The second part, consisting of Sects. 3 and 3.3, considers topological structures that aid the visualization process by presenting high dimensional data in a succinct visual form. Our focus is mainly on the Reeb graph, its extension to time-varying data, and algorithms to compute this extension.

2 Isocontour Extraction

In this section we present techniques for isocontour extraction from time-varying scalar fields. Because most research on isocontour extraction from time-varying scalar fields builds on the techniques for static fields, we briefly survey them first.

The volumetric data consists of finite point samples of a scalar function $f : \mathbb{R}^3 \to \mathbb{R}$. The sample points are connected into a mesh. We refer to the sample points as *vertices* and each mesh element as a *cell*. The function value is extended to the entire domain by interpolation. With linear or multilinear interpolation, the *min–max interval* for each cell c, is the interval $[minval(c), maxval(c)]$, where $minval(c)$, and $maxval(c)$ are the minimum and maximum scalar values at the vertices of c.

Research in isocontour extraction arose in the medical imaging field, with the use of sampled medical images to extract boundaries of organs [2, 28, 33]. In the graphics and animation community, isocontours were used to model and render implicit surfaces [8], and for modeling and animating objects [65, 66].

Lorensen & Cline [43] introduce the *marching cubes* algorithm for isocontour extraction from data sampled regularly and connected into a cubical mesh. The algorithm iterates through all cubes in the volume, and extracts a piece of the isocontour from each cube. The marching cubes algorithm has a flaw [22]; it sometimes generates isocontours with holes. Subsequent research provide algorithms that correct this problem [30, 45, 49, 51, 52]. Although simple, the marching cubes algorithm can be inefficient because it examines the entire volume, while an isocontour may intersect only a small fraction of this volume. Techniques that speed-up isocontour extraction build search structures to efficiently find cells intersecting the isocontour, and can be classified into spatial techniques, span space techniques, and topological techniques.

2.1 Spatial Techniques

Spatial techniques subdivide the volume using a octree based hierarchy [64]. Each octant is equipped with the min–max interval of function values contained in it, enabling a search through the octree hierarchy to terminate at the octant if the query isovalue does not belong to the interval. This technique works because most data sets have large regions with function values that are distributed close together and can be quickly pruned in a search. Next we look at an extension of the octree technique to time-varying data.

Temporal Branch-on-Need Tree

The Temporal Branch-on-Need Tree (T-BON) [59] extends the three-dimensional branch-on-need octree for time-varying isocontour extraction. It aims at efficient I/O while retaining the accelerated search enabled by a hierarchical subdivision. Unlike the algorithms in Sects. 2.2 and 2.3 this algorithm does not exploit temporal coherence and considers each time-step as a static volume.

Construction

The input to the T-BON construction is regularly sampled points in space–time. The output is a branch-on-need octree for each time step, stored on disk in two sections. Store information common to all trees, such as branching factor, pointers to children and siblings, once for all trees. Store the min–max intervals associated with nodes of an octree separately as a linear array, one per time-step, each packed in depth-first order.

Search and Extraction

The input is a query: Extract the isocontour for isovalue s at time t. The output is the isocontour represented as a triangulated mesh. As a first step, read the octree infrastructure from disk and re-create it in main memory. Resolve the query using the octree for t and demand-driven paging [19]. Read the min–max interval of the root from disk, and if s belongs to this interval read the child nodes. Proceed recursively, stopping at the leaf nodes. If the min–max interval of a leaf contains s add the disk block containing the corresponding cells values onto a list. After completing the octree traversal read all disk blocks from the list into memory. Repeat the tree traversal, this time extracting the isocontour using the cell data read from disk.

2.2 Span-Space Techniques

Livnat et al. [42] define the *span space* as the two-dimensional space spanned by the minimum and maximum values of the cells of the volume. A cell c with minimum value min_c and maximum value max_c maps to a point (min_c, max_c) in the span space. See Fig. 1. A variety of search structures on the span space have been used to speed-up finding cells that intersect an isocontour. Gallagher [29] uses bucketing and linked lists, Livnat et al. [42] use k-d trees [7], van Kreveld [61] and Cignoni et al. [16] use the interval tree for two- and three-dimensional data respectively. Chiang et al. [14] use a variant of the interval tree that enables out-of-core isocontour extraction, and use the algorithm of Sect. 2.2 to extend this work to time-varying data [13]. Unlike the spatial techniques that use the octree, which requires regularly gridded data, span space techniques have the advantage of being also applicable to irregularly sampled data. Next we look at an algorithm that uses span space techniques for isocontour extraction from time-varying data.

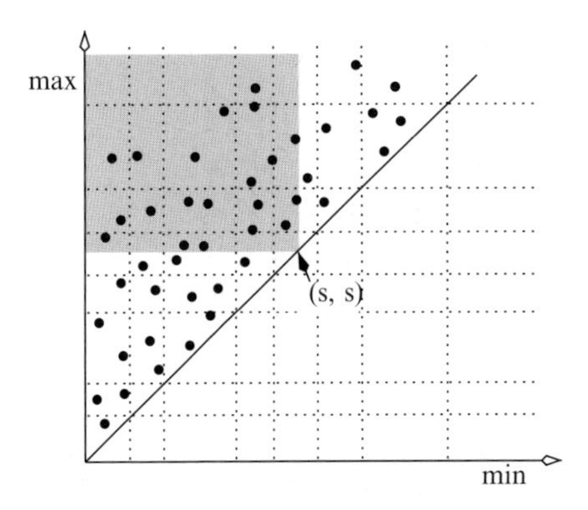

Fig. 1. Points in *shaded area* of span space correspond to cells that intersect isocontour for value s

Temporal Hierarchical Index Tree

Shen's algorithm [55] for the Temporal Hierarchical Index (THI) tree analyzes the span space of the time-varying data, and classifies and stores each cell in one or more nodes of a binary tree based on the temporal variation of its values. By placing a cell possessing a predefined small temporal variation in a single node of the THI tree, along with a conservative min–max interval of its variation over a large time span, this algorithm achieves savings in space. Cells with greater temporal variation are stored at multiple nodes of the tree multiple times, each for a short time span.

Temporal Variation

Shen uses the span space to define the temporal variation of a cell's values. The area over which the points corresponding to a cell's min–max values over time are spread out give a good measure of its temporal variation; the larger the area of spread the greater the variation. In particular, subdivide the span space into $\ell \times \ell$ nonuniformly spaced rectangles called *lattice elements* using the *lattice subdivision* scheme [56]. To perform the subdivision, lexicographically sort all extreme values of the cells in ascending order, find $\ell + 1$ values to partition the sorted list into ℓ sublists, each with the same number of cells. Use these $\ell + 1$ values to draw vertical and horizontal lines to get the required subdivision. Note that this subdivision does not guarantee that each lattice element has the same number of cells. Given a time interval $[i, j]$, a cell is defined to have low temporal variation in that interval if its $j - i + 1$ min–max interval points are located over an area of 2×2 lattice elements.

Construction

The input of the THI algorithm is a fixed mesh whose vertices are points in space. Each point has a data value for each time step in $[0, T - 1]$. Each cell has T corresponding min–max intervals. The output of the THI algorithm is a binary tree constructed as follows. In the root node N_0^{T-1}, store cells whose min–max intervals

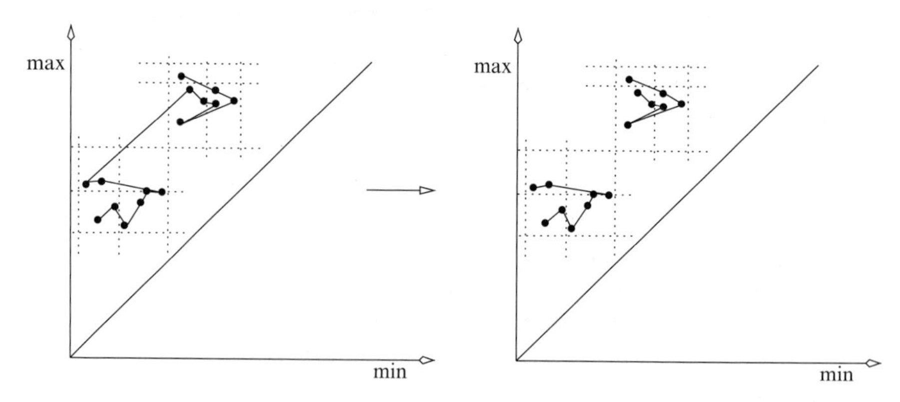

Fig. 2. The min–max intervals of a cell over a time interval are shown in the span-space as points with a path connecting them in order of time. The points on the *left* are spread outside a 2×2 lattice area. On breaking the time interval into two halves, on the *right*, the respective points fall inside a 2×2 area

have low temporal variation in the time interval $[0, T-1]$. The root has two children, $N_0^{T/2}$ and $N_{T/2+1}^{T-1}$ defined recursively on cells that are not stored in the root. Recursion stops at leaf nodes N_t^t, with $t \in [0, T-1]$. Cells that fall into leaf nodes have the highest temporal variation. See Fig. 2.

Represent each cell that falls into an internal node N_i^j by a conservative min–max interval, called the *temporal extreme values*, which contains all the cells min–max intervals for the time span $j-i+1$. Because the temporal extreme values are used to refer to a cell for more than one time step, we get a reduction in the overall index size.

Within each tree node, organize the cells using one of the span-space based techniques; Shen uses a modified ISSUE algorithm [56]. Use the lattice subdivision scheme described above, and sort cells that belong to each lattice row, excluding the cells in the diagonal lattice element, in ascending order, based on their minimum temporal extreme value. Similarly, sort the cells into another list, in descending order, based on their maximum temporal extreme value. Build an interval tree, as in [16], for cells in the lattice elements along the diagonal.

Search and Extraction

As in the T-BON algorithm, the input is a query: Extract the isocontour for isovalue s at time t, and the output is the isocontour represented as a triangulated mesh. First collect all nodes in the THI-tree whose time span contains t. Traverse the tree, starting at the root node. From the current node, visit the child N_i^j, with $i \leq t \leq j$, stopping at leaf node N_t^t.

At each node in the traversal path, use the lattice index structure to locate candidate isocontour cells. Locate the lattice element that contains the point (s, s). Because of the symmetry of the lattice subdivision this element is the kth row in the kth

column, for some integer k. The isocontour cells are contained in the upper-left corner bounded by the lines $x = s$ and $y = s$, as shown in Fig. 1. Collect these cells as follows:

- From each row $r = k + 1$ to $\ell - 1$, collect cells from the beginning of the list sorted on the minimum temporal extreme value until the cell whose minimum is greater than s.
- From row k, collect cells from the beginning of the list sorted on the maximum temporal extreme value until a cell whose maximum is lesser than s.
- From the lattice element containing (s, s), collect cells by querying the interval tree.

Recall that because we store a conservative min–max interval with cells, some collected cells may not actually intersect the isocontour. For all candidate cells, read the actual data at time t to extract the isocontour.

2.3 Topological Techniques

Topological techniques for efficient isocontour extraction typically analyze the data in a preprocess step, and compute a subset of cells called the *seed set*. A seed set contains at least one cell, called a *seed*, intersecting each component of each isocontour. To extract an isocontour first search the seed set, which is stored in an appropriate search structure, to find a seed for each connected isocontour component. To extract each component, start at the seed and visit all intersecting cells by a breadth first search of the mesh. This method of extraction, by performing a breadth first search of the mesh, is called *continuation* by Wyvill et al. [65], *mesh propagation* by Howie & Blake [34], and *contour-following* by [10]. Next we look at an algorithm that uses topological techniques for isocontour extraction from time-varying scalar fields.

Progressive Tracking

Bajaj et al. [4] extend seed set based techniques to time-varying data. They use temporal coherence to compute an isocontour at time step $t+1$ by modifying the isocontour computed at the time step t. New components at $t + 1$ are separately computed from the seed set for that time step. Unlike the T-BON and THI algorithms they accept a range of time-steps and a single isovalue as arguments, and track components over the range of time-steps.

Construction

The input can be a fixed regular or irregular mesh whose vertices are points in space. Each point has a data value for each time step in $[0, T-1]$. The output is a collection of T seed sets, one for each time step. Treat each time-step as a static scalar field and compute a seed-set using one of the algorithms proposed in [10, 62]. Organize each seed-set in an interval tree [16].

Search and Extraction

The query for this algorithm is different from the T-BON and THI algorithms. Find isocontours for isovalue s over the time range $[t_0, t_1]$. To extract an isocontour, first extract all isocontour components at time t_0 using contour propagation from the seeds for that time. Extract all isocontour components for subsequent discrete time steps, $t \leq t_1$ by a combination of modifying current isocontour components over time, and by extracting new components from the seed sets at t.

To modify components, at each time-step t, with $t_0 \leq t \leq t_1$, maintain, in one list per isocontour component, the set of intersecting cells. These cell lists are used to track the evolution of the isocontour for s over time. As in the marching cubes case, we can label each cell vertex as "above" if its value at time t is greater than s, and "below" otherwise. A cell edge joining opposite labels contains an isocontour vertex. To track an isocontour component, consider the label change for each cell edge at $t + 1$. If the labels of the end vertices of a cell edge do not change then the isocontour vertex that lies on that edge just changes position, which can be found by linear interpolation. A cell vertex label can change, in which case the isocontour experiences a local connectivity change. The changes include the contour changing its geometry but not topology, two or more components merging, or a component splitting into two or more components. All these changes can be applied to the contours by enqueueing the cell lists, extracting each cell and examining its neighborhood, and propagating the isocontour spatially. See [4] for details.

2.4 Comparison

The TBON algorithm discussed in Sect. 2.1 can be applied to regularly gridded data and is designed to be I/O-efficient by using demand driven paging; load data only when it is required. It does not exploit temporal coherence and treats each time-step as a static volumetric scalar field. The THI algorithm discussed in Sect. 2.2 can be applied to both regularly and irregularly sampled data. It reduces the storage overhead of the search structure, but it requires all data to be loaded into memory, a serious drawback for time-varying data which can be large. The progressive tracking algorithm discussed in Sect. 2.3 can also be applied to regularly and irregularly sampled data. Since this algorithm computes seed sets the storage overhead is not significant. Bajaj et al. [4] show seed set sizes of less than 2% of the total number of cells. Moreover, extracting isocontours by propagation produces coherent triangulations that are amenable to compression [35], simplification [37], and streaming [36]. The other extraction algorithms do not produce coherent triangulations.

3 Topological Structures for Supporting Visualization

In this section we review a topological structure called the Reeb graph that encodes the number of isocontour components at each isovalue, and the topological changes experienced by components. The Reeb graph can be used to display this information succinctly to aid visualization. Section 3.2 reviews an algorithm that extends the Reeb graph for visualizing time-varying scalar fields. Section 3.3 reviews a more systematic study of the evolution of Reeb graphs over time, and an algorithm to compute this evolution.

The algorithms discussed in this section are based on Morse theory [44, 48] and combinatorial and algebraic topology [1, 50]. Note that in deference to the terminology used in this mathematical literature, we sometimes use the term level set instead of isocontour. We begin with a review of smooth maps and critical points. Because algorithms work on piecewise linear data, we discuss how to translate the concepts from the smooth setting to the piecewise linear setting.

Smooth Maps on Manifolds

Let $\mathbb{M}$ be a smooth, compact d-manifold without boundary and $f : \mathbb{M} \to \mathbb{R}$ a smooth map. Assuming a local coordinate system in its neighborhood, $x \in \mathbb{M}$ is a *critical point* of f if all partial derivatives vanish at x. If x is a critical point, $f(x)$ is a *critical value*. Noncritical points and noncritical values are called *regular points* and *regular values*, respectively. The *Hessian* at x is the matrix of second-order partial derivatives. A critical point x is *nondegenerate* if the Hessian at x is nonsingular. The *index* of a critical point x is the number of negative eigenvalues of the Hessian. Intuitively, it is the number of mutually orthogonal directions at x along which f decreases. For $d = 3$ there are four types of nondegenerate critical points: the *minima* with index 0, the 1-*saddles* with index 1, the 2-*saddles* with index 2, and the *maxima* with index 3. A function f is *Morse* if:

I. All critical points are nondegenerate.
II. $f(x) \neq f(y)$ whenever $x \neq y$ are critical.

We will refer to I and II as Genericity Conditions as they prevent certain non-generic configurations of the critical points. This choice of name is justified because Morse functions are dense in $C^\infty(\mathbb{M})$, the class of smooth functions on the manifold [31, 44]. In other words, for every smooth function there is an arbitrarily small perturbation that makes it a Morse function.

The critical points of a Morse function and their indices capture information about the manifold on which the function is defined. For example, the Euler characteristic of the manifold $\mathbb{M}$ equals the alternating sum of critical points, $\chi(\mathbb{M}) = \sum_x (-1)^{\text{index}\, x}$.

Piecewise Linear Functions

A *triangulation* of a manifold $\mathbb{M}$ is a simplicial complex, K, whose underlying space is homeomorphic to $\mathbb{M}$ [1]. Given values at the vertices, we obtain a continuous function on $\mathbb{M}$ by linear interpolation over the simplices of the triangulation. The Euler characteristic of $\mathbb{M}$ can also be computed from any triangulation of $\mathbb{M}$ as the alternating sum of simplices, $\chi(\mathbb{M}) = \sum_\sigma (-1)^{\dim \sigma}$. We need some definitions to talk about the local structure of the triangulation and the function. The *star* of a vertex u, denoted $\mathrm{St}\, u$, consists of all simplices that share u, including u itself, and the *link*, denoted $\operatorname{Lk} u$, consists of all faces of simplices in the star that are disjoint from u. The *lower link*, denoted $\operatorname{Lk}_{-} u$, is the subset of the link induced by vertices with function value less than u:

$$\begin{aligned} \operatorname{St} u &= \{\sigma \in K \mid u \subseteq \sigma\}, \\ \operatorname{Lk} u &= \{\tau \in K \mid \tau \subseteq \sigma \in \operatorname{St} u,\ u \notin \tau\}, \\ \operatorname{Lk}_{-} u &= \{\tau \in \operatorname{Lk} u \mid v \in \tau \Rightarrow f(v) \leq f(u)\}. \end{aligned}$$

Banchoff [5] introduces the critical points of piecewise linear functions as the vertices whose lower links have Euler characteristic different from unity.

A classification based on the reduced Betti numbers of the lower link is finer than that defined by Banchoff. The *k-th reduced Betti number*, denoted as $\tilde{\beta}_k$, is the rank of the k-th reduced homology group of the lower link: $\tilde{\beta}_k = \operatorname{rank} \tilde{\mathsf{H}}_k$. The reduced Betti numbers are the same as the usual (unreduced) Betti numbers, except that $\tilde{\beta}_0 = \beta_0 - 1$ for nonempty lower links, and $\tilde{\beta}_{-1} = 1$ for empty lower links [50]. The first three unreduced Betti numbers are the number of connected components, the number of tunnels, and the number of voids respectively. For example, a 2-torus has unreduced Betti numbers: $\beta_0 = \beta_2 = 1$, and $\beta_1 = 2$. For $d = 3$ the link is a 2-sphere and the Betti numbers can be computed as follows: Compute the Euler characteristic χ of the lower link as the alternating sum of vertices, edges and faces in the lower link. Compute β_0, the number of connected components in the lower link, by using the union-find data structure [18]. If all the link vertices are also in the lower link then $\beta_2 = 1$ else $\beta_2 = 0$. Compute β_1 using the relation $\beta_1 = \beta_0 + \beta_2 - \chi$. The reduced Betti numbers can be computed from the definitions. For an algorithm to compute Betti numbers of simplicial complexes on the 3-sphere see [20].

When the link is a 2-sphere only $\tilde{\beta}_{-1}$ through $\tilde{\beta}_2$ can be nonzero. Simple critical points have exactly one nonzero reduced Betti number, which is equal to 1; see Table 1 and Fig. 3. The first case in which this definition differs from Banchoff's is a double saddle obtained by combining a 1- and a 2-saddle into a single vertex.

Table 1. Classification of vertices into regular and simple critical points using the reduced Betti numbers of the lower link

	$\tilde{\beta}_{-1}$	$\tilde{\beta}_0$	$\tilde{\beta}_1$	$\tilde{\beta}_2$
Regular	0	0	0	0
Minimum	1	0	0	0
1-Saddle	0	1	0	0
2-Saddle	0	0	1	0
Maximum	0	0	0	1

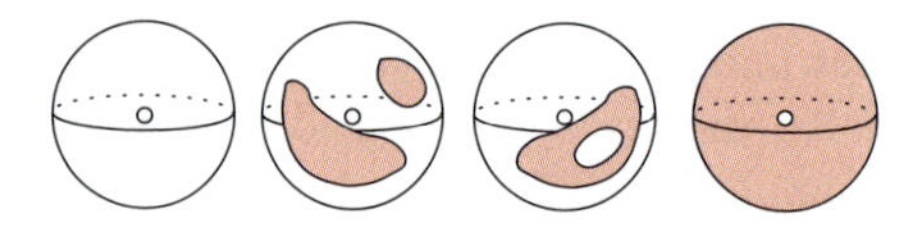

Fig. 3. Lower links, shown *shaded*, for $d = 3$. *From left to right*, a minimum, a 1-saddle, a 2-saddle, and a maximum

The Euler characteristic of the lower link is 1, which implies that Banchoff's definition does not recognize it as critical. A *multiple saddle* is a critical point that falls outside the classification of Table 1 and therefore satisfies $\tilde{\beta}_{-1} = \tilde{\beta}_2 = 0$ and $\tilde{\beta}_0 + \tilde{\beta}_1 \geq 2$. By modifying the simplicial complex, it can be unfolded into simple 1-saddles and 2-saddles as explained in [11, 25]. This allows us to develop algorithms assuming that all critical points are simple.

The Reeb Graph

A level set of a function f consists of all points in the domain whose function values are equal to a chosen real number s. A level set of f is not necessarily connected. If we call two points $x, y \in \mathbb{M}$ *equivalent* when $f(x) = f(y)$ and both points belong to the same component of the level set, then we obtain the *Reeb graph* as the quotient space in which every equivalence class is represented by a point and connectivity is defined in terms of the quotient topology [54]. Figure 4 illustrates the Reeb graph for a function defined on a 2-manifold of genus two. We call a point on the Reeb graph a *node* if the corresponding level set component passes through a critical point of f. The rest of the Reeb graph consists of arcs connecting the nodes. The *degree* of a node is the number of arcs incident to the node. A minimum creates and a maximum destroys a level set component and both correspond to degree-1 nodes. A saddle that splits one level set component in two or merges two to one corresponds to a degree-3 node. There are also saddles that alter the genus but do not affect the number of components, and they correspond to degree-2 nodes in the Reeb graph. Nodes of degree higher than three occur only for non-Morse functions.

Visualization Interfaces

Topological structures, such as the Reeb graph, provide a succinct summary of the underlying function, and are used in visualization interfaces. Consider Fig. 5. It shows a screenshot of the *contour spectrum* [3] interface; a window displays isocontour properties, such as surface area and volume, using graphs, and topological properties using the contour tree, to aid selection of interesting isocontours which are displayed separately.

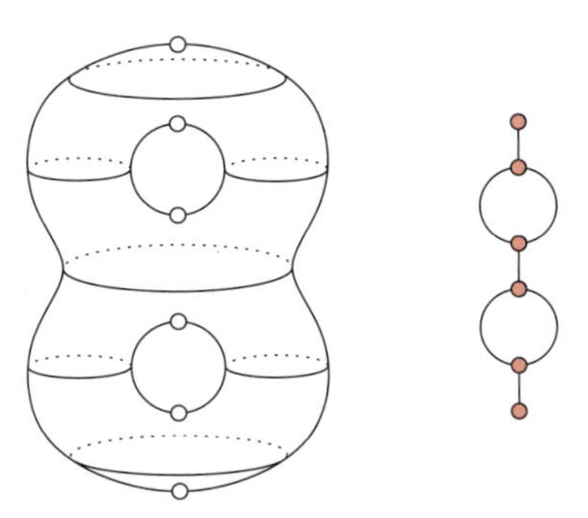

Fig. 4. The Reeb graph of the function f on a 2-manifold that maps every point of the double torus to its distance above a horizontal plane below the surface

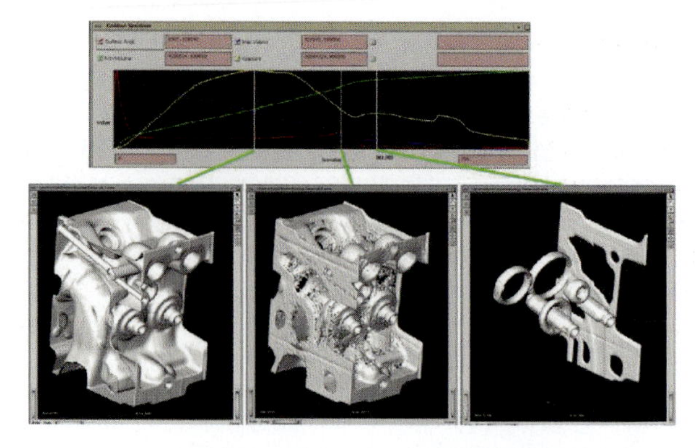

Fig. 5. The contour spectrum interface [3]. On the *top*, graphs of isocontour properties, such as surface area, and volume, vs. isovalues. On the *bottom*, three isocontours chosen using the contour spectrum

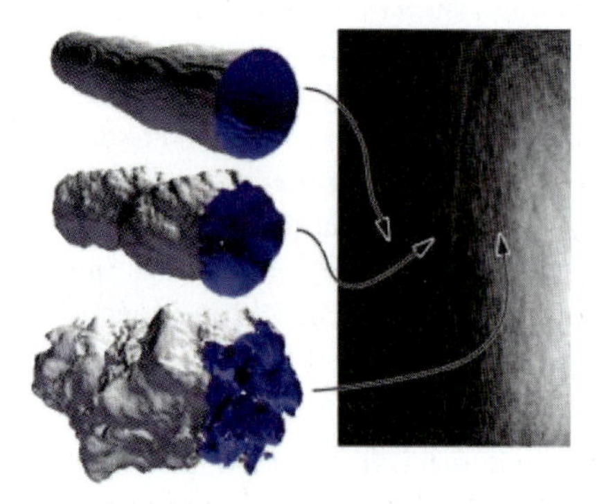

Fig. 6. A portion of the safari interface [40]. The control plane on the *right* displays the number of isocontour components for each time-step (x-axis) and isovalue (y-axis). The user selects isocontours for display by clicking on the contour plane

Figure 6 shows the safari interface [40], which extends the ideas of the contour spectrum to time-varying data, provides the user with a *(time, value)* control plane for isovalue selection, and extracts an isocontour from a time-slice for display. The control plane on the right displays the number of isocontour components for each time-step(x-axis) and isovalue(y-axis), which can be computed from the contour tree for each time step.

Recently, Carr et al. [10] have used the contour tree to compute *path seeds*, a set of edges in the triangulation that can be used to quickly find seeds for any isocontour component, and use these path seeds to create a *flexible isocontour* interface. They provide the user with an interface to select individual arcs in the contour tree and can extract chosen isocontour components for display. Building on this work, they also present an algorithm to simplify the contour tree using local geometric measures to capture the essential features of the underlying data in the presence of noise [12].

3.1 Reeb Graph Algorithms

In mathematics, the Reeb graph is often used to study the manifold $\mathbb{M}$ that forms the domain of the function. For example, the Reeb graph in Fig. 4 reveals that the function is defined on a double torus, assuming we know it is an orientable 2-manifold without boundary. In visualization, on the other hand, the Reeb graph is used to study the behavior of the function. The domain of interest is $\mathbb{R}^3$ but it is convenient to compactify it and consider functions on the 3-sphere, $\mathbb{S}^3$. All the Reeb graphs for such functions will reveal the (unexciting) connectivity of $\mathbb{S}^3$ by being trees, but the structure of the tree will tell us how the level sets of the chosen function f change topology.

History

The Reeb graph was first introduced in [54]. In the field of visualization, Boyell and Ruston [9] introduced the contour tree to summarize the evolution of contours on a map. In the interactive exploration of scientific data, Reeb graphs are used to select meaningful level sets [3] and to efficiently compute them [62]. An extensive discussion of Reeb graphs and related structures in geometric modeling and visualization applications can be found in [27].

Published algorithms for Reeb graphs take as input a function defined on a triangulated manifold. We express their running times as functions of n, the number of simplices in the triangulation. The first algorithm for functions on 2-manifolds due to Shinagawa and Kunii [57] takes time $O(n^2)$ in the worst case. Reeb graphs of simply-connected domains are loop-free, and are also known as contour trees. They have received special attention because of the practical importance of these domains, and because the algorithms to compute them are simpler. An algorithm that constructs contour trees of functions on simply-connected manifolds of constant dimension in time $O(n \log n)$ has been suggested in [11]. For the case of 3-manifolds, this algorithm has been extended to include information about the genus of the level surfaces [53]. Cole-McLaughlin et al. [17] return to the general case, giving tight bounds on the number of loops in Reeb graphs of functions on 2-manifolds and describing an $O(n \log n)$ time algorithm to construct them.

Computing a Contour Tree

We sketch the algorithm proposed by Carr et al. [11] to compute the contour tree of a function defined on a simply-connected domain. Although their algorithm works for any dimension we restrict our description to $d = 3$. This algorithm in used in Sects. 3.2 and 3.3.

The input to the contour tree algorithm is a triangulation K of a simply-connected 3-manifold, with each vertex equipped with a distinct scalar function value. The output is the contour tree of the function represented as a collection of nodes and arcs that connect them. The algorithm proceeds in two passes: the first computes the *Join tree*, and the second the *Split tree*. Since these passes are similar we describe the construction of the Join tree. The Join tree encodes the merges experienced by

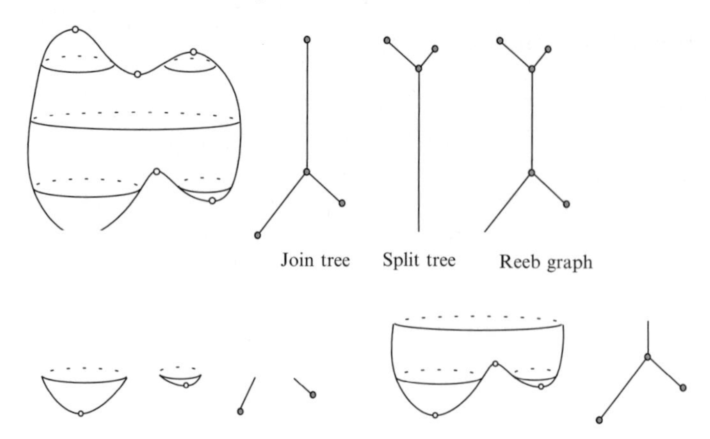

Fig. 7. In the *top row*, a 2-manifold shown with three isocontours for the height function defined on it, and the Join, Split tree and Reeb graph. In the *bottom row*, two stages during the sweep to construct the Join tree

the isocontour components as we sweep the isovalue from $-\infty$ to ∞; the Split tree does this for the sweep in the opposite direction.

To implement the sweep sort the vertices of K in ascending order of function value, and iterate through the vertices. At each step maintain the collection of vertices visited in a union-find(UF) data structure [18]. For each connected component in the collection, maintain a tree encoding the merge history of the component. Classify each vertex v based on its index and handle each case as follows. Add a regular point to the set that contains its lower link vertex. A minimum creates a new component; start a UF set, and a new Join tree arc. An index-1 critical point locally merges two components. If the components are not connected globally, then create a join node that merges two arcs and starts a new one, else create a join node that ends a single arc, and starts a new one. The latter case corresponds to the component experiencing a genus change. An index-2 critical point is handled in the split sweep. Only the global maximum appears in the Join tree; it ends the arc corresponding to the component that disappears at the maximum.

Finally, construct the contour tree by merging the Join and Split tree [11]. In Fig. 7 we see two stages during the construction of the Join tree for a function defined on a 2-manifold.

3.2 Time-Varying Contour Topology

An important problem in visualizing time-varying scalar fields is computing the correspondence of isocontour components for a fixed isovalue over time. Another problem is detecting when and how these components change topology. Sohn and Bajaj [58] address these problems by computing the correspondence of contour trees over time. They assume that the scalar field can change unpredictably between two successive time steps, and define temporal correspondence of contour tree arcs for successive time steps using a notion of an overlap between an isocontour at time t with an isocontour at time $t+1$. They develop an algorithm to compute correspondence between successive contour trees based on this definition, and use the correspondence to track isocontour components and their topology over time.

Temporal Contour Correspondence

Before we define temporal contour correspondence we need some notation. Define $\mathbb{X}$ and $\mathbb{Y}$ for the restrictions of the domain to time t and $t+1$ respectively, function f_t for the restriction of function f to $\mathbb{X}$, and isocontours $I = f_t^{-1}(s)$ and $J = f_{t+1}^{-1}(s)$. Isocontour I has connected components $I = \{I_1, \cdots, I_m\}$, and lies on the intersection of $\mathbb{X}_{\leq s} = f_t^{-1}(-\infty, s]$ and $\mathbb{X}_{\geq s} = f_t^{-1}[s, \infty)$. Identify each I_i with one component of $\mathbb{X}_{\leq s}$ and $\mathbb{X}_{\geq s}$; I_i belongs to their intersection. Sohn & Bajaj call these components the *lower object* and *upper object* of I_i, respectively. Similar definitions hold for the isocontour J.

Two contour components I_i and J_j exhibit *temporal correspondence* if their corresponding upper objects overlap and their corresponding lower objects overlap. Note, that an overlap between $\mathbb{X}_{\leq s}$ and $\mathbb{Y}_{\leq s}$ makes sense only if we assume that both f_t and f_{t+1} are defined on the same domain. See Fig. 8 for an example.

Algorithm for Contour Correspondence

Since each isocontour component corresponds to a point on an arc on the contour tree, temporal correspondence between isocontour components can be used to compute a correspondence between arcs of the contour tree at t with arcs of the contour tree at $t+1$. Sohn & Bajaj compute this correspondence by modifying

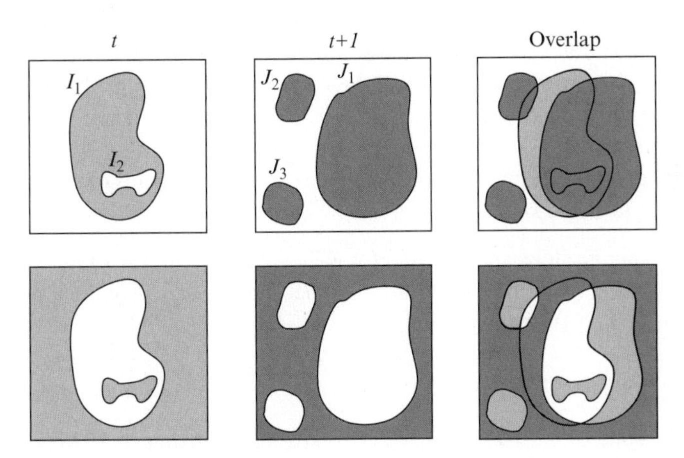

Fig. 8. Temporal correspondence between isocontours at successive time steps. In the *top row*, upper objects for each isocontour and their overlap, in the *bottom row*, lower objects and their overlap. From the definitions in [58], we get the following correspondence: $(I_1 \rightarrow J_1)$, $(I_1 \rightarrow J_2)$, $(\emptyset \rightarrow J_3)$, $(I_2 \rightarrow \emptyset)$

the contour tree algorithm of Carr et al. [11]. In the description that follows, we use JT_t, ST_t, CT_t to denote the join tree, split tree, and contour tree at time t respectively. The input is a simplicial mesh in $\mathbb{R}^3$. Each point of the mesh is equipped with T scalar values. The output is a collection of T contour trees, with each arc of CT_{t+1} labeled with arcs of CT_t. The labels indicate the temporal correspondence information. Since the join and split trees are symmetric we use the join tree for the description. For each time step t, precompute the contour tree, CT_t [11]. Label each tree arc with a unique id. The correspondence information is computed as follows:

1. Augment JT_t with the nodes that appear only in ST_t. Carr et al. [11] call the resulting join tree the augmented join tree.
2. Equip each arc a of JT_t with the ids of the corresponding arcs from CT_t. An isocontour component corresponding to a point on a, lies on the boundary of a connected component of $\mathbb{X} \leq s$ (the lower object), which may contain other isocontour components. Each of these other isocontour components correspond to an arc of CT_t. Equip a with the ids of these arcs of CT_t. This step can be done by simultaneously scanning the nodes of CT_t and JT_t in increasing order of f_t, and incrementally maintaining the arc id lists for JT_t.
3. This is the step that computes the correspondence information for the arcs of JT_{t+1}. Simultaneously sweep the functions f_t and f_{t+1} in increasing order, as if constructing the join trees JT_t and JT_{t+1}. The sweep can be thought of as incrementally generating the lower objects simultaneously in time t and $t + 1$, and detecting overlap. Recall that two isocontour components from successive time-steps exhibit temporal correspondence if their respective lower objects and upper objects overlap. This sweep tests lower object overlap, and the reverse sweep, for ST_t and ST_{t+1}, will test overlap for upper objects. During the sweep, maintain a collection of lower objects in $\mathbb{X}_{\leq s}$ and $\mathbb{Y}_{\leq s}$. Each lower object corresponds to an arc of the join tree in its respective time-step. For lower objects in $\mathbb{X}_{\leq s}$ we know their corresponding contour tree arcs from step 2. When a lower object in $\mathbb{X}_{\leq s}$ overlaps one in $\mathbb{Y}_{\leq s}$ we can create the mapping between the arcs of JT_{t+1} with the corresponding arcs of JT_t. For a more detailed description of this step see [58].
4. Map labels from the arcs of JT_{t+1} to the corresponding arcs of CT_{t+1}. This step is similar to step 2.

After the above steps are repeated for ST_{t+1}, each arc of the contour tree CT_{t+1} has two arc lists, one from JT_{t+1} and the other from ST_{t+1}. The final arc list is the intersection of the two lists.

Topology Change Graph

Consider isocontour $I = \{I_1, \cdots, I_m\}$ for isovalue s at time t. If we keep the isovalue fixed at s and proceed forward in time to $t + 1$, then the isocontour undergoes topological changes in the following possible ways: a component is created or destroyed, two components merge into one, a component splits into two, a component changes genus. The *Topology change graph(TCG)* depicts the change in topology as

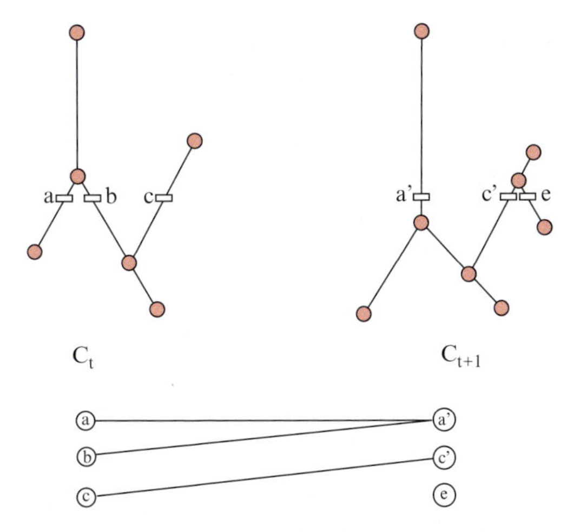

Fig. 9. In the *top row*, a contour tree at successive time steps, with arcs containing isovalue s labeled. In the *bottom row*, the topology change graph showing how contour components change topologically

a graph constructed as follows. At each time step compute the arcs of C_t that contain the isovalue s. Create a node in the TCG for each these arcs. Use the correspondence information computed for an arc of C_t to create a connection between the corresponding nodes in the TCG. See Fig. 9. Genus change can be detected by examining the Betti numbers of the component at time $t+1$, which can computed using the algorithm of Pascucci et al. [53].

3.3 Time-Varying Reeb Graph

Edelsbrunner et al. [24] have performed a systematic study of the evolution of the Reeb graph of a function over time. This evolution can be encoded in a data structure that can be used to extract the Reeb graph for any instant of time for display in a visualization interface. Moreover, the data structure can be used to track isocontour components over time, and detect changes in their topology. This information can aid the user in selecting interesting time and isovalue parameters for visualization.

Edelsbrunner et al. enumerate all the combinatorial changes experienced by the Reeb graph of a Morse function on $\mathbb{S}^3$ over time. Crucial to understanding these changes is the notion of Jacobi sets, which can be used to compute the trajectory of critical points of a function over time.

Jacobi Sets

Reeb graphs can be used to summarize a function at moments in time and Jacobi curves, as introduced in [23], to get a glimpse of their evolution through time. Edelsbrunner et al. introduce this concept for the slightly more general case of two Morse functions, $f, g\colon \mathbb{M} \to \mathbb{R}$; the specific case of a time-varying function, f, is

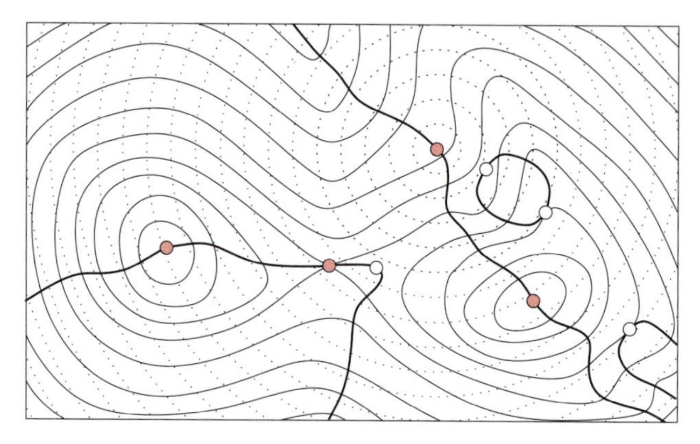

Fig. 10. The functions f and g are represented by their *dotted and solid level curves*. The Jacobi curve is drawn in *bold solid lines*. The birth–death points and the critical points of the two functions are marked by *white and shaded dots*, respectively

obtained by adding time as an extra dimension to the domain and letting g represent time. For a regular value $t \in \mathbb{R}$, consider the level set $g^{-1}(t)$ and the restriction of f to this level set $f_t : g^{-1}(t) \to \mathbb{R}$. The *Jacobi curve* of f and g is the closure of the set of critical points of the functions f_t, for all $t \in \mathbb{R}$. The closure operation adds the critical points of f restricted to level sets at critical values, as well as the critical points of g, which form singularities in these level sets. We use Fig. 10 from [23] to illustrate the definition by showing the Jacobi curve of two smooth functions on a piece of the two-dimensional plane. To understand this picture, imagine f as a cone-like mountain indicated by dotted level curves, and the solid level curves of g gliding over that mountain. On the left, we see a circle beginning at a minimum of g and expanding outwards on a slope. As this circle expands a maximum of the restriction of f moves up and a minimum moves down from the starting point.

Consider a 1-parameter family of Morse functions on the 3-sphere, $f : \mathbb{S}^3 \times \mathbb{R} \to \mathbb{R}$, and introduce an auxiliary function $g : \mathbb{S}^3 \times \mathbb{R} \to \mathbb{R}$ defined by $g(x, t) = t$. A level set has the form $g^{-1}(t) = \mathbb{S}^3 \times t$, and the restriction of f to this level set is $f_t : \mathbb{S}^3 \times t \to \mathbb{R}$. The Jacobi curve of f and g may consist of several components, and in the assumed generic case each is a closed 1-manifold. Identify the *birth–death points* where the level sets of f and g and the Jacobi curve have a common normal direction. To understand these points, imagine a level set in the form of a (two-dimensional) sphere deforming, sprouting a bud, as we go forward in time. The bud has two critical points, one a maximum and the other a 2-saddle. At the time when the bud just starts sprouting there is a point on the sphere, a birth point, where both these critical points are born. Run this in reverse order to understand a death point. Decompose the Jacobi curve into *segments* by cutting it at the birth–death points. The index of the critical point tracing a segment is the same everywhere along the segment. The indices within two segments that meet at a birth–death point differ by one:

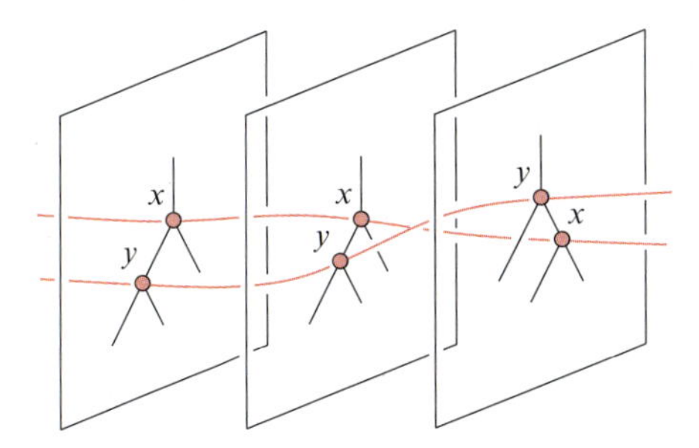

Fig. 11. Reeb graphs at three moments in time whose nodes are connected by two segments of the Jacobi curve [24]

INDEX LEMMA [24]. Let $f : \mathbb{M} \times \mathbb{R} \to \mathbb{R}$ be a 1-parameter family of Morse functions. The indices of two critical points created or destroyed at a birth–death point differ by one.

Jacobi Curves Connect Reeb Graphs

Let R_t be the Reeb graph of f_t, the function on $\mathbb{S}^3$ at time t. The nodes of R_t correspond to the critical points of f_t, and as we vary t, they trace out the segments of the Jacobi curve. The segments connect the family through time, and provide a mechanism for identifying nodes in different Reeb graphs. Figure 11 illustrates this idea.

Generically, the function f_t is Morse. However, there are discrete moments in time at which f_t violates one or both Genericity Conditions of Morse functions and the Reeb graph of f_t experiences a combinatorial change. Since time is the only varying parameter, one may assume that there is only a single violation of the Genericity Conditions at any of these discrete moments, and there are no violations at all other times. Condition I is violated iff f_t has a birth–death point at which a cancellation annihilates two converging critical points or an anticancellation gives birth to two diverging critical points. Condition II is violated iff f_t has two critical points $x \neq y$ with $f_t(x) = f_t(y)$ that form an interchange. The two critical points may be independent and have no effect on the Reeb graph, or they may belong to the same level set component of f_t and correspond to two nodes that swap their positions along the Reeb graph. We briefly sketch the changes caused by birth–death points and by interchanges.

Birth, Death, and Interchange

When time passes the moment of a birth point, we get two new critical points and correspondingly two new nodes connected by an arc in the Reeb graph. By the Index Lemma, the indices of the two critical points differ by one, leaving three possibilities: 0–1, 1–2, and 2–3. See Figs. 12 and 13, and [24] for details.

There are three similar cases when time passes the moment of a death point. Two critical points of f_t converge and annihilate when they collide, and correspondingly

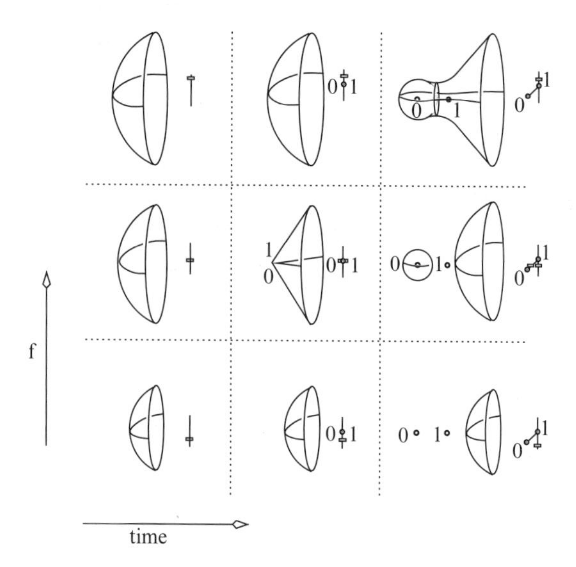

Fig. 12. Level sets and Reeb graphs around a 0–1 birth point. The 2–3 case is upside-down symmetric to this case. Time increases from left to right and the level set parameter, indicated by a *rectangular slider bar*, increases from bottom to top. Going forward in time, we see the sprouting of a bud, while going backward in time we see its retraction [24]

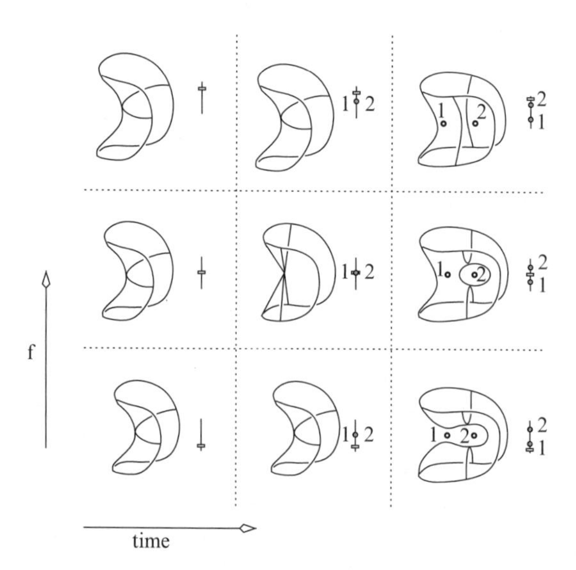

Fig. 13. Level sets and Reeb graphs around a 1–2 birth point. Time increases from left to right and the level set parameter increases from bottom to top. Going forward in time, we see a refinement of an arc in the Reeb graph and going backward we see a coarsening [24]

an arc of the Reeb graph contracts to a point, effectively removing its two nodes. The 0–1 and 2–3 cases are illustrated in Fig. 12, which we now read from right to left, and the 1–2 case is illustrated in Fig. 13, which we also read backward, from right to left.

Nodes of the Reeb graph swap position in the Reeb graph when the corresponding critical points, x and y, form an interchange and, at that moment, belong to the same level set component. Assume without loss of generality that $f_{t-\varepsilon}(x) < f_{t-\varepsilon}(y)$ and $f_{t+\varepsilon}(x) > f_{t+\varepsilon}(y)$. There are four choices for each of x and y depending on whether they add or remove a handle, merge two level set components or split a level set component. This gives a total of sixteen configurations. An analysis of possible before and after combinations and pairing them, gives the cases illustrated in Fig. 14. It is convenient to group the cases with similar starting configurations together. Edelsbrunner et al. use $+$, $-$, M, S to mean 'handle addition', 'handle deletion', 'component merge', and 'component split', respectively, and a pair of these to indicate the types of x and y. For a more detailed description of these cases see [24].

Computing Time-Varying Reeb Graphs

The input to the algorithm is a piecewise linear function defined on a triangulation K of 3-sphere cross time. The output is a collection of Reeb graphs, stored in a partially persistent data-structure [21], that captures the evolution of the Reeb graph R_t over time.

Begin by constructing the Jacobi curve as a collection of edges in K using the algorithm in [23]. Also construct the Reeb graph at time zero, R_0, from scratch, using the algorithm in [11]. Maintain R_t by sweeping forward in time, using the Jacobi curve as a path for its nodes. Implement the sweep by maintaining a priority queue of events sorted on time, repeatedly retrieving the next event, updating the Reeb graph, and deleting and inserting interchange events as arcs are removed and added. The sequence can be thought of as the evolution of a single Reeb graph. Following Driscoll et al. [21], accumulate the changes to form a single data structure representing the entire evolution, which Edelsbrunner et al. [24] refer to as the *partially persistent Reeb graph*. Adhere to the general recipe to construct it, using a constant number of data fields and pointers per node and arc to store time information and keep track of the changes caused by an update.

There are difficulties in implementing this algorithm. Properties of the Jacobi curve that are valid in the smooth setting need not necessarily hold in the piecewise-linear setting. In particular, the Jacobi curve need not be a 1-manifold; it could have vertices with degree greater than two, and edges corresponding to multiple critical points. Such vertices and edges have to be unfolded to simulate the 1-manifold property. Techniques to maintain the integrity of structural properties in the piecewise linear setting are discussed in greater detail in [26].

3.4 Comparison

We compare the algorithms of Sects. 3.2 and 3.3. The former considers a discrete set of functions, their Reeb graphs, and maps the arcs of the Reeb graph at t to the

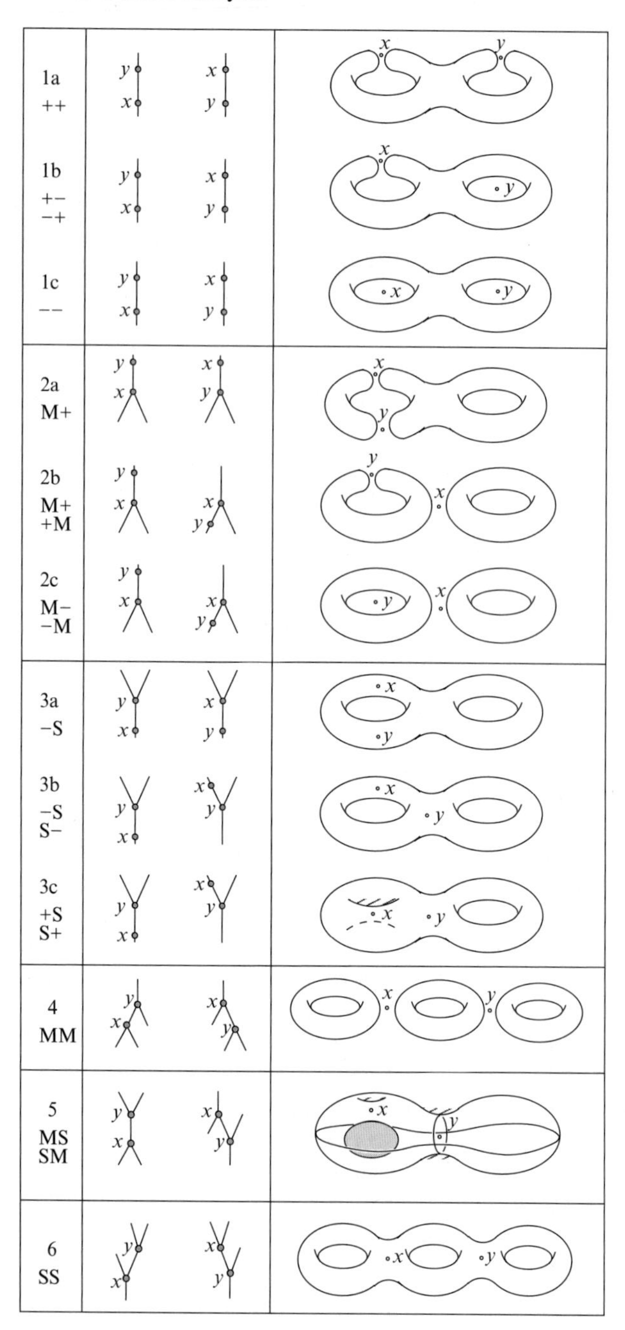

Fig. 14. On the *left*, Reeb graph portions before and after the interchange x and y. On the *right*, level sets at a value just below the function value of x and y. In each case, the index of a critical point can be inferred from whether the level set merges (index 1) or splits (index 2) locally at the critical point [24]

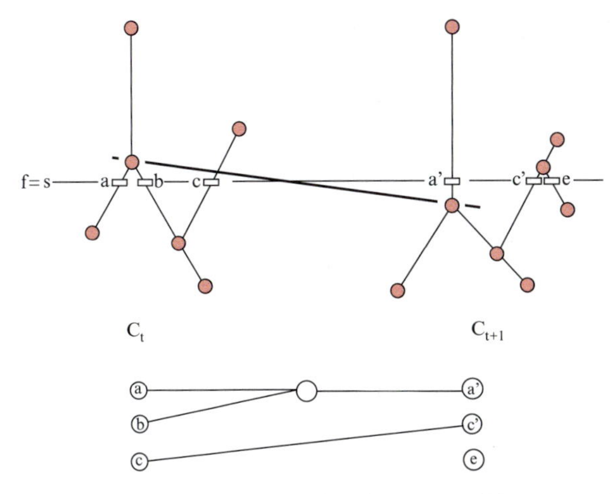

Fig. 15. In the *top row*, a contour tree at successive time steps, with arcs containing isovalue s labeled. The Jacobi segment for the upper node of the arcs containing a and b is shown. When this segment intersect the line $f = s$ the components a and b merge. Note the difference with the TCG in Fig. 9; in [24] function f is continuous in space–time and isocontour topology can change at any time

arcs of the graph at $t + 1$ using a definition of temporal coherence. This algorithm works well when the time sampling rate is high relative to the phenomenon under study so that there is good temporal coherence. Unlike the latter, it provides no understanding of how the Reeb graph changes over time, nor can it produce a Reeb graph for all continuous values of time. The algorithm of Sect. 3.3 assumes a continuous space–time function, uses Jacobi sets to connect all Reeb graphs, systematically enumerates the possible changes that the Reeb graph experiences, and captures the evolution of the Reeb graph over time. The time-varying Reeb graph can be used to compute the topological information computed in Sect. 3.2. The arc mapping information is implicit; each Reeb graph arc at time $t + 1$ has a sequence of events that maps it to an arc at time t. For the topology change graph, start with the Reeb graph R_t, examine each arc containing isovalue s and compute the time when the Jacobi segments attached to its end-nodes intersects the line $f = s$. At this time, the corresponding isocontour component experiences a topological change. See Fig. 15.

4 Conclusions

Isocontour extraction algorithms for time-varying scalar fields use three techniques to increase efficiency: spatial techniques, span space techniques, and topological techniques. Spatial techniques organize space using an octree decomposition, to detect regions that intersect, and reject regions that do not intersect the isocontour. Span space techniques organize cells in the space of function values to efficiently detect cells that intersect the isocontour. Topological techniques organize a subset of cells, called the seed set, in a search structure. The seed set contains one intersecting cell for each connected component of each isocontour, and the component can be extracted by propagation from the intersecting cell. While span space techniques and topological techniques can be used for both regularly and irregularly sampled data, spatial techniques can be used only for regularly sampled data. Unlike spatial techniques and span space techniques, topological techniques allow isocontour extraction using contour propagation and produce coherent triangulations that are amenable to compression [35], simplification [37], and streaming [36]. This feature is useful for large data set visualization when the isocontours themselves might be too large to fit in memory.

Two algorithms extend the Reeb graph to time-varying functions to aid the user in selecting interesting time and isovalue parameters for visualization. The first algorithm, by Bajaj & Sohn [58], uses a overlap heuristic to connect the Reeb graph for each time slice; it works well when the time sampling rate is high relative to the phenomenon under study so that there is good temporal coherence. The second algorithm, by Edelsbrunner et al. [24], determines the actual dynamics of the Reeb graph over all time slices, under a chosen interpolation function. The dynamics depend on the Jacobi curve which is the trajectory of the critical points over time and leads to a classification of combinatorial changes experienced by the Reeb graph over time. The algorithm constructs the Reeb graph for time $t = 0$ from scratch and sweeps forward in time, modifying the current Reeb graph to compute the evolution of the Reeb graph over time. Because the theory for this algorithm is based on smooth functions and real-world data is not smooth, special care has to be taken during implementation to ensure structural integrity of the Jacobi curve. As a consequence this algorithm is harder to implement than the first one, but it gives a more complete picture of the evolution of the Reeb graph.

Beyond the questions addressed in this paper, we see some interesting research directions that we believe will become increasingly important in visualization. Often the data is noisy and the Reeb graph is itself too large and cluttered to make any sense. We believe it is worthwhile to investigate simplification of time-varying Reeb graphs on the lines of the work of Carr et al. [12]. How can we simplify the time-varying Reeb graph, and maintain consistency with the underlying data? What meaningful measures, geometric and topological, can we devise to guide the simplification? How can we represent several levels of simplification and how can we present the user with a multiresolution view of time-varying Reeb graphs?

Acknowledgments

We thank our funding agencies: NSF grant 0128426 and subcontracts from Lawrence Livermore National Labs. Portions of this work was performed under the auspices of the US Department of Energy by University of California Lawrence Livermore National Laboratory under contract No. W-7405-Eng-48. We thank the anonymous referees for their suggestions.

References

1. P. S. Alexandrov. *Combinatorial Topology*. Dover, Mineola, NY, 1998.
2. E. Artzy. Display of three-dimensional information in computed tomography. *Computer Graphics and Image Processing*, 9:196–198, 1979.
3. C. L. Bajaj, V. Pascucci, and D. Schikore. The contour spectrum. In *IEEE Visualization*, pages 167–174, 1997.
4. C. L. Bajaj, A. Shamir, and S. Bong-Soo. Progressive tracking of isosurfaces in time-varying scalar fields. Technical report, Univ. of Texas, Austin, 2002. `http://www.ticam.utexas.edu/CCV/papers/Bongbong-Vis02.pdf`.
5. T. F. Banchoff. Critical points for embedded polyhedral surfaces. *The American Mathematical Monthly*, 77:457–485, 1970.
6. K. G. Bemis, D. Silver, P. A. Rona, and C. Feng. Case study: a methodology for plume visualization with application to real-time acquisition and navigation. In *Proc. IEEE Conf. Visualization*, pages 481–494, 2000.
7. J. L. Bentley. Multidimensional binary search trees used for associative searching. *Communications of the ACM*, 18(9):509–517, 1975.
8. J. F. Blinn. A generalization of algebraic surface drawing. *ACM Transactions on Graphics*, 1(3):235–256, 1982.
9. R. L. Boyell and H. Ruston. Hybrid techniques for real-time radar simulation. In *Proc. of 1963 Fall Joint Computer Conference (IEEE)*, pages 445–458, 1963.
10. H. Carr and J. Snoeyink. Path seeds and flexible isosurfaces: Using topology for exploratory visualization. In *Proc. of Eurographics Visualization Symposium*, pages 49–58, 285, 2003.
11. H. Carr, J. Snoeyink, and U. Axen. Computing contour trees in all dimensions. *Computational Geometry*, 24(2):75–94, 2003.
12. H. Carr, J. Snoeyink, and M. van de Panne. Simplifying flexible isosurfaces using local geometric measures. In *Proc. of IEEE Visualization 2004*, pages 497–504, 2004.
13. Y.-J. Chiang. Out-of-core isosurface extraction of time-varying fields over irregular grids. In *Proc. IEEE Visualization 2003*, pages 217–224, 2003.
14. Y.-J. Chiang, C. T. Silva, and W. J. Schroeder. Interactive out-of-core isosurface extraction. In *Proc. of the Symp. for Volume Vis.*, pages 167–174, 1998.
15. T. Chiueh and K.-L. Ma. A parallel pipelined renderer for time-varying volume data. In *Proc of Parallel Architecture, Algorithms, Networks*, pages 9–15, 1997.
16. P. Cignoni, P. Marino, C. Montani, E. Puppo, and R. Scopigno. Speeding up isosurface extraction using interval trees. *IEEE Transaction on Visualization and Computer Graphics*, 3(2):158–170, 1997.
17. K. Cole-McLaughlin, H. Edelsbrunner, J. Harer, V. Natarajan, and V. Pascucci. Loops in Reeb graphs of 2-manifolds. In *Proc. 14th Ann. Sympos. Comput. Geom.*, pages 344–350, 2003.
18. T. H. Cormen, C. E. Leiserson, and R. L. Rivest. *Introduction to Algorithms*. MIT, Cambridge, MA, 1994.
19. M. Cox and D. Ellsworth. Application controlled demand paging for out-of-core visualization. In *IEEE Proc. of Vis. '97*, pages 235–244, 1997.
20. C. J. A. Delfinado and H. Edelsbrunner. An incremental algorithm for betti numbers of simplicial complexes on the 3-sphere. *Computer Aided Geometric Design*, 12:771–784, 1995.
21. J. R. Driscoll, N. Sarnak, D. D. Sleator, and R. E. Tarjan. Making data structures persistent. *Journal of Computer and System Sciences*, 38:86–124, 1989.

22. M. J. Dürst. Additional reference to “marching cubes”. *SIGGRAPH Computer Graphics*, 22(5):243, 1988.
23. H. Edelsbrunner and J. Harer. Jacobi sets of multiple morse functions. In F. Cucker, R. DeVore, P. Olver, and E. Sueli, editors, *Foundations of Computational Mathematics*, pages 37–57. Cambridge University Press, Cambridge, 2002.
24. H. Edelsbrunner, J. Harer, A. Mascarenhas, and V. Pascucci. Time-varying Reeb graphs for continuous space-time data. In *Proc. of the 20th Ann. Sympos. on Comp. geometry*, pages 366–372. ACM Press, 2004.
25. H. Edelsbrunner, J. Harer, V. Natarajan, and V. Pascucci. Morse-smale complexes for piecewise linear 3-manifolds. In *Proc. 19th Ann. Sympos. Comput. Geom.*, pages 361–370, 2003.
26. H. Edelsbrunner, J. Harer, and A. Zomorodian. Hierarchical morse complexes for piecewise linear 2-manifolds. In *Proceedings of the 17th Annual Symposium on Computational geometry*, pages 70–79. ACM Press, 2001.
27. A. T. Fomenko and E. T. L. Kunii. *Topological Methods for Visualization.* Springer, Tokyo, 1997.
28. H. Fuchs, Z. Kedem, and S. Uselton. Optimal surface reconstruction from planar contours. *Communications of the ACM*, 20:693–702, 1977.
29. R. S. Gallagher. Span filtering: An efficient scheme for volume visualization of large finite element models. In G. M. Neilson and L. Rosenblum, editors, *Proc. of Vis. ’91*, pages 68–75, Oct 1991.
30. A. V. Gelder and J. Wilhelms. Topological considerations in isosurface generation. *ACM Transactions on Graphics*, 13(4):337–375, 1994.
31. M. Golubitsky and V. Guillemin. *Stable mappings and their singularities.* Graduate Texts in Mathematics, Vol. 14. Springer, New York, 1973.
32. P. Hanrahan. Three-pass affine transforms for volume rendering. *Computer Graphics*, 24(5):71–78, 1990.
33. G. T. Herman and H. K. Lun. Three-dimensional display of human organs from computed tomograms. *Computer Graphics and Image Processing*, 9:1–21, 1979.
34. C. T. Howie and E. H. Black. The mesh propagation algorithm for isosurface construction. In *Computer Graphics Forum 13, Eurographics ’94 Conf. Issue*, pages 65–74, 1994.
35. M. Isenburg and S. Gumhold. Out-of-core compression for gigantic polygon meshes. In *Proc. of SIGGRAPH 2003*, pages 935–942, July 2003.
36. M. Isenburg and P. Lindstrom. Streaming meshes. In *Manuscript*, April 2004.
37. M. Isenburg, P. Lindstrom, S. Gumhold, and J. Snoeyink. Large mesh simplification using processing sequences. In *Proc. of Vis. 2003*, pages 465–472, Oct 2003.
38. J. T. Kajiya and B. P. V. Herzen. Ray tracing volume densities. *Computer Graphics*, 18(3):165–174, 1984.
39. A. Kaufman and E. Shimony. 3d scan-conversion algorithms for voxel-based graphics. In *1986 Workshop on Interactive 3D Graphics*, pages 45–75, 1986.
40. L. Kettner, J. Rossignac, and J. Snoeyink. The safari interface for visualizing time-dependent volume data using iso-surfaces and contour spectra. *Computational Geometry: Theory and Applications*, 25(1-2):97–116, 2003.
41. M. Levoy. Efficient ray tracing of volume data. *ACM Transactions on Graphics*, 9(3):245–261, 1990.
42. Y. Livnat, H. W. Shen, and C. R. Johnson. A near optimal iso-surface extraction algorithm for unstructured grids. *IEEE Transaction on Visualization and Computer Graphics*, 2(1):73–84, 1996.

43. W. E. Lorensen and H. E. Cline. Marching cubes: A high resolution 3d surface construction algorithm. In M. C. Stone, editor, *Computer Graphics (SIGGRAPH '87 Proc.)*, volume 21, pages 163–169, July 1987.
44. Y. Matsumoto. *An Introduction to Morse Theory (Translated from Japanese by K. Hudson and M. Saito)*. American Mathematical Society, 2002.
45. S. V. Matveyev. Approximation of isosurface in the marching cube: ambiguity problem. In *Proceedings of the conference on Visualization '94*, pages 288–292. IEEE Computer Society Press, 1994.
46. N. Max, R. Crawfis, and D. Williams. Visualization for climate modeling. In *IEEE Computer Graphics Applications*, pages 481–494, 2000.
47. B. H. McCormick, T. A. DeFanti, and M. D. Brown. Visualization in scientific computing. *Computer Graphics*, 21(6), 1987.
48. J. Milnor. *Morse Theory*. Princeton University Press, New Jersey, 1963.
49. C. Montani, R. Scateni, and R. Scopigno. Discretized marching cubes. In *Proceedings of the conference on Visualization '94*, pages 281–287. IEEE Computer Society Press, 1994.
50. J. R. Munkres. *Elements of Algebraic Topology*. Addison-Wesley, Redwood City, CA, 1984.
51. B. K. Natarajan. On generating topologically consistent isosurfaces from uniform samples. *Visual Computer*, 11(1):52–62, 1994.
52. G. M. Nielson and B. Hamann. The asymptotic decider: resolving the ambiguity in marching cubes. In *Proceedings of the 2nd conference on Visualization '91*, pages 83–91. IEEE Computer Society Press, 1991.
53. V. Pascucci and K. Cole-McLaughlin. Parallel computation of the topology of level sets. *Algorithmica*, 38(1):249–268, 2003.
54. G. Reeb. Sur les points singuliers d'une forme de pfaff complèment intégrable ou d'une fonction numérique. *Comptes Rendus de L'Académie ses Séances, Paris*, 222:847–849, 1946.
55. H. W. Shen. Iso-surface extraction in time-varying fields using a temporal hierarchical index tree. In *IEEE Proc. of Vis. '98*, pages 159–166, Oct 1998.
56. H. W. Shen, C. D. Hansen, Y. Livnat, and C. R. Johnson. Isosurfacing in span space with utmost efficiency (issue). In *Proc. of Vis. '96*, pages 287–294, 1996.
57. Y. Shinagawa and T. L. Kunii. Constructing a Reeb graph automatically from cross sections. *IEEE Computer Graphics and Applications*, 11:44–51, 1991.
58. B.-S. Sohn and C. L. Bajaj. Time-varying contour topology. In *Manuscript*, 2004.
59. P. Sutton and C. Hansen. Isosurface extraction in time-varying fields using a temporal branch-on-need tree(t-bon). In *IEEE Proc. of Vis. '99*, pages 147–153, 1999.
60. N. Thune and B. Olstad. Visualizing 4-D medical ultrasound data. In *Proceedings of the 2nd conference on Visualization '91*, pages 210–215. IEEE Computer Society Press, 1991.
61. M. van Kreveld. Efficient methods for isoline extraction from digital elevation model based on triangulated irregular networks. In *Sixth Inter. Symp. on Spatial Data Handling*, pages 835–847, 1994.
62. M. van Kreveld, R. von Oostrum, C. L. Bajaj, V. Pascucci, and D. R. Schikore. Contour trees and small seed sets for iso-surface traversal. In *The 13th ACM Sym. on Computational Geometry*, pages 212–220, 1997.
63. L. Westover. Interactive volume rendering. In *Chapel Hill Workshop on Volume Visualization*, pages 9–16, 1989.

64. J. Wilhelms and V. Gelder. Octrees for faster isosurface generation. *ACM Transaction on Graphics*, 11(3):201–227, 1992.
65. B. Wyvill, C. McPheeters, and G. Wyvill. Animating soft objects. *Visual Computer*, 2:235–242, 1986.
66. G. Wyvill, C. McPheeters, and B. Wyvill. Data structure for soft objects. *Visual Computer*, 2:227–234, 1986.

DeBruijn Counting for Visualization Algorithms

David C. Banks[1] and Paul K. Stockmeyer[2]

[1] Florida State University
banks@csit.fsu.edu
[2] College of William and Mary
stockmeyer@cs.wm.edu

Summary. We describe how to determine the number of cases that arise in visualization algorithms such as Marching Cubes by applying the deBruijn extension of Pólya counting. This technique constructs a polynomial, using the cycle index, encoding the case counts that arise when a discrete function (or "color") is evaluated at each vertex of a polytope. The technique can serve as a valuable aid in debugging visualization algorithms that extend Marching Cubes, Separating Surfaces, Interval Volumes, Sweeping Simplices, etc., to larger dimensions and to more colors.

1 Introduction

In 2003, Banks and Linton showed that a broad range of visualization algorithms shared a common foundation [1]. These algorithms include Marching Cubes [12], Sweeping Simplices [4, 19], Marching Squares, Contour Meshing (*a.k.a.* Marching Hypercubes) [3, 11, 18, 21], Interval Volumes [14], Generalized Marching Cubes [9] [20], and Separating Surfaces [13]. In each of these algorithms, a dataset representing a mapping $M : D \rightarrow R$ is discretized into a finite set of polytopes tiling a domain D and the range R is discretized into a finite set of values that can be viewed as an abstract set of "colors" applied to the vertices of each polytope. A substitution grammar allows the polytope to be replaced by geometry, called a "substitope," that matches a particular feature (such as a level set) in the data. The task of visualizing M is accomplished by displaying the resulting substitopes.

A typical polytope, and one that we will use for many of our examples, is the square. With two colors there are clearly $2^4 = 16$ ways to apply a color to each of the four vertices of the square. However, some of these are indistinguishable from others when we consider the symmetries of the square. For example, the four colorings with one black vertex and three white vertices are all equivalent under the symmetry group of the square, and can be treated as one case. In Example 2 we illustrate our methods by confirming that these 16 colorings fall into just six equivalence classes. Moreover, in some applications the colors might be considered interchangeable, so that the coloring equivalence class of one black vertex and three white vertices would

T. Möller et al. (eds.), *Mathematical Foundations of Scientific Visualization, Computer Graphics, and Massive Data Exploration*, Mathematics and Visualization,
DOI: 10.1007/978-3-540-49926-8,

be considered the same as the class of one white vertex and three black vertices. In Example 5 we further illustrate our methods by confirming that the six classes combine to form just four classes under this additional equivalence.

The combinatorial explosion of possible colorings makes developing and debugging new substitope algorithms difficult. Banks and Linton used a tool for computational group theory, called "GAP," to automate the process of counting cases in substitope algorithms. If G is a group acting on the set C of colorings, two colorings c_1 and c_2 are in the same orbit if $c_2 = gc_1$ for some element $g \in G$. The group G is the product *shapeGroup*×*colorGroup* of the groups acting individually on the polytope and the colors. An individual case of equivalent colorings is an "orbit" of a group of symmetries acting on the set of colorings. Typically, the full symmetry group (rotations and mirror reversals) or the direct symmetry group (rotations only) is allowed to permute the vertices of the polytope, while reversal (exchanging positive and negative, for example) or full permutation is permitted on the colors. For example, two colorings c_1 and c_2 of a cube are to be considered equivalent if c_1 can be rotated and its colors reversed in order to produce c_2.

Banks and Linton used GAP to compute the cases for different choices of *shapeGroup* and *colorGroup* that have been used in various visualization algorithms for dimension $n = 1, 2, 3, 4$ and number of colors $k = 1, 2, 3, 4$. Unfortunately, the amount of memory needed for computing cases exceeded their resources, even on a Beowulf cluster of machines, for 4-cubes having four colors. This particular situation is more than a mere intellectual curiosity; it has a practical application in visualizing an unsteady (i.e., time-varying) fluid flow via displaying its vortex cores. Jiang *et al.* demonstrated that vortex cores can be classified by discretizing the velocity vector into four sets of directions (or equivalently, four colors) [10]. They assigned one of the four colors to each vertex of a three-dimensional (3D) grid of 3-cubes. If the identity group Id_4 acts on the colorings, 2,916 distinct classes of colorings arise. But if Jiang's work is extended to unsteady flows, the 4-cubes in the resulting 4D grid (in space and time) will cluster into how many distinct cases? Determining the answer exceeded the computing resources available to them.

In 2004, Banks, Linton, and Stockmeyer showed how a technique called Pólya (approximate English pronunciation: poy-*yuh*) counting can determine the number of cases that arise when the colors are not permuted, and can do so without actually computing any of the cases themselves [2]. They reported the answer to the above question: each 4-cube, with vertex colors chosen from a set of four colors, will belong to one of 22,456,756 different cases under direct symmetry and one of 11,756,666 cases under full symmetry. But Pólya counting does not answer the question of how many cases arise when a group, different from the identity group, permutes the colors. Since the permutation of the colors is essential for reducing the number of equivalence classes of colorings in substitope algorithms, one would like to know how many such cases exist.

The answer comes from an improvement on Pólya counting due to deBruijn (approximate English pronunciation: *duh*-broin or *duh*-brown). This powerful technique computes the total number of cases that arise when the vertex colors are allowed to be permuted.

2 History and Literature

The theory of Pólya counting was developed, illustrated, and explained in [16]; an English translation, with commentary, exists as [17]. The theory extending this work to include permutations acting on colors was developed by deBruijn, first in [5] and later more completely in [6]. The formulas given in these papers are complex and difficult to use, due in part to the use of differential operators. Equivalent formulas, with much more user-friendly notation, were developed by E.M. Palmer in his Ph.D. dissertation [15]. This treatment also appears in the journal article [7] and the book [8, Chap. 6] by Harary and Palmer. In the following discussion we stay close to the Harary–Palmer treatment.

2.1 The Pólya Setting

The setting for using the deBruijn method for counting cases is similar to that for Pólya's method, which we briefly review. We are given a domain set X (vertices of the cube) and a range set Y (colors), and consider functions in the set Y^X of all functions f from X to Y that assign colors to vertices.[1] The notation Y^X is fairly common in combinatorics, and reflects the fact that the number of functions from X to Y is $|Y|^{|X|}$, so we have $|Y^X| = |Y|^{|X|}$. An element in Y^X is thus an assignment of colors to the vertices of the n-cube (or more informally, a vertex-coloring of a cube) as shown below.

$$\begin{array}{c} X \\ \downarrow f \in Y^X \\ Y \end{array} \tag{1}$$

A permutation group[2] A acts on the domain set X. It is called the "shape group" in [2]. In the case of colored cubes, the group is either the group of (orientation-preserving) direct symmetries DC_n of the n-cube or the full group FC_n of symmetries of the n-cube.

$$X \xrightarrow{\alpha \in A} X \tag{2}$$

The group A acting on X induces a permutation group A' acting on the set Y^X of functions from X to Y (our set of cube colorings). Note that the groups A and A' are isomorphic as abstract groups, since a permutation of the cube also moves the colors assigned to the vertices, but they differ as permutation groups. One difference is that the degree (number of objects being permuted) of A is $|X|$ (the number of vertices), while the degree of A' is $|Y|^{|X|}$ (the number of possible colorings of the cube). For

[1] The sets X and Y are denoted D and R, respectively, by deBruijn. Pólya does not have any notation for these sets, but calls the range set the "store of figures," with individual elements called "figures."

[2] This group is also denoted G in [6] and H in [16].

any $\alpha \in A$ acting on set X, the corresponding permutation α' acting on Y^X maps a function f onto the function $\alpha' f$ defined by

$$(\alpha'(f))(x) = f(\alpha(x))$$

for all $x \in X$. This mapping is shown in the diagram below.

$$\begin{array}{ccc} X & \xrightarrow{\alpha} & X \\ f \downarrow & \swarrow \alpha' f & \\ Y & & \end{array} \tag{3}$$

Intuitively, we want to consider two functions (vertex colorings) f_i and f_j to be equivalent if one can be obtained from the other by applying some symmetry permutation on the elements of the domain space X. Thus the objects we wish to count are the orbits[3] of the group A' acting on the set Y^X of coloring functions. The colorings are equivalent (and thus lie in the same orbit) if

$$f_j = \alpha'(f_i)$$

for some α' in the group A'. Several visualization algorithms proceed by applying a geometric substitution based on the coloring of an n-simplex or an n-cube. Determining the number of equivalence classes of colorings of an n-cube by hand is tedious and error-prone because there are so many possible cases to consider. Rather than enumerate each coloring and cluster them together as we go, we can instead exploit a result due to Pólya that counts the equivalence classes based on the cycle structure of elements in the shape group.

2.2 Pólya's Theorem

Pólya counting relies heavily on a polynomial called the "cycle index." Recall that each element of a group can be written as a permutation of symbols. For example, the group element $\alpha = (1\ 4)(2\ 3)$ contains two cycles of length 2; one maps symbol 1 to symbol 4 and 4 to 1, and the other maps symbol 2 to symbol 3 and 3 to 2. See Fig. 1. The number of cases can be obtained by Pólya's theorem, which uses the cycle index $Z(A; z_1, z_2, \ \dots \ z_d)$ of the shape group A acting on the d vertices in X. The cycle index is defined by

$$Z(A;\ z_1, z_2, \ \dots \ , z_d) = \frac{1}{|A|} \sum_{\alpha \in A} z_1^{j_1(\alpha)}\ z_2^{j_2(\alpha)}\ \cdots\ z_d^{j_d(\alpha)} \tag{4}$$

where $|A|$ is the order (number of elements) of the group A, the summation is taken over each permutation α in A, the z_i are variables, and each $j_i(\alpha)$ is the number of cycles of length i in the disjoint cycle representation of α. An example clarifies the use of the cycle index.

[3] These orbits are called "patterns" in [6]; Pólya uses the term "configuration" here, but it is not clear whether his configuration refers to a single coloring or an entire equivalence class (orbit) of colorings.

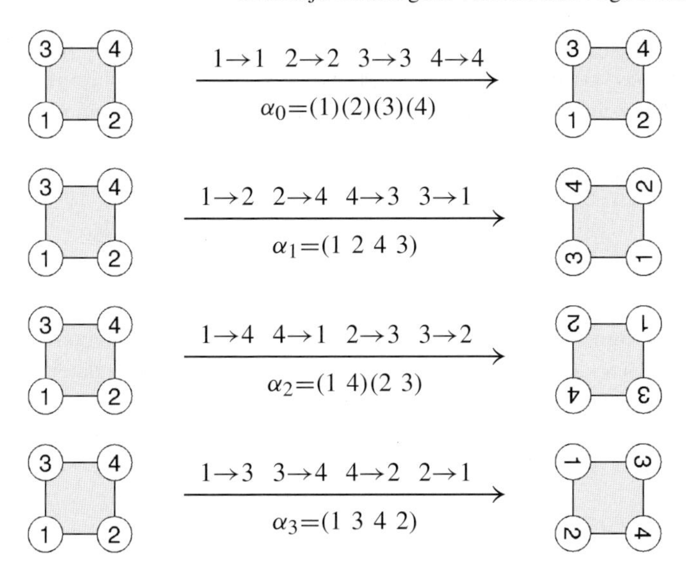

Fig. 1. The shape group DC_2 of direct symmetries acts on the 2-cube. Each of the four group elements $\{\alpha_0, \alpha_1, \alpha_2, \alpha_3\}$ permutes the vertices. Mappings of individual vertices are shown *above the arrow*; the permutation's cycle structure is shown *below the arrow*

Example 1: Cycle Index of DC_2. The (orientation-preserving) direct symmetry group of the 2-cube (i.e., the square) is DC_2. Its action on the 2-cube is shown in Fig. 1. The four elements of DC_2 can be written in permutation form as shown below, with z_α denoting the expression $z_1^{j_1(\alpha)} z_2^{j_2(\alpha)} z_3^{j_3(\alpha)} z_4^{j_4(\alpha)}$ in the summand. The non-zero cycles and non-zero exponents are indicated by boldface.

$\alpha \in A$	permutation	# cycles of length i				z_α
		$j_1(\alpha)$	$j_2(\alpha)$	$j_3(\alpha)$	$j_4(\alpha)$	
α_1	(1)(2)(3)(4)	**4**	0	0	0	$\mathbf{z_1^4}\, z_2^0\, z_3^0\, z_4^0$
α_2	(1 2 4 3)	0	0	0	**1**	$z_1^0\, z_2^0\, z_3^0\, \mathbf{z_4^1}$
α_3	(1 4)(2 3)	0	**2**	0	0	$z_1^0\, \mathbf{z_2^2}\, z_3^0\, z_4^0$
α_4	(1 3 4 2)	0	0	0	**1**	$z_1^0\, z_2^0\, z_3^0\, \mathbf{z_4^1}$

The size $|DC_2|$ of the group is 4, so the cycle index for DC_2 is the sum of the z_α's multiplied by 1/4, as shown below.

$$
\begin{aligned}
Z(DC_2;\ z_1, z_2, z_3, z_4) &= \frac{1}{4}(z_1^4 + z_4^1 + z_2^2 + z_4^1) \\
&= \frac{1}{4}(1z_1^4 + 1z_2^2 + 2z_4^1) \qquad (5)
\end{aligned}
$$

Remarkably, this polynomial characterizes the number of different cases that arise in Marching Cubes and similar visualization algorithms. We state the result that Pólya proved.

Pólya's Theorem (Constant Form):[4] The number of orbits of the permutation group A' acting on the set Y^X of functions from X to Y is obtained by replacing each variable z_i in the cycle index of the shape group A with the number $k = |Y|$ of colors. That is, when the k colors are not permuted but the vertices are, the number of cases (distinct colorings) is given by the *reduced cycle index* shown below.

$$Z(A;\ k, k, \ldots, k) = \frac{1}{|A|} \sum_{\alpha \in A} k^{j_1(\alpha)}\, k^{j_2(\alpha)} \ldots k^{j_d(\alpha)} \tag{6}$$

Example 2: Reduced Cycle Index of DC_2. The number of inequivalent ways to color the vertices of a square with k colors, with equivalence determined by the orientation-preserving group DC_2, is found using the cycle index computed in the example above, with k substituted for the variables z_i.

$$Z(DC_2;\ k, k, k, k) = \tfrac{1}{4}(1k^4 + 1k^2 + 2k) \tag{7}$$

So for $k = 1$ color the square has $(1 + 1 + 2)/4 = 1$ coloring, with $k = 2$ colors it has $(16 + 8 + 4)/4 = 6$ colorings, with $k = 3$ colors it has $(81 + 9 + 6)/4 = 24$ colorings, and so forth. These are shown in Fig. 2.

3 Permuting the Colors: The Power Group

The extension pioneered by deBruijn and Palmer involves a second permutation group B acting on the range set Y of colors. We call B the *color group*.

$$Y \xrightarrow{\beta \in B} Y \tag{8}$$

We now consider two vertex colorings to be equivalent if one can be obtained from the other by applying some permutation $\alpha \in A$ on the domain set of vertices X, and/or applying some permutation $\beta \in B$ on the range set Y of colors. Here we have two groups, A acting on the domain X and B acting on the range Y, combining to form a group of pairs $(\alpha; \beta)$, with $\alpha \in A$ and $\beta \in B$, acting on the set Y^X of functions from X to Y. Note that this new group is abstractly isomorphic to the Cartesian product $A \times B$, with order $|A| \cdot |B|$. However, the notation $A \times B$ misrepresents the nature of this group as a permutation group, since that notation usually refers to a group acting on either the set $X \times Y$ or the set $X \cup Y$. In the first case the group has degree $|X| \cdot |Y|$, and in the second case the group has degree $|X| + |Y|$. Our permutation group acts on the set Y^X and thus has degree $|Y^X| = |Y|^{|X|}$. As a permutation group, this group is thus quite different from what is usually meant by the product of two permutation groups.

[4] Historically, Polya's theorem appeared in a "weighted" version. The only previous consideration of a constant (unweighted) version was due to Palmer, who presented the constant form of his power group theorem as a warm-up for the weighted version later. Even Harary and Palmer, in their book *Graphical Enumeration*, give only the weighted form of Polya's theorem. We follow Palmer's example because of its simplicity.

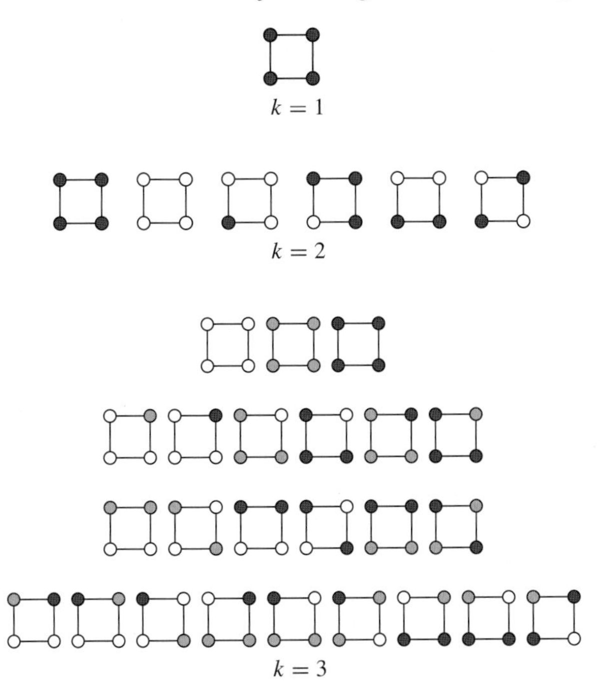

Fig. 2. Cases of k-colorings of the 2-cube permuted by the shape group DC_2 of orientation-preserving rotations

Each function f is mapped by $(\alpha;\beta)$ onto the function $(\alpha;\beta)f$ defined by

$$((\alpha;\beta)f)(x) = \beta(f(\alpha(x)))$$

for all $x \in X$. These mappings are illustrated in the diagram below.

$$\begin{array}{ccc} X & \xrightarrow{\alpha} & X \\ (\alpha;\beta)f\downarrow & & \downarrow f \\ Y & \xleftarrow{\beta} & Y \end{array} \tag{9}$$

Since the set being permuted is Y^X, Harary and Palmer call this permutation group the *power group* of groups A and B (also called the "coloring group" in [1]), and denote it B^A. The items we wish to count are the orbits of the power group B^A. The formula for counting orbits of B^A, or equivalence classes of functions (colorings) of Y^X, involves a sort of cycle index of cycle indices. It involves a sum taken over all the divisors of an integer i; recall the notation $\mathsf{s}|i$ means s divides i (divisors of i are displayed in sans-serif font). We use the Harary–Palmer formulation [8, p. 137] in stating deBruijn's counting theorem.

Power Group Enumeration Theorem (Constant Form) (deBruijn's Theorem): The number $N(A, B;\ X, Y)$ of orbits of the power group B^A acting on the set Y^X is

$$N(A, B;\ X, Y) = \frac{1}{|B|} \sum_{\beta \in B} Z(A; c_1(\beta), c_2(\beta), \ldots, c_d(\beta)) \qquad (10)$$

where

$$c_i(\beta) = \sum_{\mathsf{s}|i} \mathsf{s}\, j_{\mathsf{s}}(\beta). \qquad (11)$$

Each $j_{\mathsf{s}}(\beta)$ is the number of cycles of length s in the disjoint cycle representation of β; the sum is over all positive integers s that divide the subscript i. We work several examples to clarify the use of this theorem.

Example 3: Color Group Id_k. When the color group is the identity, the colors are not changed and so the power-group formula devolves into the ordinary Pólya formula. To see this, suppose we have k colors, so $Y = \{1, 2, 3, \ldots, k\}$, and B is the identity group Id_k consisting of just the identity permutation $\beta_0 = (1)(2)\cdots(k)$. This permutation has k 1-cycles. For all other $i > 1$ there are 0 i-cycles, as shown in the table below.

$\beta \in Id_k$	permutation	1-cycles $j_1(\beta)$	2-cycles $j_2(\beta)$	3-cycles $j_3(\beta)$	...
β_0	(1)(2)...(k)	k	0	0	...

So we have $j_1(\beta_0) = k$ and $j_i(\beta_0) = 0$ for $i \neq 1$, yielding the following values for c_i.

$$\begin{aligned} c_1(\beta_0) &= \sum_{\mathsf{s}|1} \mathsf{s}\, j_{\mathsf{s}}(\beta_0) \\ &= 1 j_1(\beta_0) && \textit{since } 1 \mid 1 \\ &= k \end{aligned}$$

$$\begin{aligned} c_2(\beta_0) &= \sum_{\mathsf{s}|2} \mathsf{s}\, j_{\mathsf{s}}(\beta_0) \\ &= 1 j_1(\beta_0) + 2 j_2(\beta_0) && \textit{since } 1,2 \mid 2 \\ &= 1k \quad + 2 \cdot 0 \\ &= k \end{aligned}$$

$$\begin{aligned} c_3(\beta_0) &= \sum_{\mathsf{s}|3} \mathsf{s}\, j_{\mathsf{s}}(\beta_0) \\ &= 1 j_1(\beta_0) + 3 j_3(\beta_0) && \textit{since } 1,3 \mid 3 \\ &= 1k \quad + 3 \cdot 0 \\ &= k \end{aligned}$$

$$
\begin{aligned}
c_4(\beta_0) &= \sum_{s|4} s\, j_s(\beta_0) \\
&= 1 j_1(\beta_0) + 2 j_2(\beta_0) + 4 j_4(\beta_0) && \textit{since } 1,2,4 \mid 4 \\
&= 1k \quad + 2 \cdot 0 \quad + 4 \cdot 0 \\
&= k \\
&\vdots
\end{aligned}
$$

We could continue calculations all the way to c_k, but it is easily seen that $c_i(\beta_0) = k$ for all i. The number of inequivalent functions is found by simply replacing each c_i in deBruijn's theorem (10) with the number k of colors. Thus we have

$$
\begin{aligned}
N(A, Id_k;\ X, \{1, 2, \ldots, k\}) &= \tfrac{1}{1} Z(A; c_1(\beta_0), c_2(\beta_0), \ldots, c_k(\beta_0)) \\
&= Z(A;\ k,\ k, \ldots,\ k)
\end{aligned}
$$

as it was in (6). Thus the power group enumeration theorem reduces to Pólya's theorem when the color group B acting on the color set Y is the trivial group that leaves colors unpermuted.

3.1 Color Groups *Rev*$_2$ and *Rev*$_3$

When the color group is nontrivial, the strength of deBruijn's power group enumeration theorem is needed in order to determine the distinct number of cases. Two nontrivial color groups have typically been used in analyzing cases of visualization algorithms like Marching Cubes: the reversal group Rev_k that swaps "opposite" colors, and the symmetric group S_k that permutes the k colors in all $k!$ possible ways. In this section we show how deBruijn counting can be applied to the reversal group acting on the colors in order to produce a closed-form polynomial in k that yields the number of distinct colorings of an n-cube. We begin with examples of reversing two or three colors (see Figs. 3 and 4), which serve as a brief tutorial in the use of deBruijn counting. We then extend these examples to reversals of any even or odd number k of colors.

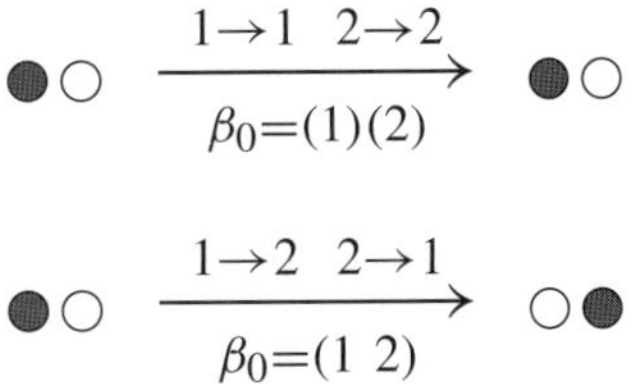

●○ $\xrightarrow[\beta_0=(1)(2)]{1\to 1\ \ 2\to 2}$ ●○

●○ $\xrightarrow[\beta_0=(1\ 2)]{1\to 2\ \ 2\to 1}$ ○●

Fig. 3. The color group $Rev_2 = \{\beta_0, \beta_1\}$ of reversals acts on colors $\{1, 2\}$. Mappings of the individual colors are shown *above the arrow*; the cycle structure of the permutation is shown *below the arrow*

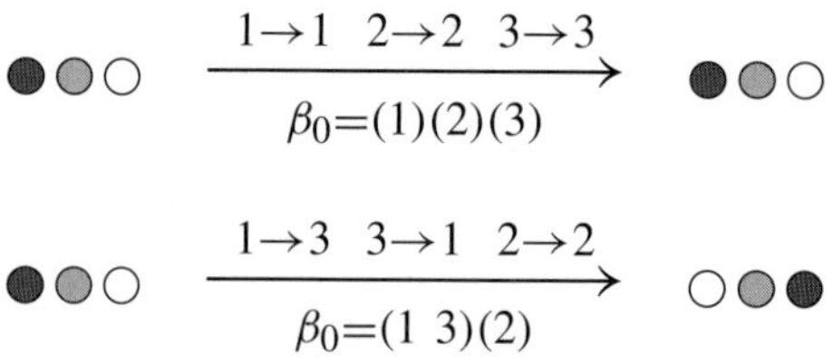

$$\bullet\bullet\circ \xrightarrow[\beta_0=(1)(2)(3)]{1\to 1\;\; 2\to 2\;\; 3\to 3} \bullet\bullet\circ$$

$$\bullet\bullet\circ \xrightarrow[\beta_0=(1\ 3)(2)]{1\to 3\;\; 3\to 1\;\; 2\to 2} \circ\bullet\bullet$$

Fig. 4. The color group $Rev_3 = \{\beta_0, \beta_1\}$ of reversals acts on colors $\{1, 2, 3\}$. Mappings of the individual colors are shown *above the arrow*; the cycle structure of the permutation is shown *below the arrow*

Example 4: Color Group *Rev*$_2$. Suppose we have two interchangeable colors (corresponding to positive and negative values at a vertex in Marching Cubes) forming the set $Y = \{\text{positive, negative}\}$. They are permuted by the group $B = \{\beta_0, \beta_1\} = Rev_2$, where the identity permutation $\beta_0 = (1)(2)$ contains 2 1-cycles and $\beta_1 = (1\ 2)$, containing 1 2-cycle, accomplishes the swap of positive with negative. The values of $j_s(\beta)$ are shown in the table below, and are used in the ensuing calculations of c_i.

$\beta \in Rev_2$	permutation	1-cycles $j_1(\beta)$	2-cycles $j_2(\beta)$	3-cycles $j_3(\beta)$	...
β_0	(1)(2)	**2**	0	0	...
β_1	(1 2)	0	**1**	0	...

We compute values of c_i for the identity permutation β_0, which contains only the pair of 1-cycles (1)(2).

$$\begin{aligned}
c_1(\beta_0) &= \sum_{s|1} s\, j_s(\beta_0) \\
&= 1 j_1(\beta_0) && \textit{since } 1 \mid 1 \\
&= 2 \\
c_2(\beta_0) &= \sum_{s|2} s\, j_s(\beta_0) \\
&= 1 j_1(\beta_0) + 2 j_2(\beta_0) && \textit{since } 1,2 \mid 2 \\
&= 1 \cdot 2 \quad + 2 \cdot 0 \\
&= 2 \\
&\vdots
\end{aligned}$$

It is evident that the subsequent terms all evaluate to 2, since 1 divides every value of i (thus contributing $1 j_1(\beta_0) = 2$ to the total) and since $j_s(\beta_0)$ evaluates to 0 for all the other $s \neq 1$. That means

$$c_i(\beta_0) = 2 \tag{12}$$

for all i. We next compute values of c_i for β_1, which contains only the single 2-cycle (1 2).

$$
\begin{aligned}
c_1(\beta_1) &= \sum_{s|1} s j_s(\beta_0) \\
&= 1 j_1(\beta_0) && \textit{since } 1 \mid 1 \\
&= 0 \\
c_2(\beta_1) &= \sum_{s|2} s j_s(\beta_0) \\
&= 1 j_1(\beta_0) + 2 j_2(\beta_0) && \textit{since } 1{,}2 \mid 2 \\
&= 1 \cdot 0 \quad + 2 \cdot 1 \\
&= 2 \\
&\vdots
\end{aligned}
$$

It is evident that the subsequent terms will evaluate to 0 whenever i is odd, since the term $2 j_2(\beta_1) = 2$ only appears in the summation when $2|i$, with all other values of j_s being 0. By similar reasoning, whenever i is even, c_i evaluates to 2. Thus

$$
c_i(\beta_1) = \begin{cases} 0 \text{ if } i \text{ is odd} \\ 2 \text{ if } i \text{ is even.} \end{cases} \tag{13}
$$

We now combine the two partial results from (12) and (13) to compute the number of inequivalent colorings

$$
\begin{aligned}
&N(A, Rev_2;\ n\text{-cube}, \{1, 2\}) = \\
&\tfrac{1}{2}\left(Z(A; 2, 2, 2, 2, 2, \dots) + Z\left(A; 0, 2, 0, 2, 0, \dots\right)\right)
\end{aligned} \tag{14}
$$

according to (10) from deBruijn's theorem. We will see in a later example that the patterns $2, 2, 2, 2, \dots$ and $0, 2, 0, 2, \dots$ generalize to $k, k, k, k, \dots$ and $0, k, 0, k, \dots$ for even values of k.

Example 5: N($DC_2, Rev_2; n$-cube, {1,2}.) Returning to the example of the colored square, we compute the number of its inequivalent colorings with shape group $A = DC_2$ and with color group $B = Rev_2$ given above. We recall the cycle index

$$
Z(DC_2;\ z_1,\ z_2,\ z_3,\ z_4) = \tfrac{1}{4}(z_1^4 + z_2^2 + 2z_4^1)
$$

that we computed in (5), evaluate it with arguments c_i from (12) and (13) above, and insert the result into (14) for $N(DC_2, Rev_2;\ 2\text{-cube}, \{1, 2\})$ above. The result is

$$
\begin{aligned}
&\tfrac{1}{2}\Big(\ Z(DC_2; 2, 2, 2, 2) + Z(DC_2; 0, 2, 0, 2)\Big) \\
&= \tfrac{1}{2}\Big(\tfrac{1}{4}(2^4 + 2^2 + 2 \cdot 2^1) + \tfrac{1}{4}(0^4 + 2^2 + 2 \cdot 2^1)\Big) \\
&= 4
\end{aligned}
$$

inequivalent 2-colorings of the square. These four colorings are shown at the top of Fig. 5. They differ from the six colorings for $k = 2$ in Fig. 2 because the 4-white case is permuted into the 4-black case and the 3-white case is permuted into the 3-black case when β_1 exchanges colors 1 and 2. This example of

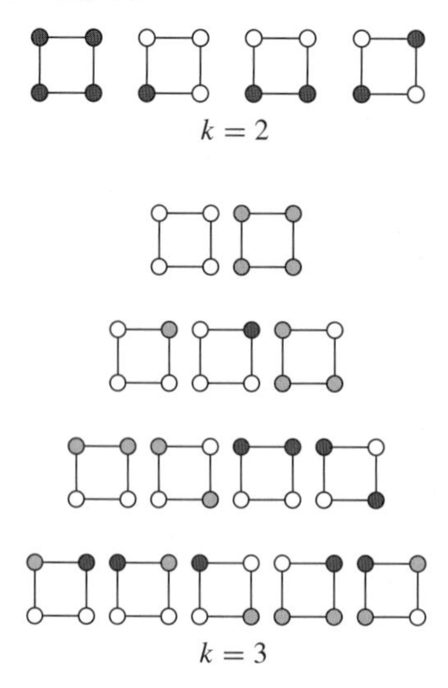

Fig. 5. Cases of k-colorings of the 2-cube permuted by the shape group DC_2 of orientation-preserving rotations, with colors permuted by Rev_k for $k = 2, 3$

$N(DC_2, Rev_2;\ n\text{-cube}, \{1, 2\})$, (and also that of the full symmetry shape group FC_2) appears in [8, p. 137]. All the later examples, unless otherwise noted, are original to this paper.

Example 6: Color Group *Rev*$_3$. Suppose we have three colors, corresponding to positive, zero, and negative values of a scalar function $h(x)$-c as used in Marching Cubes. If we allow interchanging positive and negative then we have the colors $Y = \{+, 0, -\}$ and color group $B = \{\beta_0, \beta_1\} = Rev_3$, where $\beta_0 = (1)(2)(3)$ is the identity, and $\beta_1 = (1\ 3)(2)$ swaps positive with negative. The cycle structure of Rev_3 is summarized in the table below.

$\beta \in Rev_3$	permutation	1-cycles $j_1(\beta)$	2-cycles $j_2(\beta)$	3-cycles $j_3(\beta)$	... ...
β_0	(1) (2) (3)	**3**	0	0	...
β_1	(1 3) (2)	**1**	**1**	0	...

For β_0 we have $j_1(\beta_0) = 3$ and $j_i(\beta_0) = 0$ for $i \neq 1$ (similar to the example above for Rev_2). So

$$c_i(\beta_0) = 1j_1(\beta_0) = 3 \tag{15}$$

for all i. For β_1 we have $j_1(\beta_1) = 1$, $j_2(\beta_1) = 1$, and $j_i(\beta_1) = 0$ for $i > 2$. So

$$c_i(\beta_1) = \begin{cases} 1j_1(\beta_1) & = 1 \text{ if } i \text{ is odd} \\ 1j_1(\beta_1) + 2j_2(\beta_1) & = 3 \text{ if } i \text{ is even.} \end{cases} \tag{16}$$

Then the number of inequivalent functions is

$$N(A, Rev_3;\ n\text{-cube}, \{1, 2, 3\}) = \\ \tfrac{1}{2}\Big(Z(A; 3, 3, 3, 3, 3, \ldots) + Z(A; 1, 3, 1, 3, 1, \ldots)\Big) \quad (17)$$

according to (10) from deBruijn's theorem. We will see in a later example that the patterns $3, 3, 3, 3, \ldots$ and $1, 3, 1, 3, \ldots$ generalize to $k, k, k, k, \ldots$ and $1, k, 1, k, \ldots$ for odd values of k.

For the square permuted by the shape group DC_2 we substitute into (10), using (15) and (16), to compute

$$\begin{aligned} &\tfrac{1}{2}\Big(Z(DC_2; 3, 3, 3, 3) \;+\; Z(DC_2; 1, 3, 1, 3)\Big) \\ &= \tfrac{1}{2}\Big(\tfrac{1}{4}(3^4 + 3^2 + 2\cdot 3^1) + \tfrac{1}{4}(1^4 + 3^2 + 2\cdot 3^1)\Big) \\ &= 14 \end{aligned}$$

which is less than the 24 cases shown in Fig. 2 because black and white colors (i.e., positive and negative values) can be swapped by the color group to introduce new equivalences among the colorings. Representatives of these 14 colorings are illustrated in Fig. 5. (Note that the first two colorings on the bottom-most row are mirror images of each other. Under the full symmetry group FC_2, these two become equivalent and the number of cases is reduced to 13.)

3.2 Color Groups Rev_k when k Is Even

We next extend the examples for the color group, generalizing to Rev_k for k even or k odd (which have different cycle structures, and hence different cycle indices).

Example 7: color group Rev$_k$ (k even). Suppose we have k colors, k even (so $k = 2\kappa$), with $B = \{\beta_0, \beta_1\} = Rev_{2\kappa}$, where the identity element $\beta_0 = (1)(2)\cdots(k)$ contains k 1-cycles, and element $\beta_1 = (1\ k)(2\ k-1)\cdots(\frac{k}{2}\ \frac{k+2}{2})$, containing $\frac{k}{2}$ 2-cycles, performs the swaps. The cycle structure of B is shown in the table below.

$\beta \in Rev_{2\kappa}$	permutation	1-cycles $j_1(\beta)$	2-cycles $j_2(\beta)$	3-cycles $j_3(\beta)$	...
β_0	$(1)(2)\cdots(k)$	$\boldsymbol{k}$	0	0	...
β_1	$(1\ k)(2\ k-1)\cdots(\frac{k}{2}\ \frac{k+2}{2})$	0	$\boldsymbol{\frac{k}{2}}$	0	...

For β_0 we have $j_1(\beta_0) = k$ and $j_s(\beta_0) = 0$ for all $s \neq 1$, so

$$c_i(\beta_0) = 1 j_1(\beta_0) = k \quad (18)$$

for all i. For β_1 we have $j_2(\beta_1) = \frac{k}{2}$ and $j_s(\beta_1) = 0$ for all $s \neq 2$, so

$$c_i(\beta_1) = \begin{cases} 1 j_1(\beta_1) & = 0 \text{ if } i \text{ is odd} \\ 1 j_1(\beta_1) + 2 j_2(\beta_1) & = k \text{ if } i \text{ is even.} \end{cases} \quad (19)$$

Then the number of inequivalent colorings is

$$N(A,\, Rev_{2\kappa};\ n\text{-cube}, \{1, \ldots, 2\kappa\}) = \tfrac{1}{2}\Big(\ Z(A; k, k, k, k, k,\ \ldots) + Z(A; 0, k, 0, k, 0,\ \ldots)\Big). \tag{20}$$

Example 8: Marching Cubes. We can verify the expression above with the 3-cube, for which the number of 2-colorings was found by Lorensen and Cline as 15 (for direct symmetry DC_3) and 14 (for full symmetry FC_3). The cycle indices

$$Z(DC_3; z_1, \ldots, z_d) = \tfrac{1}{24}\left(1z_1^8 + 8z_1^2z_3^2 + 9z_2^4 + 6z_4^2\right) \tag{21}$$

$$Z(FC_4; z_1, \ldots, z_d) = \tfrac{1}{48}\left(1z_1^8 + 6z_1^4z_2^2 + 8z_1^2z_3^2 + 13z_2^4 + 8z_2^1z_6^1 + 12z_4^2\right) \tag{22}$$

for the 3-cube was derived in [2, Fig. 12]. In order to make the substitution of Z into deBruijn's formula and substitution of the patterns of 0's and k's into the arguments z_i easier for the reader to follow, we highlight the patterns here in boldface. Using the formula for the number of inequivalent colorings from (20) above and letting shape group $A = DC_3$, we substitute ***k,k,k,k,***... and $\mathbf{0}, \boldsymbol{k}, \mathbf{0}, \boldsymbol{k}$,... into the arguments for the cycle index in (21) to get

$$Z(DC_3;\ \boldsymbol{k,k,k,k}, \ldots) = \tfrac{1}{24}\left(1k^8 + 8k^2k^2 + 9k^4 + 6k^2\right) = \tfrac{1}{24}\left(1k^8 + 17k^4 + 6k^2\right)$$

and

$$Z(DC_3;\ \mathbf{0}, \boldsymbol{k}, \mathbf{0}, \boldsymbol{k}, \ldots) = \tfrac{1}{24}\left(1 \cdot 0^8 + 8 \cdot 0^2 \cdot 0^2 + 9k^4 + 6k^2\right) = \tfrac{1}{24}\left(9k^4 + 6k^2\right)$$

which combine by (10) in deBruijn's theorem to produce the count

$$\begin{aligned} &N(DC_3, Rev_k;\ 3\text{-cube}, \{1..k\}) \\ &= \tfrac{1}{2}\,(Z(DC_3;\ \boldsymbol{k,k,k,k}, \ldots) \quad + Z(DC_3;\ \mathbf{0}, \boldsymbol{k}, \mathbf{0}, \boldsymbol{k}, \ldots)) \\ &= \tfrac{1}{2}\left(\frac{1}{24}\left(1k^8 + 17k^4 + 6k^2\right) + \frac{1}{24}\left(9k^4 + 6k^2\right)\right) \\ &= \tfrac{1}{48}\left(1k^8 + 26k^4 + 12k^2\right). \end{aligned} \tag{23}$$

Equation (23) gives a closed-form solution to the problem of counting cases for Marching Cubes with an even number of colors and with the orientation-preserving rotations applied to the 3-cube.

We substitute into the cycle index for FC_3 [(22) above] to get

$$Z(FC_3;\ \mathbf{k,k,k,k,}\ldots)$$
$$= \tfrac{1}{48}\left(1k^8 + 6k^4k^2 + 8k^2k^2 + 13k^4 + 8k^1k^1 + 12k^2\right)$$
$$= \tfrac{1}{48}\left(1k^8 + 6k^6 + 21k^4 + 20k^2\right)$$

and

$$Z(FC_3;\ \mathbf{0},\boldsymbol{k},\mathbf{0},\boldsymbol{k},\ldots)$$
$$= \tfrac{1}{48}\left(1\cdot 0^8 + 6\cdot 0^4k^2 + 8\cdot 0^2\cdot 0^2 + 13k^4 + 8k^1k^1 + 12k^2\right)$$
$$= \tfrac{1}{48}\left(13k^4 + 20k^2\right)$$

which combine by (10) in deBruijn's theorem to produce the count

$$N(FC_3, Rev_k;\ \text{3-cube}, \{1..k\})$$
$$= \tfrac{1}{2}\left(Z(FC_3;\ \mathbf{k,k,k,k,}\ldots) \qquad + Z(FC_3;\ \mathbf{0},\boldsymbol{k},\mathbf{0},\boldsymbol{k},\ldots)\right)$$
$$= \tfrac{1}{2}\left(\tfrac{1}{48}\left(1k^8 + 6k^6 + 21k^4 + 20k^2\right) + \tfrac{1}{48}\left(13k^4 + 20k^2\right)\right)$$
$$= \tfrac{1}{96}\left(1k^8 + 6k^6 + 34k^4 + 40k^2\right) \tag{24}$$

of distinct colorings. This is the closed-form solution for counting cases in Marching Cubes for any even number of colors where both rotations and mirror-flips are permitted on the 3-cube.

When evaluated at $k = 2$, (23) and (24) produce the numbers 15 and 14, agreeing with the number of cases found by Lorensen and Cline. When evaluated at $k = 4$, these formulas produce the numbers 1,508 and 1,036, agreeing with the number of cases found by Banks and Linton using GAP.

Example 9: Marching Hypercubes. In their table of case counts, Banks and Linton were not able to calculate the exact number of cases for $N(A, Rev_4;\ \text{4-cube}, \{1, 2\})$, corresponding to the 4-colored 4-cube [1, Fig. 7]. Counting all possible colorings through brute-force enumeration was unattainable even thought they used a high-performance cluster of computers; but with the deBruijn counting technique it becomes a simple problem that can be solved using paper and pencil. We merely evaluate the cycle index for the shape group DC_4 or FC_4, using (20) derived for k even, and evaluate at $k = 4$. The cycle indices for the 4-cube

$$Z(DC_4; z_1, \ldots, z_d) =$$
$$\tfrac{1}{192}\left(1z_1^{16} + 12z_1^4z_2^6 + 32z_1^4z_3^4 + 31z_2^8 + 32z_2^2z_6^2 + 36z_4^4 + 48z_8^2\right)$$
$$Z(FC_4; z_1, \ldots, z_d) =$$
$$\tfrac{1}{384}\left(1z_1^{16} + 12z_1^8z_2^4 + 12z_1^4z_2^6 + 32z_1^4z_3^4 + 48z_1^2z_2z_4^3\right.$$
$$\left. + 51z_2^8 + 96z_2^2z_6^2 + 84z_4^4 + 48z_8^2\right)$$

were derived in [2, Fig. 12]. The first index gives

$$Z(DC_4;\ \mathbf{k,k,k,k,}\ \ldots)$$
$$= \tfrac{1}{192}\left(1k^{16} + 12k^4k^6 + 32k^4k^4 + 31k^8 + 32k^2k^2 + 36k^4 + 48k^2\right)$$
$$= \tfrac{1}{192}\left(1k^{16} + 12k^{10} + 63k^8 + 68k^4 + 48k^2\right)$$

and

$$Z(DC_4;\ \mathbf{0,}\boldsymbol{k}\mathbf{,0,}\boldsymbol{k}\mathbf{,}\ \ldots)$$
$$= \tfrac{1}{192}\left(0 + 0 + 0 + 31k^8 + 32k^2k^2 + 36k^4 + 48k^2\right)$$
$$= \tfrac{1}{192}\left(\ 31k^8 + 68k^4 + 48k^2\right)$$

for the direct symmetry group DC_4 acting on the 4-cube. Adding them together according to (10) from deBruijn's theorem and simplifying terms, we get a closed-form solution to the number of inequivalent colorings.

$$N(DC_4,\ Rev_k;\ \text{4-cube},\ \{1,\ldots,k\})$$
$$= \tfrac{1}{2}\left(Z(DC_4;\ \mathbf{k,k,k,k,}\ \ldots) + Z(DC_4;\ \mathbf{0,}\boldsymbol{k}\mathbf{,0,}\boldsymbol{k}\mathbf{,}\ \ldots)\right)$$
$$= \tfrac{1}{384}\left(1k^{16} + 12k^{10} + 94k^8 + 136k^4 + 96k^2\right) \tag{25}$$

When $k = 2$ the result is 272 (which agrees with the table in [1]); when $k = 4$ the result is

$$N(DC_4,\ Rev_2;\ \text{4-cube},\ \{1,2,3,4\}) = 11,233,716 \tag{26}$$

answering the open question raised by Banks and Linton.

The reader can verify that evaluating (20) with $Z(FC_4; z_1,\ldots,z_d)$ produces

$$\tfrac{1}{768}\left(1k^{16} + 12k^{12} + 12k^{10} + 134k^8 + 48k^6 + 360k^4 + 96k^2\right) \tag{27}$$

as the closed-form formula for the number of inequivalent 4-colored cases. When $k = 2$ the result is 222 (agreeing with the table); when $k = 4$ the result is

$$N(DC_4,\ Rev_2;\ \text{4-cube},\ \{1,2,3,4\}) = 5,882,746 \tag{28}$$

which establishes the other missing value in the table of case-counts for the action of Rev_k on colors.

These results complete two entries of the table that were missing in [1] and [2]. They also give one possible answer to a problem raised by Weigle and Banks in 1996. Marching Cubes produces level sets $h(x, y, z) = c$ of a real-valued function; Weigle and Banks extended the algorithm to produce level sets $h(u, v) = c$ of complex functions, with $u, v, c \in \mathbb{C}$. Each variable has a real and an imaginary component, so the domain $\mathbb{C}^2$ can be treated as the 4-dimensional space $\mathbb{R}^4$ tiled by 4-cubes. The level set is invariant under negation; that is,

$$h(u, v)\text{-}c \ = \ -(h(u, v)\text{-}c) \ = \ 0$$

which permits swapping of the "colors," namely, the real part and the imaginary part of $h(u, v)$-c at each vertex. Letting β_1 be the permutation

$$\beta_1 = \Big(Re(h\text{-}c) - Re(h\text{-}c)\Big) \; \Big(Im(h\text{-}c) - Im(h\text{-}c)\Big)$$

the swaps the positive and negative real and imaginary parts of h-c, we have the color group Rev_4 acting on the 4-cube. How many cases arise? This question is answered by (26) and (28) above using deBruijn counting.

3.3 Color Groups Rev_k when k Is Odd

We next present the formula for counting cases when the colors are permuted by Rev_k for k odd. This generalizes the example for Rev_3 in Example 6 above.

Example 10: color group Rev_k (k odd). Suppose we have k colors, k odd ($k = 2\kappa + 1$), with $B = \{\beta_0, \beta_1\} = Rev_k$, where $\beta_0 = (1)(2)\cdots(k)$ is the identity and opposite colors are swapped by $\beta_1 = (1\ k)(2\ \ k\text{-}1)\cdots(\frac{k-1}{2}\ \frac{k+3}{2})(\frac{k+1}{2})$. Note that the middle color of an odd-numbered set is its own opposite, thus remaining fixed. The cycle structure of $Rev_{2\kappa+1}$ is shown in the table below.

$\beta \in Rev_{2\kappa+1}$	permutation	1-cycles $j_1(\beta)$	2-cycles $j_2(\beta)$	... ...
β_0	$(1)(2)\cdots(k)$	$\boldsymbol{k}$	0	...
β_1	$(1\ k)(2\ \ k\text{-}1)...(\frac{k\text{-}1}{2}\ \frac{k+3}{2})(\frac{k+1}{2})$	**1**	$\mathbf{\frac{k\text{-}1}{2}}$	...

For β_0 we have $j_1(\beta_0) = k$ and $j_i(\beta_0) = 0$ for $i \neq 1$, so

$$c_i(\beta_0) \;=\; 1 j_1(\beta_0) \;=\; k \tag{29}$$

for all i. For β_1 we have $j_1(\beta_1) = 1$, $j_2(\beta_1) = \frac{k-1}{2}$, and $j_i(\beta_1) = 0$ when $i > 2$, so

$$c_i(\beta_1) = \begin{cases} 1 j_1(\beta_1) & = \ 1 \text{ if } i \text{ is odd} \\ 1 j_1(\beta_1) + 2 j_2(\beta_1) & = \ k \text{ if } i \text{ is even.} \end{cases} \tag{30}$$

Then the number of inequivalent functions is

$$\begin{aligned} & N\,(A, Rev_k;\ n\text{-cube}, \{1..k\}) \\ & = \tfrac{1}{2}(Z(A; k, k, k, k, k, \ldots) + Z(A; 1, k, 1, k, 1, \ldots)) \end{aligned} \tag{31}$$

when k is odd.

The number of cases arising from the actions of color groups DC_n and FC_n and shape groups Id_k and Rev_k can therefore be written explicitly as polynomials in k. This completely solves the question of how many cases arise for colorings of n-cubes in visualization algorithms that traverse cube-shaped tiles and evaluate functions that can be reversed (or negated) on a feature set of interest. In order to complete the case-counts that are still missing from the taxonomy in [2], all that remains is to develop the deBruijn formulas for the symmetric group. We have already begun this process, which will fill the two unsolved case-counts still remaining in the table in [2].

Taking the polynomials $Z(A, Rev_k;\ n\text{-cube}, \{1, 2, \cdots, k\})$ derived in [2], substituting them into (10), and evaluating them with formulas (20) and (31) for even and odd k, one can derive the complete set of polynomials that generate all 32 of the Rev_k entries in the table by Banks and Linton, and can extend the table arbitrarily far to the right (in the k direction). Figure 6 presents this collection of case-count polynomials, which the reader can derive from the principles and examples found in this paper together with the cycle indices in [2].

n	$N(DC_n, Rev_k;\ n\text{-cube}, \{1, 2, \ldots, k\})$	
1	$\frac{1}{2}\left(1k^2\right)$	k even
	$\frac{1}{2}\left(1k^2 + 1\right)$	k odd
2	$\frac{1}{8}\left(1k^4 + 2k^2 + 4k\right)$	k even
	$\frac{1}{8}\left(1k^4 + 2k^2 + 4k + 1\right)$	k odd
3	$\frac{1}{48}\left(1k^8 + 26k^4 + 12k^2\right)$	k even
	$\frac{1}{48}\left(1k^8 + 26k^4 + 12k^2 + 9\right)$	k odd
4	$\frac{1}{384}\left(1k^{16} + 12k^{10} + 94k^8 + 0k^6 + 136k^4 + 96k^2\right)$	k even
	$\frac{1}{384}\left(1k^{16} + 12k^{10} + 94k^8 + 12k^6 + 136k^4 + 96k^2 + 33\right)$	k odd

n	$N(FC_n, Rev_k;\ n\text{-cube}, \{1, 2, \ldots, k\})$	
1	$\frac{1}{4}\left(1k^2 + 2k\right)$	k even
	$\frac{1}{4}\left(1k^2 + 2k + 1\right)$	k odd
2	$\frac{1}{16}\left(1k^4 + 2k^3 + 6k^2 + 4k\right)$	k even
	$\frac{1}{16}\left(1k^4 + 2k^3 + 6k^2 + 6k + 1\right)$	k odd
3	$\frac{1}{96}\left(1k^8 + 6k^6 + 34k^4 + 40k^2\right)$	k even
	$\frac{1}{96}\left(1k^8 + 6k^6 + 34k^4 + 46k^2 + 9\right)$	k odd
4	$\frac{1}{768}\left(1k^{16} + 12k^{12} + 12k^{10} + 134k^8 + 48k^6 + 360k^4 + 96k^2\right)$	k even
	$\frac{1}{768}\left(1k^{16} + 12k^{12} + 12k^{10} + 134k^8 + 60k^6 + 420k^4 + 96k^2 + 33\right)$	k odd

Fig. 6. Closed-form expressions for the case-counts of k-colored n-cubes, where colorings within a case are equivalent under the action of the shape group A (either the direct symmetry DC_n or the full symmetry FC_n) and the color group Rev_k

4 Summary

In this paper we showed how the number of equivalence classes (cases) that arise in visualization algorithms can be counted without explicitly generating their members. The strategy is to use deBruijn's extension of Pólya's counting technique. A shape group A acts on the vertices of an n-cube tiling a domain, while a color group B acts on the k abstract "colors" assigned to each vertex. The two groups together act on the colorings of the figures. Two colorings are equivalent if one can be mapped to the other via elements of A and B. In practice, determining the size of the equivalence classes is a combinatorial problem that becomes much too large to accomplish by hand when $n > 3$, which limits the creation of, or extensions of, visualization algorithms that depend on substituting geometry for polytopes based on coloring.

The deBruijn counting technique makes this class of enumeration problems tractable. In fact, we can write closed-form expressions for the case counts when $B = Rev_k$ (the group that reverses k colors). This allows us to determine the exact number of cases for situations that previously required too much time and memory to compute by brute force. Knowing the exact number is a necessary first step for creating a look-up table that allows a visualization algorithm to replace grid cells with features.

References

1. David C. Banks and Stephen A. Linton. Counting cases in marching cubes: Toward a generic algorithm for producing substitopes. In *Proceedings of Visualization 2003*, pages 51–58. IEEE, 2003.
2. David C. Banks, Stephen A. Linton, and Paul K. Stockmeyer. Counting cases in substitope algorithms. *IEEE Transactions on Visualization and Computer Graphics*, 10(4):371–384, 2004.
3. Praveen Bhaniramka, Rephael Wenger, and Roger Crawfis. Isosurfacing in higher dimensions. In *Proceedings of IEEE Visualization 2000*, pages 267–273. IEEE, 2000.
4. Jules Bloomenthal. Polygonization of implicit surfaces. In *Computer Aided Geometric Design*, volume 5, pages 341–355, 1988.
5. N. G. deBruijn. Generalization of pólya's fundamental theorem in enumerative combinatorial analysis. *Nederl. Akad. Wetensch. Proc. Ser. A* **62** = *Indag. Math.* **21**, pages 59–69, 1959.
6. N. G. deBruijn. Pólya's theory of counting. In Edwin F. Bechenbach, editor, *Applied Combinatorial Mathematics*, pages 144–184. Wiley, New York, 1964.
7. Frank Harary and Ed Palmer. The power group enumeration theorem. *Journal of Combinatorial Theory*, 1:157–173, 1966.
8. Frank Harary and Edgar Palmer. *Graphical Enumeration*. Academic, New York, 1973.
9. Hans-Christian Hege, Martin Seebass, Detlev Stalling, and Malte Zöckler. *A Generalized Marching Cubes Algorithm Based on Non-Binary Classifications*. Konrad-Zuse-Zentrum für Informationstechnik Berlin, 1997. Technical Report SC 97-05.
10. Ming Jiang, Raghu Machiraju, and David Thompson. A novel approach to vortex core region detection. In *Proceedings of the symposium on data visualization 2002*, pages 217–225. Eurographics Association, 2002.

11. Stefan F. Kirchberg. *Marching Hypercubes – ein Verfahren zur Konstruktion von Hyperflächen aus 4D-Rasterdaten.* Universität Dortmund, 1993. Doplomarbeit am Lehrstuhl 7.
12. William E. Lorensen and Harvey E. Cline. Marching cubes: A high resolution 3D surface construction algorithm. In *Proceedings of SIGGRAPH 1987*, pages 163–169. ACM, 1987.
13. Gregory M. Nielson and Richard Franke. Computing the separating surface for segmented data. In *Proceedings of IEEE Visualization 1997*, pages 229–233. IEEE, 1997.
14. Gregory M. Nielson and Junwon Sung. Interval volume tetrahedrization. In *Proceedings of IEEE Visualization 1997*, pages 221–228. IEEE, 1997.
15. E. M. Palmer. *Graphical Enumeration and the Power Group (Ph.D. dissertation).* University of Michigan, 1965.
16. G. Pólya. Kombinatorische anzahlbestimmungen für gruppen, graphen und chemische verbindungen. *Acta Mathematica*, 68:145–254, 1937.
17. G. Pólya and R. C. Read. *Combinatorial Enumeration of Groups, Graphs, and Chemical Compounds.* Springer, New York, 1987.
18. Jonathan C. Roberts and Steve Hill. Piecewise linear hypersurfaces using the marching cubes algorithm. In Robert Erbacher and Alex Pang, editors, *Visual Data Exploration and Analysis VI, Proceedings of SPIE*, pages 170–181. IS&T and SPIE, January 1999.
19. Han-Wei Shen and Christopher R. Johnson.Sweeping simplices: A fast iso-surface extraction algorithm for unstructured grids. In *Proceedings of IEEE Visualization 1995*, pages 143–151. IEEE, 1995.
20. Detlev Stalling, Malte Zöckler, O. Sander, and Hans-Christian Hege. *Weighted Labels for 3D Image Segmentation.*Konrad-Zuse-Zentrum für Informationstechnik Berlin, 1998. Technical Report SC 98-39.
21. Chris Weigle and David C. Banks. Complex-valued contour meshing. In *Proceedings of IEEE Visualization 1996*, pages 173–180. IEEE, 1996.

Topological Methods for Visualizing Vortical Flows

Xavier Tricoche[1] and Christoph Garth[2]

[1] Computer Science Department, Purdue University, West Lafayette, IN, USA
xmt@purdue.edu
[2] Department of Computer Science, University of California, Davis, USA
cgarth@ucdavis.edu

Summary. The paper describes the application of topological methods to the visualization of vortical flow patterns that arise in simulations from Computational Fluid Dynamics. Two techniques are presented: the first is concerned with the exploration of complicated, instantaneous flow structures while the second one permits the visualization of their temporal evolution in large-scale transient simulations. In both cases the mathematical framework is derived from the notion of parametric topology. This yields a unified formalism that permits to efficiently address the challenges raised by typical flow problems. The benefits of this approach are demonstrated in the analysis and visualization of transient vortical flows that undergo the phenomenon of vortex breakdown.

1 Introduction

Scientific computing is an important tool for the development of new prototypes in the design of modern aircrafts. While the basic theoretical principles of aerodynamics are well established, they are applicable to large scale problems only and do not describe the increasingly important details on small scales. The quality of numerical models has risen to a point where simulations can fill this gap. As the demand for faster aircrafts and improved security is high, they have established themselves as an extremely valuable alternative to physical experiments. Aside from the validation of prototypes, simulations can help to increase our understanding of the dynamics of some of the more complex flow patterns that keep appearing in aviation-related problems.

A prominent example is vortex breakdown. This phenomenon has stood in the way of a wide application of delta-wing type aircrafts as it limits the controllability in critical flight situations and causes damage to the aircraft through the induced pressure differences. In order to understand the origin of this phenomenon and avoid its occurrence in future designs, it has been reproduced and is now investigated in numerical simulations. In this case like in Computational Fluid Dynamics problems in general, these simulations facilitate complicated flow experiments and provide accurate measurements of multiple quantities over the whole 3D domain considered.

T. Möller et al. (eds.), *Mathematical Foundations of Scientific Visualization, Computer Graphics, and Massive Data Exploration*, Mathematics and Visualization,
DOI: 10.1007/978-3-540-49926-8,

However, this comes along with a hindrance for analysis at the post-processing stage. Since detailed models require fine resolutions, the amount of generated data is enormous which is especially true for time-dependent problems. This obstacle must be properly addressed by visualization methods, as they are essential to assist and improve the evaluation of the resulting numerical datasets.

For the analysis of planar flows, flow topology has proven valuable in distilling a complete structural picture of the prevalent structures by an analysis of the critical points and separatrices of the flow vector field. Parametric topology has extended this methodology to time-dependent flows. The resulting visualization is expressive, while the algorithms are efficient. Therefore, planar flow topology can be regarded as mostly complete as a flow analysis tool. Unfortunately, the extension to three dimensions is far from having achieved the same quality of visualization. This can be in part attributed to the fact that the elements involved (e.g., separating surfaces) are inherently 3D. A full display of nontrivial 3D topology is very complicated at best and suffers from mutual obstruction of the corresponding primitives. Approaches exist for simplified depictions of 3D topology, however, they remain rather unsatisfying in comparison to their 2D counterparts. Therefore the aim of this work is to provide a visualization approach for complex 3D flows that inherits the appealing properties of planar flow topology. In that way we are able to complement well established feature extraction methods in a unified framework built upon rigorous mathematical notions.

More precisely, the key idea behind the visualization methods introduced in this paper is the notion of parametric topology. Depending on the considered application the corresponding parameter can be interpreted as the time underlying a transient evolution or as the distance reached along a particular curve that traverses a region of interest. Practically, to obtain accurate and intuitive depictions of intricate flow structures we transform traditional cutting planes into flexible and powerful tools for exploring flow volumes in a continuous way. These moving cutting planes smoothly travel along trajectories that can be either obtained automatically by standard feature extraction schemes or directly provided by the user to explore a particular region. We accurately track the parametric vector field topology captured on the cutting planes. This allows us to dissect the 3D flow, detect and visualize essential properties of the flow, especially for recirculation bubbles which are key features of vortex breakdown. While understanding of this phenomenon is still incomplete, it is known that it is characterized by the appearance of stagnation points (critical points of the flow velocity field) on the vortical axis. To gain insight from the temporal behavior of the stagnation points, the critical point tracking from 2D parametric topology is extended to 3D vector fields defined over tetrahedral grids. For visualization, the four-dimensional trajectories are reduced to two dimensions by using the symmetry inherent to the vortical structures. To further enhance the understanding of the full 3D flow pattern, we also incorporate stream surfaces into the representation.

The paper is structured as follows. Section 2 summarizes previous and related work. In Sect. 3, we recall essential theoretical notions of steady and parametric flow topology. In this context we also provide a detailed discussion of the Poincaré index. The tracking of vector field critical points with respect to a parameter change is discussed in Sect. 4 along with the corresponding algorithm. Next, Sect. 5 introduces the moving cutting plane approach. We complete our presentation with our visualization results for two CFD datasets in Sect. 6 and conclude in Sect. 5.

2 Related Work

The importance of topology for depicting flow fields was first recognized by Helman and Hesselink [7] and resulted in a 2D visualization method. Complete 3D topology has not been attempted yet, however there are authors that examine subsets, such as Globus et al. [6] and Theisel et al. [17] using saddle connectors. Tricoche et al. [18] describe how the time-tracking of singularities and the corresponding topological variations can be investigated for instationary 2D vector fields. Theisel and Seidel also propose a method for the tracking of critical points in more general settings by integrating streamlines of the derived feature flow field [16]. However, the construction of this field is prohibitively expensive for large datasets.

Concerning the temporal variation of features, there are approaches that detect features in several time steps and perform a matching procedure to extract their evolution (e.g., Silver and Wang [14] and Samtaney et al. [12]). Making explicit use of the temporal interpolation, Weigle and Banks [19] extract features in the form of four-dimensional isosurfaces. A similar course is followed by Bauer and Peikert [2]. They incorporate a scale-space approach into their method for the tracking of vortex cores. As to the interrelations among multiple features over time, Silver et al. [3] have developed the *Feature Tree* that is remotely related to the much simpler structural graph we establish here.

In our development of a critical point tracking algorithm on tetrahedral grids, we make use of the Poincaré Index concept, which was described earlier by Mann and Rockwood [9]. They explain its basic premise and show how it can be applied to the study of critical points and other types of singularities. Their work is however limited to the study of analytical vector fields and is not directly applicable to our work.

From the viewpoint of fluid mechanics, vortex breakdown (or vortex burst) has concerned many authors due to its relevance for a large number of applications (see, e.g., [10]). In the field of visualization, Kenwright and Haimes [8] were among few to write about the detection and visualization of vortex breakdown. They already emphasized its importance in aeronautics. However, their interpretation of vortex breakdown is not in accordance with modern theories.

3 Theoretical Aspects of (Parametric) Topology

We introduce in this section basic notions of vector field topology both in the 2D and 3D settings as required by the visualization methods discussed in the paper. The emphasis is put on linear structures induced by piecewise linear interpolation over simplicial grids. This choice is justified by the fact that arbitrary grid types can be decomposed into simplices.

3.1 Phase Portrait, Limit Sets, and Separatrices

The essential idea behind topology analysis in the steady case is to characterize the nature of a flow with respect to the asymptotic behavior of its streamlines. For that purpose, one associates the domain of definition of the flow with its *phase portrait* that consists of the set of all streamlines. This corresponds to introducing an equivalence relation that groups all the points located on the path of the same streamline into a single class.

The topological structure of the flow is essentially a subdivision of the phase portrait into regions where all streamlines are asymptotically equivalent, thus forming a uniform flow. More specifically, all streamlines belonging to such a region converge toward the same so-called *limit sets* both forward and backward. Limit sets have a general mathematical definition but for the needs of this presentation we are only interested in critical points and closed orbits. The boundaries of the different topological regions are called *separatrices* and can be either streamlines or *stream surfaces*.

3.2 Critical Points

The *critical points* (or singular points) of a vector field are the positions where the field magnitude is zero. These points play a fundamental role in the field structure because they are the only locations where streamlines can meet. In the linear case the classification of critical points is based on the eigenvalues of the Jacobian matrix.

In planar fields, depending on the real and imaginary parts of these eigenvalues, there exist several basic configurations, some of which are shown in Fig. 1. The *saddle points* are of particular interest since the *separatrices* start or end at their location along the eigenvectors. Note that for every other critical point type, the sign of the real parts of both eigenvalues is either positive or negative, corresponding to a repelling (source) or an attracting (sink) nature, respectively. Thus separatrices emanate from saddle points and end at sources or sinks.

In the 3D case, the Jacobian matrix has three eigenvalues, and more combinations exist. If all eigenvalues have a positive (resp. negative) real part the corresponding critical point is a source (resp. a sink). Other cases correspond to different types of saddle points. As in the 2D setting, separatrices of the topology start originate 3D saddle points along their eigenvectors. However these separatrices are either one-dimensional (streamlines) or two-dimensional (stream surfaces). The latter are spanned by both eigenvectors associated with the eigenvalues whose real parts have same sign. The various cases are illustrated in Fig. 2.

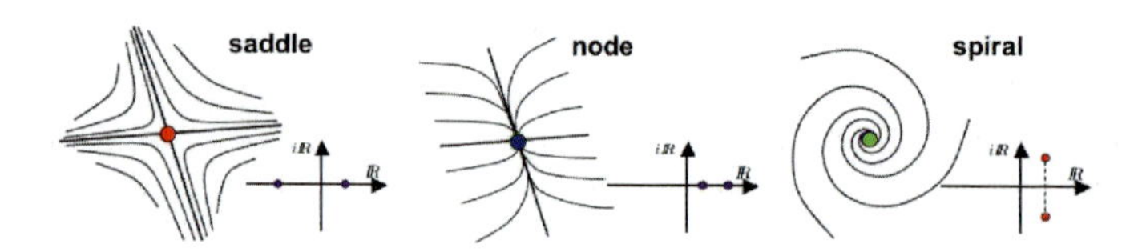

Fig. 1. Linear critical points in the plane

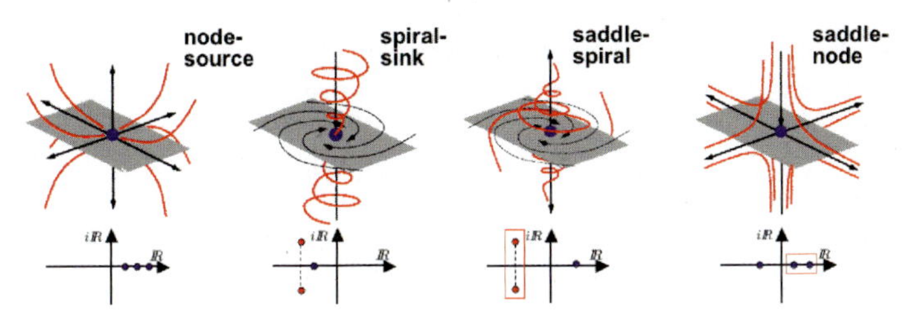

Fig. 2. Linear critical points in three dimensions

3.3 Poincaré Index

Definition and Properties

A fundamental concept in planar vector field topology is the so-called *Poincaré index* of a simple (i.e., non self-intersecting) closed curve. It measures the number of rotations of the vector field while traveling along the curve in positive direction (also called *winding number*). In a more mathematical way, one gets the following definition for the index of a simple curve γ:

$$\mathrm{ind}_\gamma = \frac{1}{2\pi} \oint_\gamma \mathrm{d}\phi, \quad \text{where } \phi = \arctan \frac{v_y}{v_x}.$$

(ϕ is the angle coordinate of the vector field $\mathbf{v}(v_x, v_y)$.) Remark that the index is always an integer.

Similarly one defines the index of a critical point as the index of a simple closed curve around the critical point enclosing no other singular point. For linear critical points, the possible index values are $+1$ and -1. A saddle point has index -1 whereas every other critical point has index $+1$. Following properties of the Poincaré index are essential in practice [1]:

1. A simple closed curve that encloses no critical point has index 0.
2. The index of a simple closed curve that encloses several critical points is the sum of the respective indices of those critical points.

This notion can be generalized to 3D vector fields. First, one defines the winding number $\#_x(\mathrm{S})$ of a closed surface S with respect to a point x as

$$\#_x(\mathrm{S}) := \frac{1}{4\pi} \int_\mathrm{S} \frac{y - x}{|y - x|^3} dS(y).$$

It is an integral value as in the planar case and corresponds to the number of times S wraps around x. For example, the x-centered sphere $S_\epsilon(x)$ of radius $\epsilon > 0$ has the canonical winding number 1. Now, to define the index of a closed surface S with respect to a three-dimensional vector field, one introduces the notion of *Gauss map*

$$\gamma : \mathbb{R}^3 \backslash \{0\} \to S^2, x \mapsto \frac{x}{||x||},$$

that maps any nonzero vector to its (normalized) direction. The index of a closed surface S is then defined as the number of times the vector field directions on S cover the origin as we move around all of S. In other words, it is the winding number of the Gauss map of **v** restricted to S with respect to the origin. Mathematically speaking, we have

$$4\pi \operatorname{ind}_S = \#_0(\gamma(\mathbf{v}|_S)) = \int_S \gamma(\mathbf{v}(x)) dS(\gamma(\mathbf{v}(x))). \quad (1)$$

Note that the winding number can be read as an oriented area integral of $\gamma(\mathbf{v}|_S)$. Hence, the sign of ind_S depends on the orientation of S relative to $\mathbb{R}^3$. We are able to define $\operatorname{ind}_z(v)$ of a singularity z via

$$\operatorname{ind}_z(\mathbf{v}) := \lim_{\epsilon \to 0} \#_0(\gamma(v|_{S_\epsilon(z)})). \quad (2)$$

The properties mentioned previously for the planar case hold in the three-dimensional case too. Let S be a closed surface that encloses the vector field singularities z_i. Then

$$\sum_i \operatorname{ind}_{z_i}(\mathbf{v}) = \#_0(\gamma(v|_S)).$$

As in the 2D case positive orientation is assumed for all closed surfaces under consideration. From the last equation, we find that the index vanishes if S does not enclose any singularity in its interior. Observe that the converse holds only in the linear case.

Computation

In the piecewise linear setting the critical points that may be encountered in the interior domain of each linearly interpolated triangle or tetrahedron cell are of first order and have therefore either index $+1$ or -1. We consider 3D critical points first and show how the 2D setting can be seen as a special case. To compute the index of an isolated linear 3D critical point z we can use a simple approach that is based on the Jacobian J of the corresponding linear vector field. Indeed, assuming a nondegenerate case, J has full rank which implies that $|\operatorname{ind}_z(\mathbf{v})| = 1$. Hence, the index is $+1$ if J is orientation-preserving and 1 otherwise. In other words, the index of a linear critical point is determined by the sign of its determinant. Therefore, if we consider the types of linear critical points mentioned previously, a source has index $+1$ and a sink has index -1. Concerning saddle points, their index depends on the particular type. If the dimension associated with the 1D separatrix corresponds to a source, the index is $+1$, otherwise -1. Refer to Fig. 2. If we now consider two dimensional critical points, we easily see that a similar result applies: sink and sources correspond to two eigenvalues of same sign and both types have index $+1$ while saddle points have index -1, see Fig. 1.

3.4 Parameter Dependent Topology

In the case of a parameter-dependent (e.g., instationary) flow, parameter changes entail transformations of the topology. Despite the unlimited variety of such transformations they always preserve qualitative consistency. In particular, the Poincaré index acts as a topological invariant.

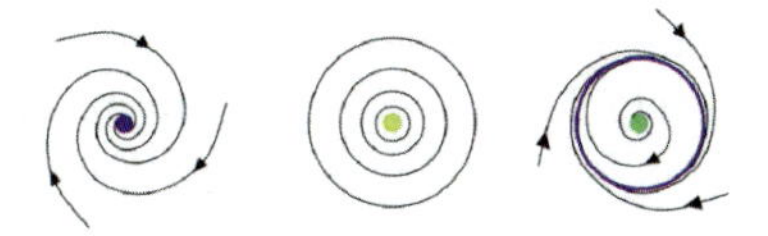

Fig. 3. Hopf bifurcation

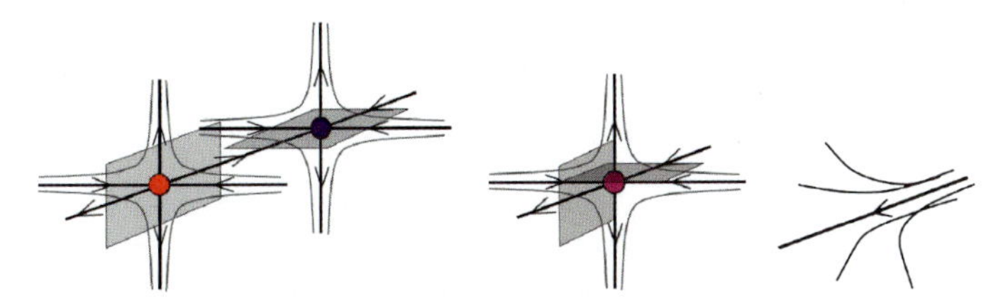

Fig. 4. Three-dimensional fold bifurcation

For the needs of our method we only mention two types of local bifurcations in the 2D case and derive a similar example for the 3D case. The first local bifurcation in 2D involves pairs of critical points, namely a saddle point and a sink or a source. When both critical points become progressively closer, merge and eventually vanish the bifurcation is a pairwise annihilation. The reverse phenomenon is called pairwise creation. The common terminology for both evolutions is fold bifurcations. The second type of planar bifurcation affects a single spiral critical point, either a sink or a source, and is known as Hopf bifurcation. The corresponding evolution for a planar field is pictured in Fig. 3. The corresponding critical point (a center) is an unstable configuration and any change in the parameter value will transform it into a source surrounded by a closed orbit that behaves as a sink. The reverse evolution is possible too, as well as swapped roles for sinks and sources. Similar transformations occur in the 3D case. A simple example can be obtained by adding a one-dimensional source behavior to a saddle point and a source involved in a fold bifurcation. This creates two 3D saddle points that merge and vanish in the very same way. An illustration is proposed in Fig. 4. Observe that as in the 2D case the basic ingredient of this fold bifurcation is the fact that the overall index of both critical points is 0 (two saddle points of opposite indices) which corresponds to a neighborhood without singularity. Consequently, the local value of the Poincaré index is preserved throughout the corresponding transformation. This remark also explains why a fold bifurcation in 3D can concern a source (index $+1$) and a sink (index -1).

4 Topology Tracking

In the following we describe a simple algorithmic solution to track the *continuous* evolution of the topology and detect the associated bifurcations. The method was originally designed for time-dependent 2D vector fields [18] and has been recently extended to the 3D case [5]. We focus the description hereafter on the latter case, which is in essence very similar to the planar case. In particular, bifurcations on the common boundary of two neighboring cells, tracking through successive entry and exit points, as well as the ability to filter out insignificant details or interpolation artifacts in postprocessing are common features of both the 2D and 3D implementation. Section 5.1 explains how the 2D method can be applied to steady 3D vector fields. Observe that in both cases, and in contrast to the original method, we do not explicitly track the separatrices of the topology.

Setting

The objective is to determine the paths of isolated critical points of a time-dependent piecewise-linear vector field, given on a tetrahedral grid. Let $p_i \in \mathbb{R}^3$ be a set of points and v_i^j the vector values associated with the p_i at discrete times $t_j \in \mathbb{R}$. Let T_k be a set of tetrahedra defined on the points p_i. Then every tetrahedron T_k gives rise to a vector field $v(x,t)$ that is linear in both space and time: if $x \in T_k$ and $t \in [t_j, t_{j+1}]$, then set

$$v(x,t) = \sum_{l=0}^{3} \beta_l(x) \left(\frac{t - t_j}{t_{j+1} - t_j} v_l^{j+1} + \frac{t_{j+1} - t}{t_{j+1} - t_j} v_l^j \right),$$

where β_l are the barycentric coordinates w.r.t. T_k and l refers to the vertices of T_k. We will next examine the paths of singularities in a single tetrahedron T_k.

Bifurcations

Due to the inherent limitations imposed on the singularities by the piecewise linear nature of the vector field, we can conclude that fold bifurcations that involve two critical points can only occur on the common face of two neighboring tetrahedra. Bifurcations on an edge or a vertex are special cases that are numerically highly unstable. Therefore there are not addressed here.

Assume we have two tetrahedra T_1 and T_2 that share a common face on which we find a bifurcation at some time t. Since the field is linear in both tetrahedra, from the two singularities involved, one is located in T_1 and the other in T_2. Moreover, due to index conservation, the overall index must be zero. Hence the indices of the critical points must be $+1$ and -1, respectively.

Paths in a Single Cell

We first consider a single tetrahedron T and determine what possibilities exist for the path of a singularity z. To simplify the notations, we assume that the vector field in T is given in the form

$$v(x) = \sum_{i=0}^{3} \beta_i(x) \left((1-t) v_i^n + t v_i^{n+1} \right), \qquad x \in T, t \in [0,1]$$

and that v is nondegenerate, i.e., it contains exactly one isolated zero at all times. For fixed t we can solve for the position of the singularity of this field in barycentric coordinates. For example, with $w_i(t) = (1-t)v_i^n + tv_i^{n+1}$ we write (omitting the parameters)

$$v = w_0 + \beta_1(w_1 - w_0) + \beta_2(w_2 - w_0) + \beta_3(w_3 - w_0)$$

and apply Cramer's rule to find

$$\beta_1(t) = \frac{\det(-w_0,\ w_2 - w_0,\ w_3 - w_0)}{\det(w1 - w0,\ w2 - w0,\ w3 - w0)} =: \frac{b_1(t)}{q(t)}.$$

The same can be done for all β_i. Brief computation shows that the resulting $b_i(t)$ and $q(t)$ are polynomials of degree 3 in t. We required that v be nondegenerate, this reflects in $q(t) \neq 0$ for all $t \in [0, 1]$. Naturally, if $\beta_i(t) < 0$ for some i, the singularity of v is outside the tetrahedron for this specific t. In other words, we have found an explicit representation for the location of z. Taking a closer look at b_i, we find that the zeros of these polynomials allow us to determine when z crosses one of T's faces. If for $\hat{t} \in [0, 1]$ we find $\beta_i(\hat{t}) = 0$ and $\beta_j(\hat{t}) >= 0$ for $j \neq i$, then the singularity is located on the face of T opposite the vertex p_i (its barycentric coordinate is zero). For this case, by evaluating the sign of the derivative

$$\beta_i'(\hat{t}) = \left(\frac{b_i}{q}\right)'(\hat{t}) = \frac{b_i'(\hat{t})}{q(\hat{t})} \qquad (\text{since } b_i(\hat{t}) = 0)$$

we can tell if the singularity enters or leaves the tetrahedron at $\hat{t}$. We will say that T has an entrance/exit on face F at $\hat{t}$. This information is important to determine in which neighboring tetrahedron (if one exists for F) the singularity path continues.

For fixed $t \in [0, 1]$ there can be at most one singularity inside T, hence we can conclude that if there is a singularity in T at some $t \in (0, 1)$, it must either have entered T at an earlier time $0 < \hat{t} < t$ or remained in T since $t = 0$ (in this case we will say that z enters at $t = 0$). In complete analogy, it must either exit T at $t < \tilde{t} < 1$ or remain in T until $t = 1$ (read z exits at $t = 1$). In other words, a singularity path always connects an entrance to an exit, and exits and entrances always come in pairs. Moreover an entrance is always connected to the closest exit (in time).

When z passes from T to a neighbor T' through the face F at $\hat{t}$, there is a singularity on F in both T and T' at $\hat{t}$. Two possibilities exist: either we find an exit/entrance combination in T and T', in which case the path continues in T', or we find an exit/exit or entrance/entrance combination. In the latter case, the vector field has a fold bifurcation on F at $\hat{t}$, and the paths of both singularities involved start or end on F.

4.1 Tracking Algorithm

Using previous results, we give in the following a simple scheme for tracking a singularity between two time steps $t = 0$ and $t = 1$. It works by simply connecting entrance/exit path segments over tetrahedron boundaries. Observe that the iterative nature of our scheme allows to restrict to two consecutive time steps the amount of data that is needed at once for processing.

Assume that a singularity z is present in T at $t \in (0, 1)$. Then, to compute the path forward in time:

1. Compute the b_i and q for T, and determine entrances and exits.
2. If there is no exit later than t, z exits T at $t = 1$; the path is complete.
3. If there are exits in T, then z leaves T at the earliest exit later than t; determine the neighbor tetrahedron T' corresponding to the exit face F and compute b'_i, q' for T'.
4. If T' has an exit on F corresponding to the exit on T ($\rightarrow$ bifurcation), the path of z ends on F.
5. Otherwise, T' has an entrance on F corresponding to the exit on T; z is now in T'. Set $T = T'$ and restart at 1.

Following the path of z backwards in time can be achieved in a completely analogous manner. Both directions are completely equivalent. We use this procedure as a building block for computing the paths of all singularities present in two given time steps between $t = 0$ and $t = 1$:

1. Find the sets of tetrahedra S_0 and S_1 that contain a singularity at $t = 0$ and $t = 1$ respectively. Let $B = \{\}$ be the set of bifurcations encountered in between $t = 0$ and $t = 1$.
2. For every $T \in S_0$: follow the path of z forward in time
   (a) If it ends in T' at $t = 1$, eliminate T' from S_1.
   (b) If it ends at a bifurcation, add it to B.
3. For every $T \in S_1$ (singularities not reached by paths from $t = 0$): follow the path of z backward in time
   (a) It must end at a bifurcation; add it to B.
4. For all bifurcations in B: check if B has two paths connecting to it; if it does not, there must be another singularity involved. Follow its path forward or backward in time depending on whether the bifurcation is a creation of singularities or an annihilation.
   (a) The path must end at a bifurcation; add it to B; goto 4.

The algorithm avoids multiple tracing of the same path by using the equivalence between forward and backward tracing (i.e., if a path extends from $t = 0$ to $t = 1$, we only need to trace it forward). The test in step 4 is required because nonintuitive situations can occur (see Fig. 5). The final result is a set of paths that completely describe the continuous structural variation of the vector field between the two time

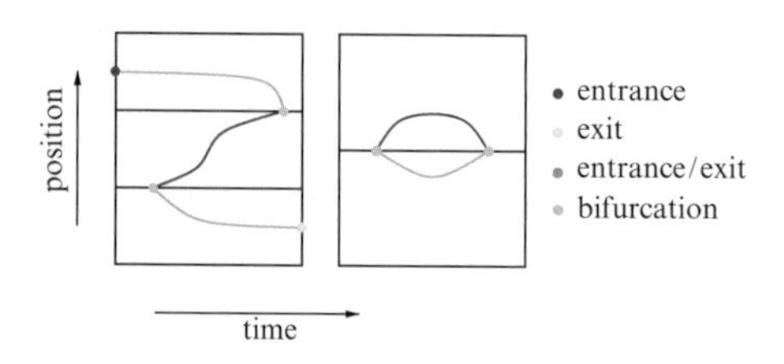

Fig. 5. Special cases of cell-wise singularity tracking

steps. Going to several time steps from here is easy as it only involves connecting the paths from different time steps according to which singularity they start/end at.

Observe that some cases are not covered by the given algorithm. If two bifurcations that create and annihilate a pair of singularities lie between two time steps, neither of the singularities will show up in either time step, and hence their paths will not be discovered by the algorithm (see Fig. 5). However they do not interact with other singularities and therefore they do not play an important role in understanding the structural changes in between the time steps. Moreover, it is often desirable to ignore small-scale local behavior.

4.2 Structural Graph Filtering

To obtain a complete picture of the structural evolution of a given dataset, the interaction of the various singularities form a *structural graph* with bifurcations as vertices and paths as edges (see Fig. 7 for an example). We describe here how this graph can be manipulated at the postprocessing stage.

The method described above is restricted to tetrahedra which implies that arbitrary input datasets must be tetrahedrized before application. Although the tracking algorithm could be extended to nontetrahedral grid cells, this would result in a number of special cases that complicate the simple structure of the algorithm. In its present form, implementation is straightforward and fast. However, the price to pay is that tetrahedrization of arbitrary grids can result in the creation of singularities that are not in the original dataset. It is possible that a cell of index 0 is split up such that the resulting tetrahedra have nonzero indices. These "artificial" singularities are not an issue since they are always created pair-wise and typically only last for a short period of time.

Numerical datasets are often subject to noise, especially if the computations involve some kind of differentiation. It is common practice to apply smoothing operators to datasets in order to account for this limitation. Numerical noise usually reflects in short-lived pairs of artificial singularities that exist in isolation and are not part of the datasets structural evolution over time. It can also occur that a path is "interrupted" by a pair of artificial bifurcations that enclose a path segment of very short duration (Fig. 5 (left) gives an example).

What seems a drawback at first can be turned into an advantage: instead of smoothing the dataset we filter the resulting set of singularity paths by removing paths that last less than, e.g., one time step. Filtering can be applied on the structural graph directly and can be implemented in an efficient way by first removing edges that represent paths with short duration and successively removing all isolated vertices. In our experiments, we found this method to be very effective in treating noisy datasets. It must be mentioned here that conventional smoothing does not significantly reduce the number of artificial singularities. Moreover it affects the structure of the dataset in such a way that the structural evolution is obscured or changed (this is especially true for minimum/maximum tracking as described in the next para- graph). Consequently a filtering based directly on the topological structure of the flow offers a much more accurate control over the complexity of the structural information.

4.3 Algorithm Performance

The tracking algorithm itself is of linear complexity in both the number of singularities and the number of time steps. The most time-intensive part is the precomputation of all singularities in a time step, for which each cell has to be considered individually. This is not a drawback of our algorithm but rather a limitation inherent in this class of tracking algorithms (cf. [16, 18]) If this information is assumed given, the running times for our examples are on the order of very few seconds. Since the algorithm only needs two successive time steps to do its work, it is possible to integrate it directly into the CFD simulation. The structural graph for all time steps can then be completed in postprocessing. This would also allow for online supervision of simulations that are still in progress.

5 Planar Topology Tracking for Volume Exploration

As mentioned previously the approach developed in this section consists in using the framework of planar topology tracking to explore steady 3D flow structures. More precisely, the 3D (steady) flow is now investigated through the parametric topology of its 2D projection onto a plane that is swept along a given curve across the volume of interest. In other words, the curve provides the third dimension and is interpreted as the parameter space for topology tracking. Essential algorithmic aspects of this method are discussed next.

5.1 Moving Cutting Planes

Trajectories

By definition, the choice of a particular trajectory to explore a flow volume is essential to ensure the quality and the usefulness of the extracted topology and must therefore be carefully chosen with respect to the considered application. The general idea followed in our implementation is to use any inherent symmetry of the dataset to yield a natural way to split the physical space. Since the application of our technique is focused on vortical flows and vortex breakdown we selected the curves described next. Refer to Fig. 6 for an illustration.

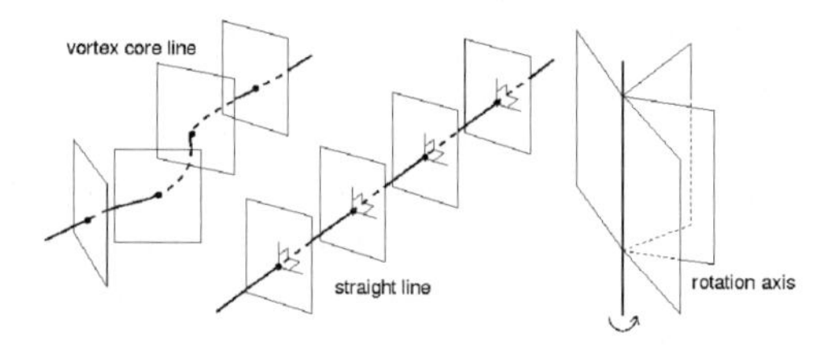

Fig. 6. Different types of moving cutting planes

- *Vortex Core Lines* are a natural choice to explore vortical regions. We extract them using an implementation of the standard method by Sujudi and Haimes [15] based on the Parallel Operator [11]. A smoothing step applied in preprocessing permits to improve the results.
- *Straight Line* across the grid are a straightforward alternative. They are mainly used to explore large structures whose overall orientation is known.
- *Recirculation Bubble Axis:* to explore the recirculation bubbles induced by vortex breakdown, the medium axis usually exhibited by these structures can be used. More specifically, as mentioned previously a recirculation bubble is delimited by two stagnation points and the line connecting them plays the role of a rotation axis in our method.

Cutting Plane Orientation

A robust computation of the cutting plane orientation is mandatory for our flow exploration technique. It can be seen on Fig. 6 that choosing the recirculation bubble axis as exploratory curve fully determines the plane orientation. Similarly we also used the straight line as plane normal when it is selected to capture large-scale features. On the contrary, when dealing with a vortex core line the inaccuracy of the extraction method results in an approximated position of the actual vortex core which can have a negative impact on the resulting normal value. The same holds true when approximating the curved, possibly complex path of a vortex by a straight line segment. In both cases we need an automatic way to compute a suitable normal at each point along the discrete path according to the local flow orientation. Practically, the quality of a normal is evaluated with respect to the amount of flow crossing the plane over a small region around the considered point. We use a simple iterative scheme to maximize this quantity which consists in assigning in every step the mean direction of crossing flow to the current approximation of the normal.

Planar Resampling

The remaining task consists in resampling the 3D vector field on the cutting plane while ensuring consistency of the coordinate frames between consecutive positions along the followed curve. This is mandatory to obtain meaningful results during the topology tracking procedure. To do so, it is sufficient to assign a single basis vector to each plane, the second one being readily obtained by cross product with the normal. Practically we select an arbitrary vector in the first plane and we iteratively transport this vector from one plane to the next by successive projections and normalization, similar to, e.g., [13]. Once provided with the grid resolution (i.e., sampling rate) we only need to control the spatial extent of the sampled region around the curve on the plane. This is to ensure that the sampling will not include data points corresponding to positions lying outside the structure of interest. Practically we either assume a constant size of the feature or we apply the technique described in a recent paper [4] to determine the boundary of the vortex core region.

5.2 Topology Tracking

The previous step collects the successive values of the projected vector field as the cutting plane moves through the volume. As mentioned previously we can now abstract these data from their original embedding in three-space and treat them as the successive states of a parameter-dependent planar vector field. In that way we can apply the two-dimensional tracking scheme proposed in [18] and whose extension to three dimensions was discussed in the previous section. In essence the setting of the original method corresponds here to the computational space. One difference is that the results obtained (singularity paths, bifurcations) must be mapped back into physical space after tracking for visualization and interpretation. Moreover we need to account for the lack of smoothness of the vector field projected on the moving cutting plane along a curve. Specifically this may cause spiraling critical points to oscillate between sink and source behavior, creating numerous Hopf bifurcations. We correct this effect by filtering out small-scale features like pairs of critical points vanishing shortly after their creation or type swap between sources and sinks. The latter is handled by assigning the type *center* to the critical point. Although this is an unstable structure in planar topology, this may be monitored in cutting plane topology when inspecting a vortex whose spiraling flow neither converges nor diverges with respect to its core line.

6 Results

We now show the results of both topology tracking methods applied to two CFD simulations specifically designed to investigate the impact of vortex breakdown on a vortical flow.

6.1 Datasets

Delta Wing: This simulation describes a sharp-edged delta wing at subsonic speed (0.2 Mach) with the characteristic vortical systems above the wing. The angle of attack increases over time, eventually leading to vortex breakdown in later time steps. The viscous simulation of the full configuration was performed with- out the assumption of symmetry and was carried out using the DLR Tau Code solver. The grid consists of 11.1 million unstructured grid cells based on about 3 million vertices. At these, a number of variables is given (velocity, pressure, kinetic energy, etc.) for each of the 90 time steps. Among the significant physical features are secondary and tertiary vortices on the wing and corresponding separation and attachment structures.

Can Dataset: The aim behind this simulation of a cylindrical container filled with an incompressible and highly viscous liquid was to study vortex breakdown under ideal conditions, created by the viscosity of the fluid and the high symmetry in the problem that lead to numerically very accurate and smooth data. The bottom cylinder cap rotates, creating a vortex on the symmetry axis of the cylinder. A variation in the rotation speed leads to the appearance and successive vanishing of vortex breakdown during the 500 time steps. The dataset is given in the same form as the delta wing dataset, with the grid containing approx. 750,000 elements.

6.2 3D Critical Point Tracking

We have employed the critical-point tracking described in Sect. 4.1 for the analysis of both datasets. It is already known that vortex breakdown is associated with the occurrence of (pairwise) stagnation points, therefore we have applied the tracking algorithm to the velocity fields first. Furthermore, there are speculations that both acceleration and signed helicity (i.e., dot product of velocity and vorticity) play an important role in this context. We have computed these fields for those datasets and applied tracking to them as well, in the case of signed helicity minimum tracking was performed. Since these computations involve derivative computation, we observe strong numerical noise in both helicity and acceleration yielding many artificial singularities. Using structural graph filtering we are still able to obtain meaningful results.

For the can dataset, the results are of almost analytical quality (see Fig. 7). The simulation actually shows two occurrences of vortex breakdown (and two corresponding pairs of stagnation points) and it is interesting to observe how primary and secondary vortex breakdown successively merge and split again. Acceleration zeros and helicity minima show a strong correlation with the onset of the breakdown process and the bifurcation that creates the two stagnation points. It is also obvious that the structural graph helps locate interesting time steps quickly.

In treating the delta wing dataset, we focus on two regions that correspond to breakdown on both sides of the wing. After a coordinate transformation consisting in a projection onto the vortical axis, the structural graph of the right region (cf. Fig. 8) clearly shows the evolution of the stagnation points as they move towards the wing.

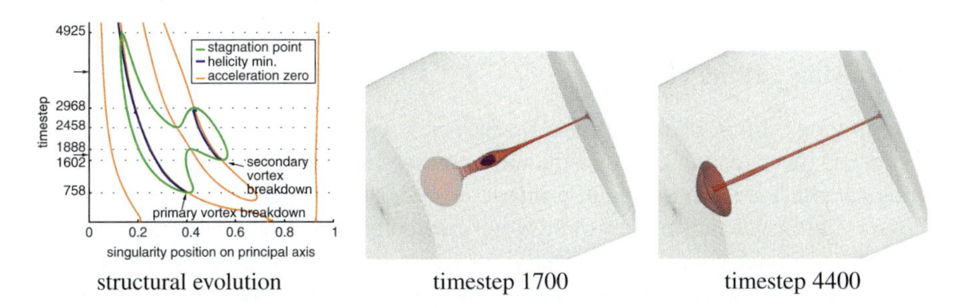

Fig. 7. *Left:* Structural graph of the can dataset. The *green paths* represent the stagnation points in the velocity field. Primary and secondary breakdown each create a pair of stagnation points. Around time step 1,888, the two phenomena join, only to re-split at time step 2,458 and successively decay. The *blue and orange paths* belong to helicity minima and acceleration zeros. Note the strong interrelation between the three quantities. *Middle and right:* Two snapshots from the can dataset. Separation stream surfaces are started at the singularity positions. Time step 1,700 shows both breakdowns, whereas the second breakdown has already vanished in time step 4,000 and the first breakdown shows the typical "mushroom" structure

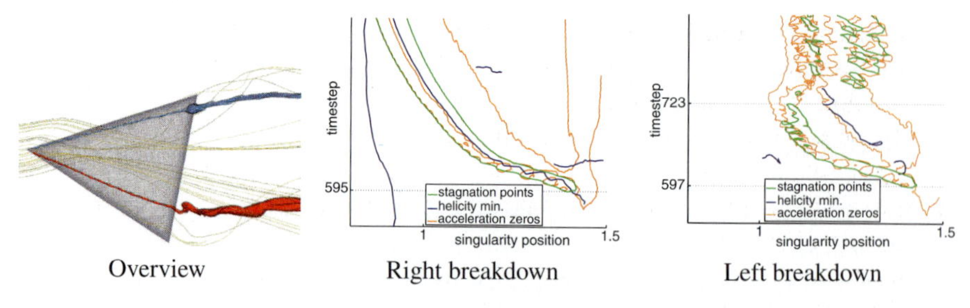

Fig. 8. *Left:* Overview of the delta wing dataset with its two primary vortices above the wings. Stream surfaces wrap around the vortices and are eventually distorted by vortex breakdown. Note the asymmetrical breakdown structure. *Middle and right:* Structural graphs for right and left breakdown. Again a connection between various quantities involved in vortex breakdown can be observed for the right breakdown. In the left breakdown, several oscillating breakdown structures are visible in the later time steps

Again, acceleration zeros and a helicity minimum seem to play a role in formation of breakdown, although the correlation is not as obvious as in the can dataset. This is, in part to be blamed upon the lack of resolution and numerical noise. Filtering of the structural graph for the helicity gradient field reduces the number of meaningful paths from roughly 1.000 to 4, effectively eliminating all artificial singularities. The left region is even more complex, and it is clearly visible how the stagnation points begin to oscillate and disappear around time step 730, to be followed by what appears to be a sequence of short-lasting vortex breakdowns in different places. In this case, the structural graph makes identification of multiple breakdown bubbles possible by grouping the velocity field singularities according to their common origin in a bifurcation. The stream surfaces shown are separation surfaces originating in the separation planes of the (saddle) stagnation points. Although this picture conveys the basic structure of the breakdown bubbles, for an accurate interpretation the structural graph is necessary.

6.3 Moving Cutting Planes

The moving cutting planes scheme from Sect. 5 was applied to both datasets with the aim of investigating the flow structures obtained by the simulations. For the delta wing dataset, the reproduction of primary, secondary, and tertiary vortices is crucial. Figure 9 left gives an overview of the wing created with parallel cutting planes along the wing symmetry axis. The primary vortices are presented prominently, and the vortex axis results from the tracking of the corresponding singularities. Using the cutting plane orientation scheme described in Sect. 5.1 with the vortex core as input curve for the plane generation, both secondary and tertiary vortices are visible. Moreover, the planar cut reveals interactions between the three vortices that are hard to determine by other means. This includes the separation surface between the primary and secondary vortices and the so-called primary separation, i.e., the flow sheet that emanates from the wing edge and divides the flow above the wing from the surrounding flow. Both appear as a separatrix in the plane.

The dataset had been examined for the presence of the vortical system before, using the method of Sujudi and Haimes [15]. However, this scheme requires careful computation of derivatives and involves smoothing. The result is a set of disconnected line segments and is hard to interpret. In comparison, the approach employed here was easily applied. This can be attributed in part to the fact that the approximate location of the sought features was a priori known, which is usually the case in the verification of datasets.

Application of the planar topology to the can dataset has revealed a peculiarity. The simulation exhibits vortex breakdown, hence a so-called breakdown bubble is visible. Over time, this bubble grows, merges and successively re-splits with a second bubble, and shrinks until it vanishes as the breakdown is resolved.

Aside from the strict validation of datasets, parametric planar topology can also serve as a feature extraction method for vortex core lines under limited circumstances. For example, the primary vortex axes in the delta wing dataset can be extracted in this manner (cf. Fig. 9). Although it is in this case equivalent to other algorithms, it excels in the extraction of recirculation cores. As the vortex breakdown bubble encloses a mostly rotation symmetric region of recirculation, there is essentially a bent vortex inside the bubble. Its core appears as a singularity in the section planes revolving around the original vortex axis. Then, tracking provides a connection between different planes and thus constructs the core of the recirculation vortex. Figures 10 and 11 show these recirculation rings.

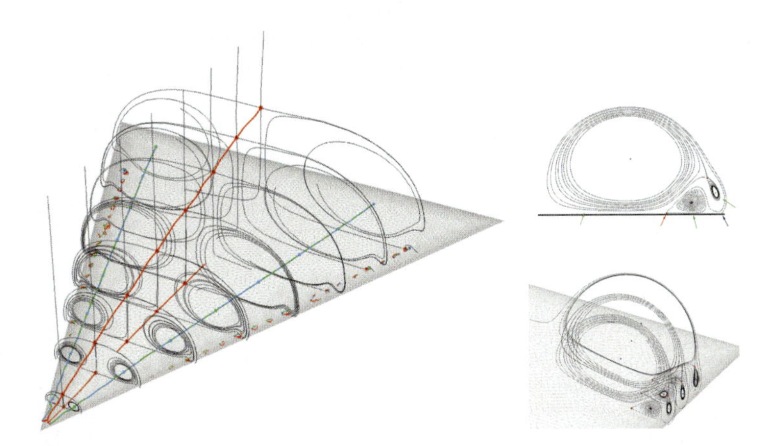

Fig. 9. *Left:* An overview of the delta wing dataset: parametric topology visualizes the primary vortices. The planes are computed along the symmetry axis of the wing and are parallel. Each planes shows two sinks/sources (primary vortices) and a number of saddle points (separation from the wing). Note how the separatrices end in cycles. This indicates very weak attracting/repelling behavior of the vortices. *Right:* Primary, secondary and tertiary vortices visualized by planar topology. Here, the planes are on the primary vortex core and oriented to the flow. Note how plane orientation affects the resulting structures. *Green arrows* indicate the three vortices in the *top image*. The *red arrow* shows the separation sheet between primary and secondary vortex. The primary separation at the wing edge is indicated by the *blue arrow*. All three vortices are present as expected

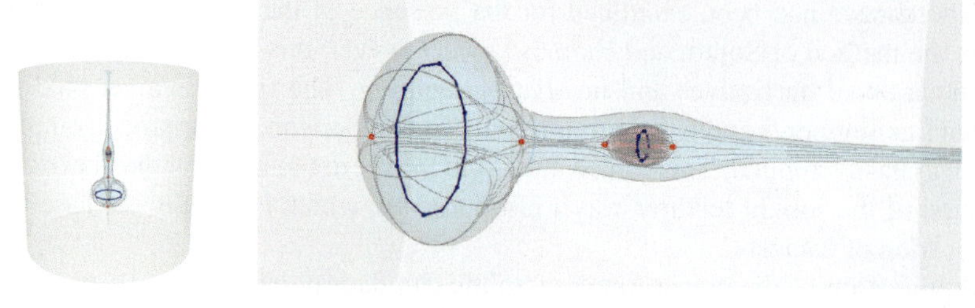

Fig. 10. *Left:* An overview of the can dataset. *Right:* Parametric topology shows the essentials of vortex breakdown including the recirculation ring (*blue*) and a secondary vortex breakdown. To show that the separatrices accurately model the flow behavior, the breakdown bubbles are surrounded with transparent stream surfaces [4] (*light blue/light red*) originating at the upstream stagnation points that are reproduced as saddle points in the topology of the planes (*red*)

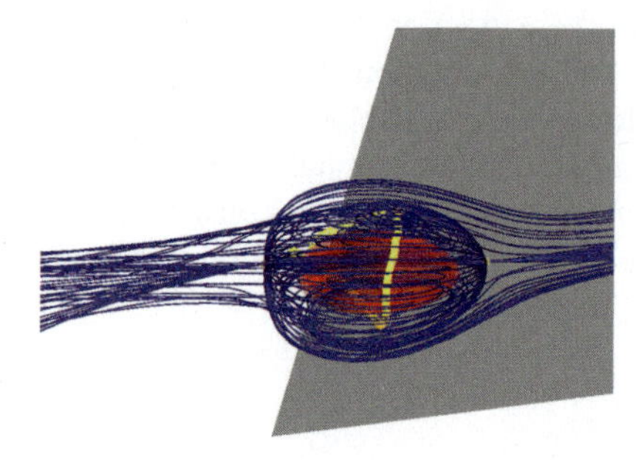

Fig. 11. Enlargement of the right recirculation bubble in the delta wing dataset. Continuous tracking on the projected topology onto a plane rotating around the axis connected the 3D stagnation points. The resulting parametric topology exhibits numerous Hopf bifurcations that are smoothed out to yield the center type critical point (*yellow*) corresponding to the typical closed vortex core

7 Conclusion

We have introduced a unified algorithmic framework to address the challenging task of analyzing and visualizing the flow structures exhibited by typical CFD simulations of complex vortical flows. Building upon the central idea of parametric topology we investigate 3D flow patterns like vortical systems and recirculation bubbles to yield intuitive and accurate representations. Moreover we extend a topology tracking scheme originally designed for 2D transient vector fields to the three-dimensional setting and show how to leverage it to efficiently explore large time-dependent datasets and understand the temporal evolution of features of interest. The corresponding algorithms are easily implemented and suitable for the processing of typical CFD datasets, both online during numerical simulations and at the postprocessing stage.

References

1. A. A. Andronov. *Qualitative Theory of Second-Order Dynamic Systems.* Wiley, New York, 1973.
2. D. Bauer and R. Peikert. Vortex tracking in scale-space. In *Data Visualization 2002. Proc. VisSym '02*, 2002.
3. J. Chen, D. Silver, and L. Jiang. The feature tree: visualizing feature tracking in distributed AMR datasets. In *IEEE Symposium on Parallel and Large-Data Visualization and Graphics*, 2003.
4. C. Garth, X. Tricoche, T. Salzbrunn, and G. Scheuermann. Surface techniques for vortex visualization. In *Proceedings Eurographics – IEEE TCVG Symposium on Visualization*, May 2004.
5. C. Garth, X. Tricoche, and G. Scheuermann. Tracking of vector field singularities in unstructured 3D time-dependent datasets. In *IEEE Visualization Proceedings '04*, October 2004.
6. A. Globus, C. Levit, and T. Lasinski. A tool for visualizing the topology of three-dimensional vector fields. In *IEEE Visualization Proceedings*, pages 33–40, October 1991.
7. J. L. Helman and L. Hesselink. Visualizing vector field topology in fluid flows. *IEEE Computer Graphics and Applications*, 11(3):36–46, May 1991.
8. D. N. Kenwright and R. Haimes. Vortex identification – applications in aerodynamics: a case study. In R. Yagel and H. Hagen, editors, *IEEE Visualization '97*, pages 413–416, Los Alamitos, CA, 1997.
9. S. Mann and A. Rockwood. Computing singularities of 3D vector fields with geometric algebra. In *IEEE Visualization Proceedings '03*, 2003.
10. T. Mullin, J. J. Kobine, S. J. Tavener, and K. A. Cliffe. On the creation of stagnation points near straight and sloped walls. *Physics of Fluids*, 12(2), 2000.
11. R. Peikert and M. Roth. The "Parallel vectors" operator – a vector field visualization primitive. In *IEEE Visualization Proceedings '00*, pages 263–270, 2000.
12. R. Samtaney, D. Silver, N. Zabusky, and J. Cao. Visualizing features and tracking their evolution. *IEEE Computer*, 27(2):20–27, 1994.
13. W. J. Schroeder, R. Volpe, and W. E. Lorensen. The stream polygon: a technique for 3D vector field visualization. In *IEEE Visualization Proceedings*, 1991.
14. D. Silver and X. Wang. Tracking and visualizing turbulent 3D features. *IEEE Transactions on Visualization and Computer Graphics*, 3(2), 129–141, 1997.
15. D. Sujudi and R. Haimes. Identification of swirling flow in 3D vector fields. Technical Report AIAA Paper 95–1715, American Institute of Aeronautics and Astronautics, 1995.
16. H. Theisel and H.-P. Seidel. Feature flow fields. In *Joint Eurographics-IEEE TVCG Symposium on Visualization*, 2003.
17. H. Theisel, T. Weinkauf, H.-C. Hege, and H.-P. Seidel. Saddle connectors – an approach to visualizing the topological skeleton of complex 3D vector fields. In *IEEE Visualization '03*, 2003.
18. X. Tricoche, T. Wischgoll, G. Scheuermann, and H. Hagen. Topology tracking for the visualization of time-dependent two-dimensional flows. *Computers and Graphics*, 26(2):249 – 257, 2002.
19. C. Weigle and D. C. Banks. Extracting iso-valued features in 4-dimensional scalar fields. In *Proc. Symposium on Volume Visualization*, 1998.

Stability and Computation of Medial Axes: A State-of-the-Art Report

Dominique Attali[1], Jean-Daniel Boissonnat[2], and Herbert Edelsbrunner[3]

[1] Gipsa-lab, ENSE[3], Domaine Universitaire, BP 46, 38402 Saint Martin d'Hères, France
Dominique.Attali@gipsa-lab.inpg.fr
[2] INRIA, 2004 Route des Lucioles, BP 93, 06904 Sophia-Antipolis, France
Jean-Daniel.Boissonnat@sophia.inria.fr
[3] Department of Computer Science, Duke University, Durham, and Raindrop Geomagic, Research Triangle Park, NC, USA
edels@cs.duke.edu

Summary. The medial axis of a geometric shape captures its connectivity. In spite of its inherent instability, it has found applications in a number of areas that deal with shapes. In this survey paper, we focus on results that shed light on this instability and use the new insights to generate simplified and stable modifications of the medial axis.

1 Introduction

In this paper, we survey what is known about the medial axis of a geometric shape. To get an intuitive feeling for this concept, consider starting a grass fire along a curve in the plane, like the outer closed curve in Fig. 1. The fire starts at the same moment, everywhere along the curve, and it grows at constant speed in every direction. The medial axis is the set of locations where the front of the fire meets itself. In mathematical language: it is the set of points that have at least two closest points on the curve. If we start the fire along the boundary of a geometric shape in $\mathbb{R}^k$ we generically get a medial axis of dimension $k - 1$, one less than the dimension of the space.

In the plane, the medial axis is a (one-dimensional) graph whose branches correspond to regions of the shape it represents. Its structure has found applications in image analysis for shape recognition [52] and in robotics for motion planning [42]. The distance-to-boundary recorded at points of the medial axis provides information about local thickness, which can be used to segment the shape, separating it into large regions with relatively narrow connections [21, 25, 45]. In reverse engineering, the medial axis appears naturally as a tool to characterize the sampling density needed to reconstruct a curve in the plane and a surface in space [1, 2]. Other applications include domain decomposition in mesh generation [46,50], feature extraction in geometric design [38, 39], and tool-path creation in computer-aided manufacturing [36].

In this paper, we make no attempt to cover the large amount of work on medial axes in digital image processing and instead refer to texts in the area [35, 40, 41].

T. Möller et al. (eds.), *Mathematical Foundations of Scientific Visualization, Computer Graphics, and Massive Data Exploration*, Mathematics and Visualization,
DOI: 10.1007/978-3-540-49926-8,

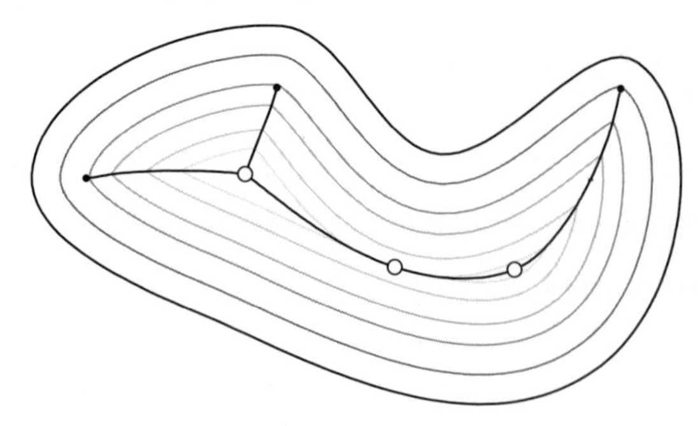

Fig. 1. Medial axis of shape whose boundary is the outer closed curve. The distance-to-boundary function has three critical points, one saddle and two maxima. One of the maxima coincides with a branch point

Whenever possible, we state the definitions and results for $\mathbb{R}^k$, where k is an arbitrary but fixed constant, but sometimes we need to limit ourselves to $\mathbb{R}^2$ and $\mathbb{R}^3$. Since most applications are in two and three dimensions, this limitation implies only a minor loss of relevance. The style adopted in this paper is not that of a typical survey paper. Rather than aiming at a broad coverage of the literature, we focus on a small number of results that we deem important. Those are centered around questions of stability and computation of the medial axes. We encapsulate the various topics in a relatively large number of small and by-and-large independent sections. Starting with fundamental properties, we slowly progress towards more advanced results. In Sect. 2, we define medial axes and skeletons. Sections 3 and 4 state properties of the medial axis that concern its finiteness and its homotopy type. Sections 5, 6, and 12 discuss the stability of the medial axis under various notions of distance. Section 7 recalls that computing the medial axis exactly runs into obstacles except for certain classes of shapes, and Sect. 8 introduces the approximation paradigm designed to circumvent these obstacles. Sections 9, 10, 11 and 13 describe steps and aspects of this paradigm. We find that topics of stability and computation are related, which is the reason for interleaving the sections as enumerated.

2 Medial Axis and Skeleton

There is no generally agreed upon definition for either the medial axis or the skeleton of a shape; the precise meaning of these terms changes from one author to another. The medial axis has been introduced by Blum [12] as a tool in image analysis. In this paper, we adopt the definitions given in [43]. Let X be a bounded open subset of the k-dimensional Euclidean space, $\mathbb{R}^k$. The *medial axis*, $\mathcal{M}[X]$, is the set of points that have at least two closest points in the complement of X. We call an open ball $B \subseteq X$ *maximal* if every ball that contains B and is contained in X equals B. The *skeleton* is the set of centers of maximal balls. The two notions are closely related but not the same. Specifically, the medial axis is a subset of the skeleton which is, in turn, a subset of the closure of the medial axis [44, Chap. 11]. In the most general case, the skeleton is not necessarily closed and the last inequality is strict. Examples of shapes in $\mathbb{R}^2$ whose skeletons are composed of an infinite number of curves and which are not closed can be found in [16, 22]. A simple example of a skeleton in $\mathbb{R}^3$ that is not closed is given in [20]. Even though medial axes are not necessarily the same as skeletons, the two concepts are too similar to warrant a balanced treatment of both. The rest of this paper will therefore focus exclusively on the medial axis.

If we weight each point x of the medial axis with the radius $\rho(x)$ of the maximal ball centered at x, then we have enough information to reconstruct the shape. In other words, the medial axis together with the map ρ provides a reversible coding of shapes. This coding is not necessarily minimal and some shapes, such as finite unions of balls, can be reconstructed from proper subsets of their weighted medial axes.

3 Finiteness Properties

There are cases in which the medial axis has infinitely many branches, even if the shape is bounded and its boundary is C^∞-smooth [22]. To construct such an example, let $f : \mathbb{R} \to \mathbb{R}$ be a C^∞-smooth function defined by

$$f(x_1) = \begin{cases} 1 - e^{-\frac{1}{x_1^2}} \left(\sin \frac{\pi}{x_1}\right)^2 & \text{if } |x_1| < 1, \\ 1 & \text{if } |x_1| \geq 1. \end{cases}$$

Consider the set of points $X \subseteq \mathbb{R}^2$ above its graph, as shown in Fig. 2. The medial axis of X consists of infinitely many branches, one for each oscillation of f. To obtain a bounded shape, we apply inversion, mapping every point $x \in X$ to $\iota(x) = x/\|x\|^2$. The inversion preserves circles and incidences between curves. It follows that the medial axis of $\iota(X)$ has the same structure and number of branches as the medial axis of X. More specifically, if we compactify $\mathbb{R}^2$ and join all branches of $\mathcal{M}[X]$ at the added point at infinity, then we have a homeomorphism between $\mathcal{M}[X]$ and $\mathcal{M}[\iota(X)]$. The point at infinity maps to the center of the circle that is the image

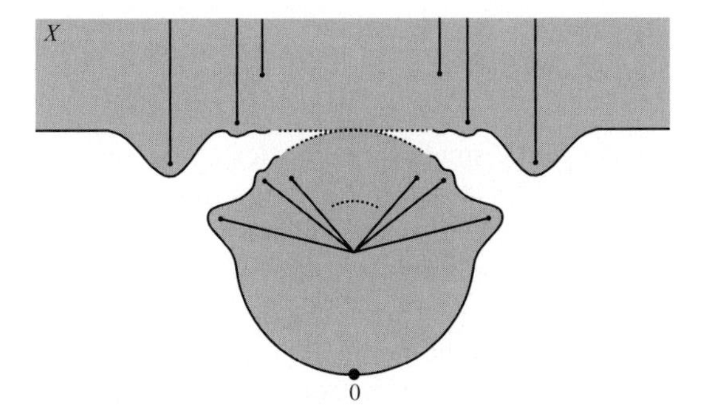

Fig. 2. The upper "half-plane" X bounded by a smooth curve and its image under inversion. Both shapes have medial axes with infinitely many branches

of the line $x_2 = 1$. We note, however, that the homeomorphism between the two medial axes is different from ι, which does not preserve centers of circles. The above construction can be extended to produce medial axes with infinitely many branch points and similar pathological examples in higher dimension.

In the plane, Choi et al. [22] establish that the medial axis of a bounded shape whose boundary is piecewise real analytic is a finite graph. Chazal and Soufflet [20] extend this result to *semianalytic* bounded open sets, which are bounded and open subsets $X \subseteq \mathbb{R}^k$ for which every point of $\mathbb{R}^k$ has a neighbourhood U with $X \cap U$ defined by a finite system of analytic equations and inequalities. They prove that such sets have medial axes which admit stratifications and satisfy finiteness properties. Specifically, the medial axis can be decomposed into a finite number of strata, each a connected i-manifold with boundary, for $i < k$. Furthermore, the medial axis has finite j-dimensional volume, where j denotes the largest dimension of any stratum. In addition, in $\mathbb{R}^3$ the boundary of the medial axis consists of a finite union of points and curves of finite length. For shapes in $\mathbb{R}^3$, a classification of points in the closure of the medial axis can be found in [4, 33].

4 Homotopy Equivalence

In [43], Lieutier proves that any bounded open subset $X \subseteq \mathbb{R}^k$ is homotopy equivalent to its medial axis. Unlike earlier works [51,53], he assumes no regularity condition on the boundary of X. Intuitively, this result implies that the medial axis and the shape are connected the same way, ignoring local dimensionality. To be formal, we say that two maps f and g from X to Y are *homotopic* if there exists a continuous map $H : X \times [0, 1] \to Y$ with $H(x, 0) = f(x)$ and $H(x, 1) = g(x)$. Using this definition, two spaces X and Y are *homotopy equivalent* if one can find two continuous maps $f : X \to Y$ and $g : Y \to X$ such that $g \circ f$ is homotopic to the identity on X and $f \circ g$ is homotopic to the identity on Y. To establish the homotopy equivalence between a shape X and its medial axis, Lieutier considers the distance-to-boundary function, ρ, which associates to each point $x \in X$ its distance to the complement of X and defines the vector field $\nabla\rho : X \to \mathbb{R}^k$ by

$$\nabla\rho(x) = \frac{x - c(x)}{\rho(x)}, \tag{1}$$

where $c(x)$ is the center of the smallest ball enclosing the set of points in the complement of X closest to x. Used before for the purpose of surface reconstruction in [29, 34], this vector field extends the gradient of ρ to points on the medial axis, where the gradient is not defined. The extended vector field is not continuous but induces flow lines used in the proof to map the shape to its medial axis. Specifically, each point x is mapped to the point $f(x)$ it occupies after flowing along the vector field $\nabla\rho$ for a sufficiently long but constant amount of time. The proof uses $f : X \to \mathcal{M}[X]$ and the inclusion map $g : \mathcal{M}[X] \to X$ to establish the homotopy equivalence.

5 Instability and Semi-continuity

We think of $\mathcal{M}$ as a transform that maps the shape X to its medial axis, $\mathcal{M}[X]$. As emphasized in [43], geometric shapes are usually not known exactly and represented by approximations of one kind or another. For example, the boundary of a shape may be approximated by a triangulation obtained by software for surface reconstruction or segmentation. Under these circumstances, it would be important that the transform be continuous. In other words, one should be able to compute an arbitrarily accurate approximation of the output for a sufficiently accurate approximation of the input. Most commonly, one would use the Hausdorff distance to quantify the difference between two inputs and two outputs and this way define what it means for the transform to be continuous. Unfortunately, the medial axis transform is not continuous under this notion of distance: small modifications of the input shape can induce large modifications of its medial axis. This effect is illustrated in Fig. 4, where we compare the medial axis of an oval on the left with the medial axis of a set whose Hausdorff distance to the oval is bounded from above by $\epsilon > 0$. The difficulty of approximating the medial axis due to its instability with respect to the Hausdorff distance is a well-known but until recently not well-understood problem.

One can observe experimentally that small modifications of a shape do not affect the entire medial axis. Typical effects for shapes in $\mathbb{R}^2$ are fluctuating branches that leave the rest of the medial axis unchanged. Similarly, for shapes in $\mathbb{R}^3$ we notice fluctuating spikes, added to or removed from the otherwise stable structure. This observation is consistent with the fact that the medial axis is semicontinuous with respect to the Hausdorff distance [44, Chap. 11]. To explain this concept, we let A and B be subsets of $\mathbb{R}^k$ and write $d_H(A \mid B) = \sup_{x \in A} d(x, B)$ for the *one-sided Hausdorff distance* of A from B, where $d(x, B)$ is the infimum of the Euclidean distances between x and points y in B. Observe that $d_H(A \mid B) < \epsilon$ if and only if A is contained in the parallel body $B^{+\epsilon} = \{x \in \mathbb{R}^k \mid d(x, B) < \epsilon\}$. The *Hausdorff distance* between A and B is $d_H(A, B) = \max\{d_H(A \mid B), d_H(B \mid A)\}$. We write A^c and B^c for the complements of A and B and note that the Hausdorff distance between A^c and B^c is generally different from that between A and B. Indeed, $d_H(A^c, B^c)$ is forgiving for small islands of A far away from B, while $d_H(A, B)$ is forgiving for small holes of A far away from B^c. With this notation, we are ready to define the concept of semicontinuity. Specifically, a transform $\mathcal{T}$ is *semicontinuous* if for every bounded open subset $X \subseteq \mathbb{R}^k$ and for every $\delta > 0$, there exists $\epsilon > 0$ such that for every open subset Y of $\mathbb{R}^k$,

$$d_H(X^c, Y^c) < \epsilon \implies d_H(\mathcal{T}[X] \mid \mathcal{T}[Y]) < \delta. \tag{2}$$

Note that ϵ depends on X. In words, small Hausdorff distance between the complements of X and Y implies that $\mathcal{T}[X]$ is contained in a tight parallel body of $\mathcal{T}[Y]$. As mentioned earlier, this condition is satisfied for $\mathcal{T} = \mathcal{M}$.

6 Stability Under C^2-Perturbations

In [20], Chazal and Soufflet prove that the medial axis transform is continuous when C^2-perturbations are applied to shapes in $\mathbb{R}^3$. To define what this means, we call two multilinear maps *ϵ-close* if the norm of their differences is less than ϵ. A map $f : \mathbb{R}^3 \to \mathbb{R}^3$ is an *ϵ-small C^m-perturbation* if:

(1) $f(x) = x$ outside some compact subset of $\mathbb{R}^3$.
(2) f is a C^m-diffeomorphism.
(3) The i-th derivatives of f and f^{-1} are ϵ-close to the i-th derivative of the identity map, for all points $x \in \mathbb{R}^3$ and all i from 0 to m.

Let X be an open subset of $\mathbb{R}^3$ whose boundary is a C^2-smooth manifold [37]. Chazal and Soufflet [20] prove that a small C^2-perturbation f implies a small Hausdorff distance between the medial axes of X and $f(X)$. Formally, for every $\delta > 0$, there exists $\epsilon > 0$ such that for every ϵ-small C^2-perturbation f, $d_H(\mathcal{M}[X], \mathcal{M}[f(X)]) < \delta$. This result is optimal for m since the medial axis of a shape is already instable under C^1-perturbations [20]. Therefore, for approximating the medial axis of X with the medial axis of Y, the boundary of Y must be close to the boundary of X both in position, normal direction and curvature. Unfortunately, effective implementations of exact alorithms for the medial axis are known only for restricted families of shapes, such as polyhedra, unions of balls and complements of discrete point sets, whose boundaries are generally not C^2. In other words, it is unlikely that the positive approximation result for C^2-perturbations can be turned into a practical algorithm.

7 Exact Computation of Medial Axes

A fairly general class of shapes for which it is possible, in principle, to compute the medial axis exactly are the *semi-algebraic sets*, each the set of solutions of a finite system of algebraic equations and inequalities. The medial axis of such a set is itself semi-algebraic and can be computed with tools from computer algebra. To describe this, let X be a shape in $\mathbb{R}^3$ whose boundary is a C^1-smooth manifold. We introduce the *symmetry set* of X, consisting of the centers of spheres tangent to the boundary of X at two or more points. It contains all points of the medial axis but possibly additional points since the spheres are not constrained to bound balls contained in X. Suppose now the boundary of X is defined by the algebraic equation $f(x) = 0$ and 0 is a regular value of f. It follows that the gradient for all points of the boundary is non-zero, $\nabla f(x) \neq 0$. In this case, the symmetry set is the closure of the set of points z for which there exists points x and y that satisfy the following system of algebraic equations:

$$\begin{cases} f(x) \; = \; 0, \\ f(y) \; = \; 0, \\ (x-z) \times \nabla f(x) \; = \; 0, \\ (y-z) \times \nabla f(y) \; = \; 0, \\ \|x-z\|^2 \; = \; \|y-z\|^2, \\ t\|x-y\|^2 \; = \; 1 \end{cases}$$

In the last condition, t is an additional free variable that ensures that x and y are distinct. If 0 is not a regular value of f, we need to add $\nabla f(x)\nabla f(y)s = 1$ as yet another equation, with s a free variable. Finally, the medial axis is obtained by imposing the additional conditions that $\|u - z\|^2 \geq \|x - z\|^2$, for all points u on the boundary, and z be contained in X. Considering u a new free variable, we thus remove points from the solution, namely the points z for which $f(z) < 0$ or for which there exists u with $f(u) = 0$ and $\|u - z\|^2 < \|x - z\|^2$. This new set is still semi-algebraic since it is the difference between two semi-algebraic sets.

Although possible in principle, we are not aware of an implementation that effectively constructs the exact medial axes of general semi-algebraic sets. The most advanced effective implementations are limited to the planar case, to piecewise linear shapes, and to shapes constructed from finitely many balls. Even for shapes bounded by simple curves in the plane, the algebraic difficulties in computing medial axes are significant and satisfactory implementations are rare and far in between. Piecewise linear curves involve the comparison of expressions with two nested square roots [17] and efficient and fully robust implementations are few [36]. Ramamurthy and Farouki tackle the case of algebraic curve segments whose bisectors have rational parametrizations [47]. An exact algorithm for not-necessarily convex polyhedra in $\mathbb{R}^3$ can be found in [24]. For the complement of a union of balls in $\mathbb{R}^k$, the medial axis can be derived from the Apollonius diagram of the corresponding spheres or from convex hulls of finitely many points in $\mathbb{R}^{k+2}$ [11, 14]. Perhaps suprisingly, the medial axis of the union of finitely many balls is simpler than that of the complement. As first described in [5], it is piecewise linear and can be constructed from the Voronoi diagram of a finite set of points. As discussed in more detail shortly, the cells of dimension less than k in this diagram may be interpreted as the medial axis of a punctured Euclidean space, a case that permits particularly simple exact algorithms.

8 Approximation Paradigm for Medial Axes

The difficulty of computing medial axes exactly motivates a serious look at approximation algorithms. We describe a framework that captures a common line of attack to approximating the medial axis, as sketched in Fig. 3. First, we find Y approximating X that belongs to a class of shapes for which the medial axis can be constructed exactly. Second, we construct the medial axis of Y. Third, we prune the medial axis of Y to get a subset $\mathcal{P}[\mathcal{M}[Y]] \subseteq \mathcal{M}[Y]$ that approximates the medial axis of X.

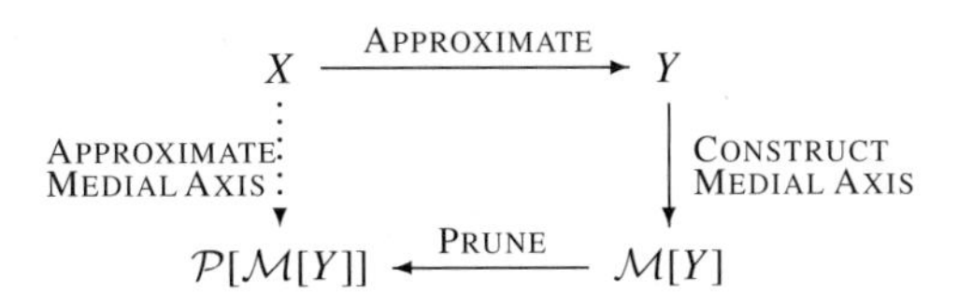

Fig. 3. An approximation $\mathcal{P}[\mathcal{M}[Y]]$ of the medial axis of a shape X can be found as part of the medial axis of a shape Y approximating X

The composition of the three steps provides the approximation of the medial axis of X. The most challenging step in this paradigm is the extraction of a subset $\mathcal{P}[\mathcal{M}[Y]]$ of $\mathcal{M}[Y]$ that indeed approximates $\mathcal{M}[X]$. Recent mathematical results that rationalize this approach are discussed shortly.

The notion of approximation used in the first step varies between different implementations of the approximation paradigm. It either means that Y is the image of X under a small C^m-perturbation, or that the Hausdorff distance between the complements of X and Y is small, as in [19]. Other notions of approximation are conceivable.

9 Punctured Euclidean Spaces

We start by identifying a class of shapes for which the medial axis can be constructed exactly and efficiently. We obtain shapes in this class by puncturing the k-dimension real space at a discrete set of locations. Equivalently, we consider the complement of a discrete set of points P in $\mathbb{R}^k$. The medial axis of this space is the *Voronoi graph* of P, which we define as the union of all cells in the Voronoi diagram of dimension $k-1$ or less. Algorithms for constructing the Voronoi graph are well-studied in computational geometry and implementations are available from the geometric software library CGAL (`http://www.cgal.org/`). For a set P of n points in $\mathbb{R}^k$, the graph can be constructed in time $O(n^{\lceil k/2 \rceil} + n \log n)$, which is optimal in the worst case because the graph can consist of a constant times $n^{\lceil k/2 \rceil}$ faces. In most practical applications, the number of faces, F, is much less and the output-sensitive algorithm in [18] constructs the graph in $\mathbb{R}^3$ in time $O((n + F) \log^2 F)$. Examples of point sets with provably small Voronoi graphs are so-called κ-light ϵ-samples of compact smooth generic surfaces in $\mathbb{R}^3$, with $F = O(n \log n)$ [8], and κ-light ϵ-samples of polyhedral surfaces in $\mathbb{R}^3$, with $F = O(n)$ [7]. Such samples will be studied in more detail shortly.

Consider a finite point set P whose Hausdorff distance to the boundary of a shape X is less than ϵ and write $\text{Vor}[P]$ for the Voronoi graph of P. Using the semicontinuity of the medial axis expressed in (2), we can show that the subset of $\text{Vor}[P]$ inside X contains an approximation of the medial axis of X. In the approximation paradigm for medial axes, this subset can be interpreted as part of the medial axis of a shape Y close to X. Following [19], we let Y be the parallel body $X^{+\epsilon}$ of X minus the points in P; see Fig. 4. Since the Hausdorff distance between P and the boundary of X is less than ϵ, the same is true for the complements of X and the thus constructed space: $d_H(X^c, Y^c) < \epsilon$. In summary, we have $\mathcal{M}[Y] \cap X = \text{Vor}[P] \cap X$.

10 Voronoi Graph and Medial Axis

We now consider results that focus on the detailed relationship between the Voronoi graph of a finite point set and the medial axis of the shape whose boundary the points sample. We need precise notions. A *sample* of the boundary of a shape X is a finite

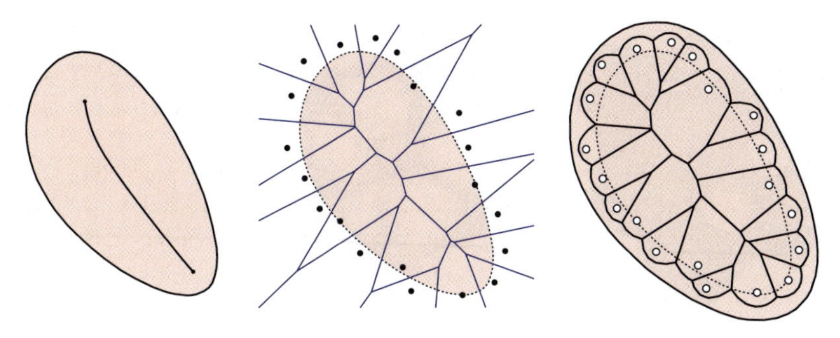

Fig. 4. On the *left*, a shape X and its medial axis. In the *middle*, a finite set of points P whose Hausdorff distance to the boundary of X is less than ϵ and its Voronoi graph. On the *right*, $X^{+\epsilon} - P$ and its medial axis

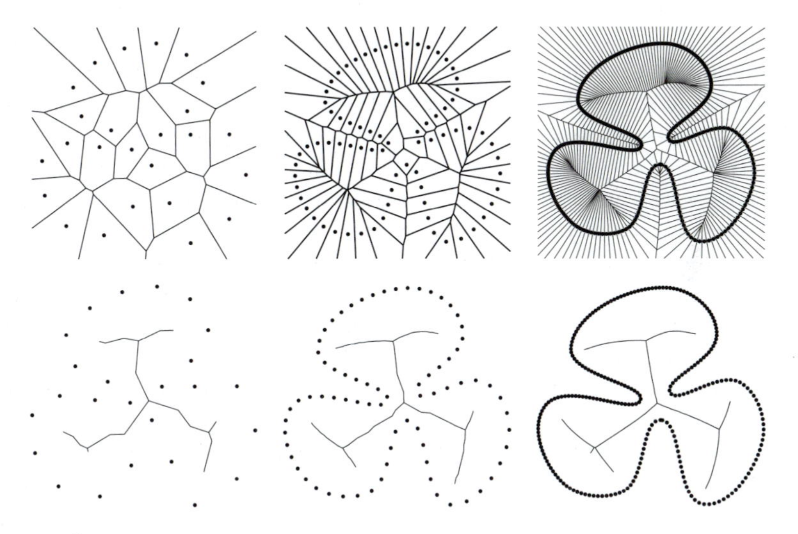

Fig. 5. In $\mathbb{R}^2$, vertices and edges lying inside a shape and extracted from the Voronoi graph of an ϵ-sample of the boundary approximate the medial axis (courtesy of Attali and Montanvert [10])

set of points (exactly and not just approximately) on that boundary. An ϵ*-sample* is a sample whose Hausdorff distance to the boundary of X is less than ϵ. In other words, every point of the boundary is less than distance ϵ away from a point in the ϵ-sample. The ϵ-sample is κ*-light* if the number of sample points within distance ϵ is never more than κ. The ϵ-sample is *noisy* if points are not necessarily on the boundary but at Hausdorff distance less than ϵ to the boundary.

An early result on the connection between the Voronoi graph and the medial axis is due to Brandt [15]. Given a shape in $\mathbb{R}^2$, he takes an ϵ-sample on the boundary curve and considers the Voronoi edges and vertices that are completely contained in the shape; see Fig. 5. He then proves that under some technical conditions on the boundary curve, the portion of the Voronoi graph defined by these edges and

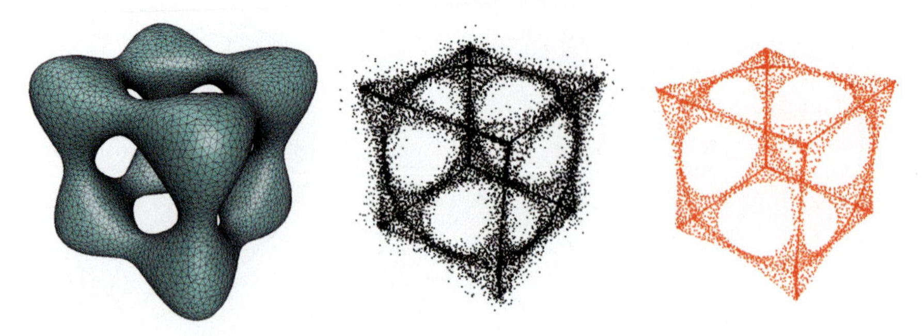

Fig. 6. On the *left* we see a triangulation of the boundary of a shape in $\mathbb{R}^3$. Its vertices determine a Voronoi diagram whose vertices inside the shape are shown in the *middle*. The subset of poles inside the shape is shown on the *right*

vertices approximates the medial axis. Amenta and Bern [1] point out that the direct extension of this result to shapes in $\mathbb{R}^3$ does not hold; see Fig. 6. The validity of the extension is spoiled by the existence of slivers in three-dimensional Delaunay triangulations, which occur for ϵ-samples with arbitrarily small $\epsilon > 0$. Roughly, a *sliver* is a tetrahedron whose four vertices are almost cocircular. The location of the Voronoi vertex corresponding to the sliver depends on the four vertices but is generally unrelated to any feature of the surface and does not necessarily lie near the medial axis. As a first step to cope with slivers, Amenta and Bern eliminate all but a few Voronoi vertices they refer to as poles. Every sample point p generates a Voronoi polyhedron and the vertices furthest away from p on the two sides of the surface are the *poles* of p. Clearly, there are at most $2n$ poles for a sample of n points. As proved in [3], for a shape whose boundary is a smooth C^1-manifold, the poles tend to the medial axis of the shape and its complement as ϵ goes to zero.

To extend the result of Brandt to $\mathbb{R}^3$, we need more than just points (the poles) near the medial axes, we also need to connect them to form a geometric structure approximating the medial axis. In [3], Amenta, Choi and Kolluri use simplices of the (weighted) Delaunay triangulation of the poles. To avoid the construction of this weighted Delaunay triangulation and connect the poles directly inside the Voronoi graph, we need to know about its local distance from the medial axis. Bounds on this distance can be found in [6, 13, 23]. Assuming the boundary of the shape is a smooth C^1-manifold and using these bounds, among other things, Dey and Zhao [27] give an algorithm that identifies a subgraph of the Voronoi graph that approximates the medial axis for the Hausdorff distance. We note that the above results are limited to smooth surfaces and to samples of points that lie on that surface. The next two sections deal with more general data.

11 Pruning in the Presence of Noise

Assuming the medial axis of a shape Y approximating X has been constructed, we prune $\mathcal{M}[Y]$ to retrieve an approximation of $\mathcal{M}[X]$. In this paragraph, the terms shape, medial axis, and stable part refer to Y, $\mathcal{M}[Y]$, and $\mathcal{P}[M[Y]]$, respectively.

Pruning methods shorten peripheral branches of the medial axis, trying to capture its stable part. Typically, points on the border are successively removed until a stopping condition is satisfied. This condition may be a threshold on the difference between the initial shape and the shape reconstructed from the simplified medial axis [16, 23, 28, 48, 49], or it may be based on an estimate of the stability of portions of the medial axis [9, 10, 26, 31, 32, 45]. We present experimental results due to Attali and Montanvert [10] that shed light on the latter approach. To each point $y \in Y$, we associate the distance to Y^c, the complement of Y, and the largest angle formed by points in Y^c closest to y:

$$\rho(y) \;=\; d(y, Y^c) \qquad \text{and} \qquad \theta(y) \;=\; \max_{a,b \in \Pi(y)} \angle ayb,$$

where $\Pi(y) = \{x \in Y^c \mid d(y, Y^c) = \|y - x\|\}$. We obtain the *parameter graph* by collecting, for all points y of the medial axis, the points $(\theta(y), \rho(y))$ in the two-dimensional parameter space [9, 10]. Points in this graph lie on curves associated with branches of the medial axis, as illustrated in Fig. 7. When noise is added to the boundary, new branches appear on its medial axis; see Fig. 8. The corresponding

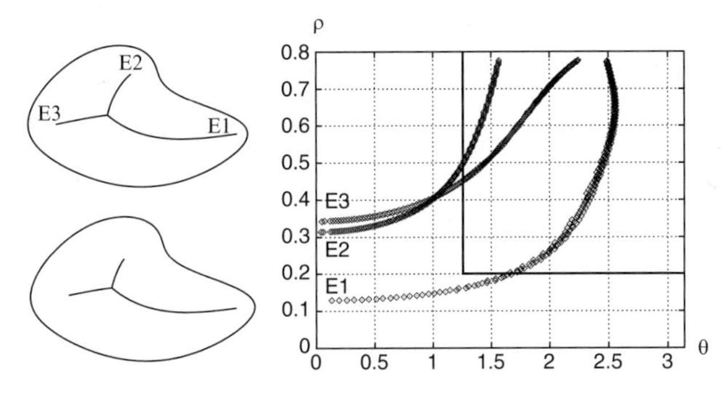

Fig. 7. Medial axis, parameter graph and simplified medial axis obtained by keeping points in the *upper right quadrant* of the parameter graph (courtesy of Attali and Montanvert [10])

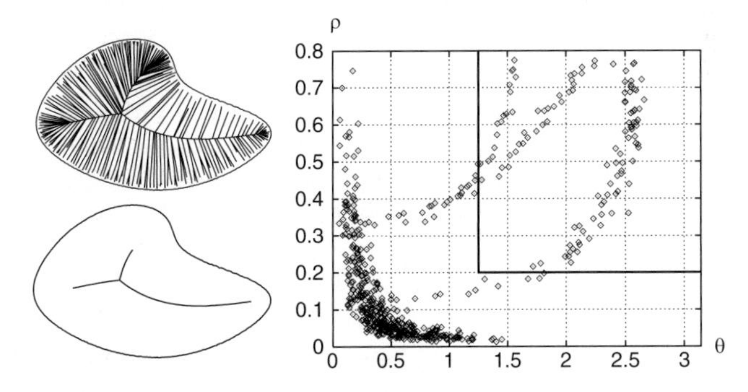

Fig. 8. A small amount of noise added to the boundary of a shape suffices to dramatically change its medial axis. The simplified medial axis is defined by points in the *upper right quadrant* of the parameter graph (courtesy of Attali and Montanvert [10])

effect on the parameter graph is the appearance of a hyperbola-like point cloud located near the coordinate axes. This experimental observation suggests a method for recognizing points on the medial axis that owe their existence to noise in the input data. As a first cut, we remove points y for which $\rho = \rho(y)$ is smaller than a first threshold or $\theta = \theta(y)$ is smaller than a second threshold. In order to refine the method, let us introduce $h(y) = \sup_{a,b\in\Pi(y)} \|a - b\|/2$. The new quantity, $h = h(y)$ is related to the previous ones by

$$\rho = \frac{h}{\sin(\theta/2)}.$$

For each fixed h we get ρ as a function of θ. By plotting a few of these functions, we can experimentally find a value of h for which the graph of this function approximates the hyperbola-like point cloud induced by the noise. This suggests that the stable part of the medial axis corresponds to points y for which h exceeds a given threshold. The next section describes a theoretical justification of this experimental finding.

12 Stability of the λ-Medial Axis

Chazal and Lieutier [19] define the λ-medial axis of a bounded open subset X of $\mathbb{R}^k$ and prove its stability under the Hausdorff distance, for regular values of λ. Remember that this property is not shared by the medial axis transform. To describe their results, let $r(x)$ be the radius of the smallest ball enclosing $\Pi(x)$, the set of points in the complement of X with minimum distance to x. By definition, the *λ-medial axis* of X is

$$\mathcal{M}_\lambda[X] = \{x \in X \mid r(x) \geq \lambda\}.$$

For $\lambda > 0$, the λ-medial axis is a subset of the medial axis and the Hausdorff distance between the two tends to zero when λ goes to zero. We say that λ is a *regular* value of X if the function that maps $\mu \in \mathbb{R}$ to $\mathcal{M}_\mu[X]$ in $\mathbb{R}^k$ is continuous under the Hausdorff metric at $\mu = \lambda$. In other words, a small modification of a regular value λ implies a small modification of the λ-medial axis. Typical non-regular values are radii of locally largest maximal balls. We are now ready to give a precise statement of the result in [19]: if λ is a regular value of a shape X, the λ-medial axis transform is continuous at X for the Hausdorff distance. In other words, for every $\delta > 0$, there exists $\epsilon > 0$ such that for every open subset Y of $\mathbb{R}^k$,

$$d_H(X^c, Y^c) < \epsilon \implies d_H(\mathcal{M}_\lambda[X], \mathcal{M}_\lambda[Y]) < \delta. \tag{3}$$

Note the similarity with (2), which expresses the same property using one-sided instead of two-sided Hausdorff distance. As part of the approximation paradigm for medial axes, this result sheds new light on the pruning method described above, which is now seen as approximating the λ-medial axis. Furthermore, an approximation of the medial axis can be obtained by forcing λ to decrease as Y gets closer to X.

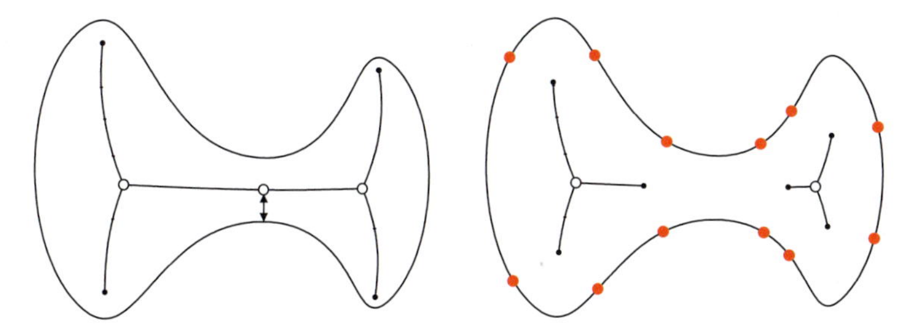

Fig. 9. *Left:* a shape with weak feature size indicated by the *arrow*. *Right:* the λ-medial axis for a value of λ greater than the weak feature size. Each endpoint has two closest points on the boundary, whose distance from each other is 2λ

This idea appears in [27] and can also be found in [19]. Specifically, we consider a sequence of shapes Y_ϵ whose Hausdorff distance to X is at most ϵ. Writing D for the diameter of X and introducing $g(\epsilon) = 10\sqrt{3}D^{3/4}\sqrt[4]{\epsilon}$, we get

$$\lim_{\epsilon \to 0} d_H(\mathcal{M}[X], \mathcal{M}_{g(\epsilon)}[Y_\epsilon]) = 0, \tag{4}$$

[private communication with André Lieutier]. Unlike the medial axis, the λ-medial axis is not necessarily homotopy equivalent to the shape (Fig. 9). To shed light on this phenomenon, Chazal and Lieutier [19] call a point $x \in X$ *critical* if the vector field $\nabla \rho$ defined in (1) vanishes at x; see Fig. 1. The *weak feature size* of X is the smallest distance between a critical point and the boundary of X. As proved in [19], the λ-medial axis is homotopy equivalent to X if λ is smaller than the weak feature size.

13 What Now?

How do we best harness the power of the new insights, in particular the stability of the λ-medial axis? In this section, we speculate how this stability can be used to obtain improved implementations of the approximation paradigm for medial axes. We also mention some of the open issues that are still obstacles in our quest for a satisfactory solution in the absence of any knowledge on the shape X other than a possibly noisy finite sample of its boundary.

Given a finite point set P, we call the λ-medial axis of the complement the *λ-Voronoi graph* of P. The *λ-complex* consists of all simplices in the Delaunay triangulation that can be enclosed by a sphere of radius less than λ. The relation between the two is one of duality and complementarity: a Voronoi cell of dimension less than k belongs to the λ-Voronoi graph iff the dual Delaunay simplex does not belong to the λ-complex. To derive an alternative description, let B be the set of open balls with radius λ whose centers are points in P. The λ-complex consists of all Delaunay simplices spanned by points in P whose balls have a non-empty intersection. This is similar to but slightly weaker than the condition for the simplex to belong to

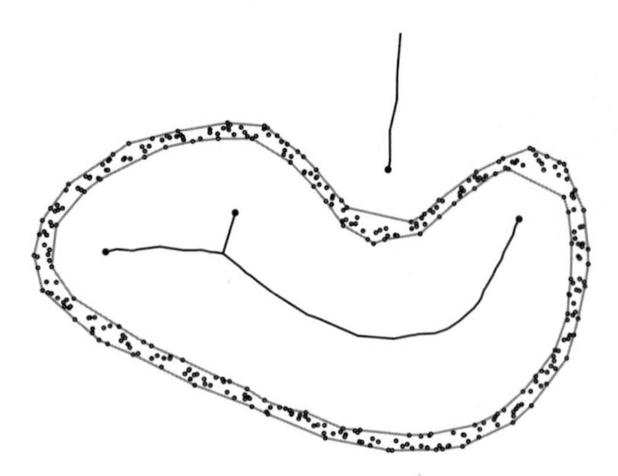

Fig. 10. The λ-complex and the λ-Voronoi graph of a noisy sample of a simple closed curve in the plane

the α-shape [31], which requires that the balls and the (k-dimensional) Voronoi cells have a non-empty intersection. Indeed, it is not difficult to prove that for $\lambda = \alpha$, the λ-complex and the α-complex are homotopy equivalent. To construct the λ-Voronoi graph, we simply select the Delaunay simplices that belong to the λ-complex and collect the dual Voronoi cells of the remaining Delaunay simplices as the pieces of the λ-Voronoi graph. An example of this construction is shown in Fig. 10. Assume now that P is a noisy sample of the boundary of an open set $X \subseteq \mathbb{R}^k$. If we know that P is an ϵ-sample of that boundary, we may set $\lambda = g(\epsilon)$ and compute the λ-Voronoi graph. The results presented in Sects. 10 and 12 assure that as ϵ goes to zero, the λ-Voronoi graph restricted to X is an approximation of the medial axis of X. To recapitulate, we go through the following steps to obtain an approximation of the medial axis:

1. Determine how small an ϵ is needed
2. Obtain an ϵ-sample of the boundary of X
3. Construct the λ-Voronoi graph, with $\lambda = g(\epsilon)$
4. Select the part of the graph inside X

In practice, we rarely have enough knowledge about X to know what $\epsilon > 0$ is sufficiently small, and even if we knew, we might not have the means to obtain an ϵ-sampling of the boundary. In exceptional cases, the boundary of X is defined mathematically, e.g. as the zero-set of an algebraic function $f : \mathbb{R}^k \to \mathbb{R}$, and we can determine sufficiently fine ϵ-samples and therefore λ-Voronoi graphs that approximate the medial axis, as in Fig. 11. This approach to medial axes thus suffers from the same difficulties as the α-shape approach to surface reconstruction: it is usually not clear which value of λ (or α) is most appropriate, and in many cases there is no such most appropriate value. This suggests we re-trace some of the developments aimed at fixing this drawback for α-shapes, namely looking at the filtration (nested sequence) of λ-Voronoi graphs and use topological persistence [30] to select and combine pieces of λ-Voronoi graphs for different values of λ in different portions of X.

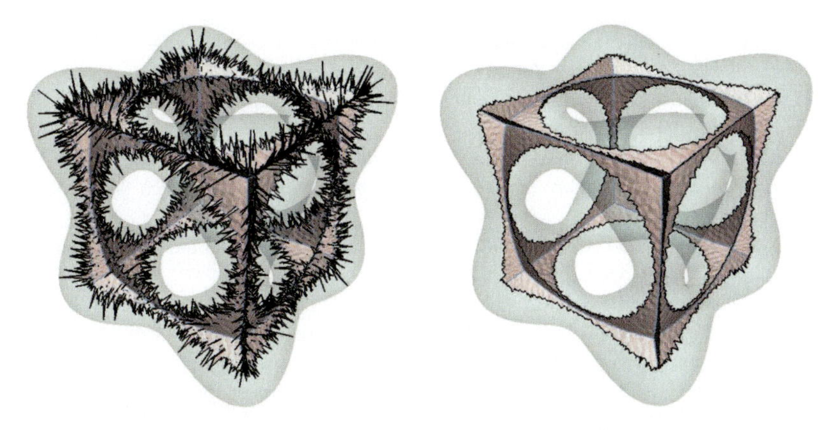

Fig. 11. Two λ-medial axes of the same shape, with λ increasing from *left* to *right*, constructed as a subset of the λ-Voronoi graph of a sample of the boundary

Acknowledgement

We thank Frédéric Chazal and André Lieutier for communications related to questions discussed in this paper. We thank David Cohen-Steiner for fruitful discussions. The first two authors acknowledge the support of the EU through the Network of Excellence AIM@SHAPE Contract IST 506766. The third author acknowledges the support of the NSF through grant CCR-00-86013.

References

1. N. Amenta and M. Bern. Surface reconstruction by Voronoi filtering. *Discrete Comput. Geom.*, 22:481–504, 1999.
2. N. Amenta, M. Bern, and D. Eppstein. The crust and the beta-skeleton: combinatorial curve reconstruction. *Graph. Model. Image Process.*, 60:125–135, 1998.
3. N. Amenta, S. Choi, and R. K. Kolluri. The power crust, unions of balls, and the medial axis transform. *Comput. Geom. Theory Appl.*, 19:127–153, 2001.
4. E. V. Anoshkina, A. G. Belyaev, O. G. Okunev, and T. L. Kunii. Ridges and ravines: a singularity approach. *Internat. J. Shape Modeling*, 1:1–11, 1994.
5. D. Attali. *Squelettes et graphes de Voronoi* 2-*d et* 3-*d*. PhD thesis, Univ. Joseph Fourier, Grenoble, 1995.
6. D. Attali and J.-D. Boissonnat. Complexity of the Delaunay triangulation of points on polyhedral surfaces. *Discrete Comput. Geom.*, 30:437–452, 2003.
7. D. Attali and J.-D. Boissonnat. A linear bound on the complexity of the Delaunay triangulation of points on polyhedral surfaces. *Discrete Comput. Geom.*, 31:369–384, 2004.
8. D. Attali, J.-D. Boissonnat, and A. Lieutier. Complexity of the Delaunay triangulation of points on surfaces: the smooth case. In *Proc. 19th Ann. Sympos. Comput. Geom.*, pages 201–210, 2003.
9. D. Attali and J.-O. Lachaud. Delaunay conforming iso-surface. *Skeleton Extraction and Noise Removal*, 19:175–189, 2001.

10. D. Attali and A. Montanvert. Modeling noise for a better simplification of skeletons. In *Proc. Internat. Conf. Image Process.*, volume 3, pages 13–16, 1996.
11. F. Aurenhammer and H. Imai. Geometric relations among Voronoi diagrams. *Geom. Dedicata*, 27:65–75, 1988.
12. H. Blum. A transformation for extracting new descriptors of shape. In W. Wathen-Dunn, editor, *Models for the Perception of Speech and Visual Form*, pages 362–380, MIT, Cambridge, MA, 1967.
13. J.-D. Boissonnat and F. Cazals. Natural neighbor coordinates of points on a surface. *Comput. Geom. Theory Appl.*, 19:155–173, 2001.
14. J.-D. Boissonnat and M. Karavelas. On the combinatorial complexity of Euclidean Voronoi cells and convex hulls of d-dimensional spheres. In *Proc. 14th ACM-SIAM Sympos. Discrete Alg.*, pages 305–312, 2003.
15. J. W. Brandt. Convergence and continuity criteria for discrete approximations of the continuous planar skeletons. *CVGIP: Image Understanding*, 59:116–124, 1994.
16. J. W. Brandt and V. R. Algazi. Continuous skeleton computation by Voronoi diagram. *CVGIP: Image Understanding*, 55:329–337, 1992.
17. C. Burnikel. *Exact computation of Voronoi diagrams and line segment intersections*. PhD thesis, Universität des Saarlandes, March 1996.
18. T. M. Chan, J. Snoeyink, and C. K. Yap. Primal dividing and dual pruning: Output-sensitive construction of 4-d polytopes and 3-d Voronoi diagrams. *Discrete Comput. Geom.*, 18:433–454, 1997.
19. F. Chazal and A. Lieutier. Stability and homotopy of a subset of the medial axis. In *Proc. 9th ACM Sympos. Solid Modeling Appl.*, 2004.
20. F. Chazal and R. Soufflet. Stability and finiteness properties of medial axis and skeleton. *J. Control Dyn. Syst.*, 10:149–170, 2004.
21. C. H. Chen, L. F. Pau, and P. S. Wang, editors. *Segmentation tools in Mathematical Morphology*. World Scientific, Singapore, 1993.
22. H. I. Choi, S. W. Choi, and H. P. Moon. Mathematical theory of medial axis transform. *Pacific J. Math.*, 181:57–88, 1997.
23. S. W. Choi and H.-P. Seidel. Linear one-sided stability of MAT for weakly injective 3D domain. In *Proc. 7th ACM Sympos. Solid Modeling Appl.*, pages 344–355, 2002.
24. T. Culver. *Computing the medial axis of a polyhedron reliably and efficiently*. Depart. comput. sci., Univ. North Carolina, Chapel Hill, NC, 2000.
25. T. K. Dey, J. Giesen, and S. Goswami. Shape segmentation and matching with flow discretization. In F. Dehne et al., editor, *Proc. Workshop Alg. Data Structures*, pages 25–36, 2003.
26. T. K. Dey and W. Zhao. Approximate medial axis as a Voronoi subcomplex. In *Proc. 7th ACM Sympos. Solid Modeling Appl.*, pages 356–366, 2002.
27. T. K. Dey and W. Zhao. Approximating the medial axis from the Voronoi diagram with a convergence guarantee. *Algorithmica*, 38:179–200, 2004.
28. A. R. Dill, M. D. Levine, and P. B. Noble. Multiple resolution skeletons. *IEEE Trans. Pattern Anal. Mach. Intell.*, 9:495–504, 1987.
29. H. Edelsbrunner. Surface reconstruction by wrapping finite point sets in space. In B. Aronov, S. Basu, J. Pach, and M. Sharir, editors, *Discrete and Computational Geometry – The Goodman-Pollack Festschrift*, pages 379–404. Springer, Berlin, 2004.
30. H. Edelsbrunner, D. Letscher, and A. Zomorodian. Topological persistence and simplification. *Discrete Comput. Geom.*, 28:511–533, 2002.
31. H. Edelsbrunner and E. P. Mücke. Three-dimensional alpha shapes. *ACM Trans. Graphics*, 13:43–72, 1994.

32. M. Foskey, M. Lin, and D. Manocha. Efficient computation of a simplified medial axis. In *Proc. 8th ACM Sympos. Solid Modeling Appl.*, pages 96–107, 2003.
33. P. Giblin and B. B. Kimia. A formal classification of 3D medial axis points and their local geometry. *IEEE Trans. Pattern Anal. Mach. Intell. (PAMI)*, 26:238–251, 2004.
34. J. Giesen and M. John. Surface reconstruction based on a dynamical system. In *Proc. 23rd Ann. Conf. European Association for Computer Graphics (Eurographics), Computer Graphics Forum*, pages 363–371, 2002.
35. R. Haralick and L. Shapiro. *Computer and Robot Vision*, volume 1. Addison-Wesley, New York, 1992.
36. M. Held. VRONI: An engineering approach to the reliable and efficient computation of Voronoi diagrams of points and line segments. *Comput. Geom. Theory Appl.*, 18:95–123, 2001.
37. M. W. Hirsch. *Differential Topology*. Springer, New York, 1988.
38. M. Hisada, A. G. Belyaev, and T. L. Kunii. A skeleton-based approach for detection of perceptually salient features on polygonal surfaces. In *Computer Graphics Forum*, volume 21, pages 689–700, 2002.
39. C. Hoffmann. *Geometric and Solid Modeling*. Morgan-Kaufmann, San Mateo, CA, 1989.
40. B. Jähne. *Digital Image Processing*, 4th edition. Springer, Berlin, 1997.
41. L. Lam, S.-W. Lee, and C. Y. Suen. Thinning methodologies – a comprehensive survey. *IEEE Trans. on PAMI*, 14(9):869–885, 1992.
42. J.-C. Latombe. *Robot Motion Planning*. Kluwer Academic, Boston, 1991.
43. A. Lieutier. Any open bounded subset of $\mathbb{R}^n$ has the same homotopy type as its medial axis. In *Proc. 8th ACM Sympos. Solid Modeling Appl.*, pages 65–75. ACM, 2003.
44. G. Matheron. Examples of topological properties of skeletons. In J. Serra, editor, *Image Analysis and Mathematical Morphology, Volume 2: Theoretical Advances*, pages 217–238. Academic, London, 1988.
45. R. Ogniewicz. A multiscale MAT from Voronoi diagrams: the skeleton-space and its aplication to shape description and decomposition. In C. Arcelli et al., editors, *Aspects of Visual Form Processing*, pages 430–439. World Scientific, Singapore, 1994.
46. M. A. Price and C. G. Armstrong. Hexahedral mesh generation by medial surface subdivision: Part II. solids with flat and concave edges. *Int. J. Numer. Methods Eng.*, 40:111–136 1997.
47. R. Ramamurthy and R. T. Farouki. Voronoi diagram and medial axis algorithm for planar domains with curved boundaries, I and II. *J. Comput. Appl. Math.*, 102:119–141 and 253–277, 1999.
48. G. Sanniti di Baja and E. Thiel. A multiresolution shape description algorithm. In D. Chetverikov et al., editor, *Lecture Notes in Computer Science*, volume 719, pages 208–215. Springer, Berlin, 1993.
49. D. Shaked and A. M. Bruckstein. Pruning medial axes. *Comput. Vis. Image Underst.*, 69:156–169, 1998.
50. A. Sheffer, M. Etzion, A. Rappoport, and M. Bercovier. Hexahedral mesh generation using the embedded Voronoi graph. *Engineering Comput.*, 15:248–262, 1999.
51. E. Sherbrooke, N. M. Patrikalakis, and F.-E. Wolter. Differential and topological properties of medial axis transforms. *Graph. Model. Image Process.*, 58:574–592, 1996.
52. M. Sonka, V. Hlavac, and R. Boyle. *Image Processing, Analysis and Machine Vision*, 2nd edition. PWS, Pacific Grove, 1999.
53. F.E. Wolter. Cut locus and medial axis in global shape interrogation and representation. Technical Report Design Laboratory Memorandum 92-2, MIT, 1992.

Local Geodesic Parametrization: An Ant's Perspective

Lior Shapira[1] and Ariel Shamir[2]

[1] Tel Aviv-University
lior@liors.net
[2] The Interdisciplinary Center
arik@idc.ac.il

Summary. Two-dimensional parameterizations of meshes is a dynamic field of research. Most works focus on parameterizing complete surfaces, attempting to satisfy various constraints on distances and angles and produce a 2D map with minimal errors. Except for developable surfaces no single map can be devoid of errors, and a parametrization produced for one purpose usually doesn't suit others.

This work presents a different viewpoint. We try and acquire the perspective of an ant living on the surface. The point on which it stands is the center of its world, and importance diminishes from there onward. Distances and angles measured relative to its position have higher importance than those measured elsewhere. Hence, the local parametrization of the geodesic neighborhood should convey this perspective by mostly preserving geodesic distances from the center. We present a method for producing such overlapping local-parametrization for each vertex on the mesh. Our method provides an accurate rendition of the local area of each vertex and can be used for several purposes, including clustering algorithms which focus on local areas of the surface within a certain window such as Mean Shift.

1 Introduction

Contrasted with the modern evidence that the Earth is round, the ancient belief that the Earth is flat is reasonable from the point of view of a person standing on it. Similarly, an Ant standing on a manifold surface in 3D has a particular perspective concerning the shape of the surface it lives on. The distances and angles to points which lie on the surface relative to the Ant's position create a perspective map of the neighborhood surface. We call this map local-geodesic parametrization and in this work we present a method to create such maps for any position on a manifold mesh.

A manifold is a topological space that is locally Euclidean. This means that around every point, there is a neighborhood that is topologically the same as the open unit ball in $\mathbb{R}^n$ (taken from [13]). In $\mathbb{R}^3$ a manifold surface-mesh is locally homeomorphic to an open unit circle. The basic problem of parametrization is finding a map from 2D Euclidean space to the surface-mesh or to sub-parts of the mesh with minimum distortion of lengths and/or angles between points on the mesh.

T. Möller et al. (eds.), *Mathematical Foundations of Scientific Visualization, Computer Graphics, and Massive Data Exploration*, Mathematics and Visualization,
DOI: 10.1007/978-3-540-49926-8,

Numerous solution have been proposed focusing on different aspects of the field, achieving results which suit different objectives, and using different methods. In several important papers Floater presented barycentric coordinates as a way to allow vertices in triangulations to be expressed as convex combinations of their neighbors [3]. This was later enhanced by the mean value coordinates in [4]. Such constraints can be solved as a sparse linear system of equations. Levy and Mallet presented a method to solve the same convex combination problem as a minimization problem [8]. Several papers such as Sheffer and Sturler [12], and Zigelman et al. [14], concentrate on parametrization with free boundaries. Lee et al. [7] present parametrization with virtual boundaries. In this method a surface homeomorphic to a disc which is to be parameterized is surrounded with one to several layers of virtual vertices and edges. These virtual vertices lie on a convex boundary, allowing more flexibility in relaxation of the parameterization, achieving better results. Other works have focused on processing the mesh to make it suitable for parameterizations and improve the parametrization results. For example, by introducing seams to an arbitrary mesh, the mesh is modified to be homeomorphic to a disc and the parametrization distortion is reduced [11] and [2]. For a thorough review of the field the reader is referred to a recent survey by Floater and Hormann [5].

In fact, most previous papers concentrate on minimizing the angle, length or area distortions on a global scale. In contrast, in this work we are interested in minimizing the distortion with a bias towards a specific position (the Ant's viewpoint) and locally in its neighborhood. This gives raise to a different type of parametrization which can be seen as a perspective geodesic map.

Figure 1 gives an overview of our method. The algorithm starts from a vertex which is the central position of the map. Using a front marching method a surface

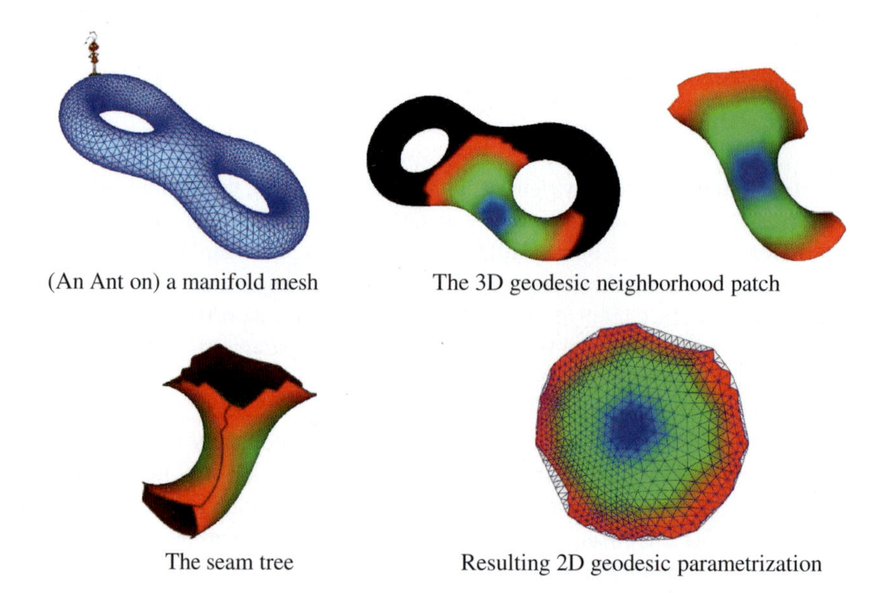

Fig. 1. Overview of local geodesic parametrization

patch with a given geodesic distance is built around the vertex. If the patch is not homeomorphic to a disk, seams are introduced in order to have only one boundary loop. Next, the vertices of the boundary loop are placed on a 2D circle according to their angle and then moved to their true geodesic distance from the center. Filler vertices and triangles are added to create a new convex boundary. Lastly, mean-value coordinates are used to calculate the parametrization inside the patch.

The algorithm is described in details in Sect. 2. Next, in Sect. 3 we discuss several implementation issues. We give an example for the use of such maps for geodesic mean shift in Sect. 4, and conclude in Sect. 5.

2 The Algorithm

2.1 Building the Neighborhood Patch

The first step in the algorithm is to define the geodesic neighborhood of the vertex. This is done using a front marching algorithm similar to the one used in [6]. The algorithm starts from the vertex and expands along a breadth first front. At all times the algorithm keeps a set of vertices for which geodesic distance has been found (*fixed*), a set of vertices which are on the advancing front (*close*) and a set of vertices whose distance is yet unknown (*far*). The front advances until it reaches the geodesic distance defined as maximum radius (Fig. 2a–g). At this stage, all fixed vertices are added to the local geodesic neighborhood. Note that if the front meets itself, it merges to produce a new unified front. This will mean that there is a need to cut seams in the patch in later stages (Fig. 2d–f).

The output of this step is a collection of vertices and edges which comprise a vertex's local geodesic neighborhood within a given radius. Even though the mesh is manifold, the computed geodesic neighborhood will not, in general, be homeomorphic to a disk. This is the case when the neighborhood radius is large or the mesh is complex (e.g., fingers of a hand). To create a patch which is homeomorphic to a

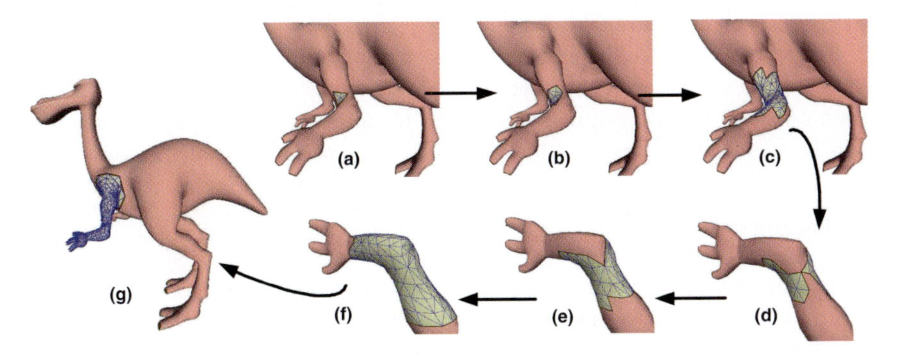

Fig. 2. Building the geodesic neighborhood using the front marching method

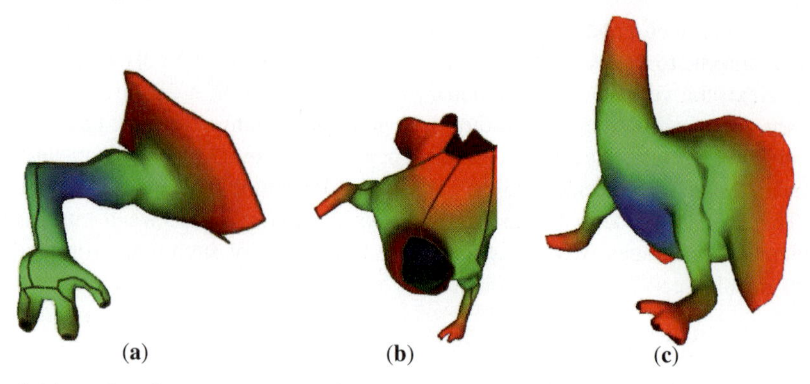

Fig. 3. Examples of seam-trees cuts: (**a**) for the geodesic neighborhood of Fig. 2, and (**b**) and for larger geodesic neighborhood defined in (**c**). The color represents the geodesic distance from the center (from *blue* to *red*)

disk, we wish to ensure that it includes only one boundary loop. This boundary loop will become the boundary of the patch's local geodesic map, and define the patch's shape.

Based on [11] we attempt to connect the different boundary loops using two steps. First, we build a set of essential vertices which must be in the seams. This set includes all the vertices sitting on the different boundary loops, but also specific vertices which were marked during the front marching algorithm. These are called *maxima vertices* and are vertices where the front merged with itself (see Fig. 2d). Next, we assign weights to all edges in the patch. Boundary edges are assigned a small weight of $\epsilon > 0$ and other edges are given weight relative to the distance from the center vertex. We use all-pairs-shortest path to find the shortest paths between the subset of vertices. In the second step, a minimum cost spanning tree of these paths is created. This tree connects all the vertices we marked as essential in the previous step creating the seams needed to cut the patch (Fig. 3). It will include all the boundary loops, and by including maxima vertices, this process guarantees that the seams occur at "natural" boundaries of the patch, i.e., along maxima ridges of the geodesic distance from the center.

To create a disk-like shape, the patch is cut along the branches of the seam tree. Vertices from which several branches in the seam tree grow are duplicated and maintain a reference to their original vertex. The resulting patch has only one boundary loop, hence it is homeomorphic to a disc and can be parameterized.

2.2 Parametrization

Once the patch has a single boundary, we can order the vertices on this boundary to form a single loop. We place the vertices in order on a 2D circle, whose initial radius is the geodesic neighborhood radius. We calculate the loop length and assign an *importance* measure to each vertex based on the ratio of the lengths of its adjacent boundary edges to the total length. The angle between the vertices is chosen relative to the importance of each vertex.

After this stage we have several options for defining the boundary of the patch to be parameterized. We can keep the boundary vertices on the initial circle (which is convex) and fill the rest of the patch using a barycentric method. This method is quick and simple but fairly inaccurate in terms of preserving the geodesic distance (Fig. 4a).

If we want to preserve the shape of the patch as defined by its boundary, we pull towards the center all the vertices whose distance to the center is smaller than the neighborhood-radius, and position them in their true distance. This forms a non-convex shape, which is harder to parameterize. This can be solved in two possible ways. The first alternative is to calculate the convex hull of the shape, and pull the

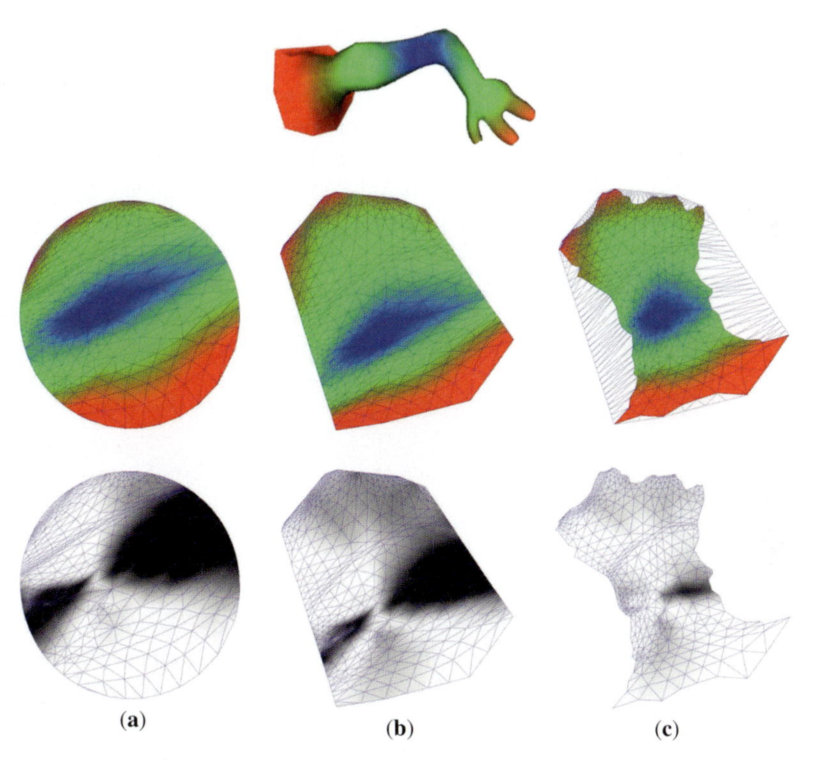

Fig. 4. Different parametrization alternatives for the 3D patch of the Dino arm (*top*): (**a**) Placing the patch boundary loop on a circle. (**b**) Pulling the boundary vertices to their true geodesic distance and back to the convex hull. (**c**) Pulling the boundary vertices to their true geodesic distance and adding filler vertices and triangles. The *top row* shows a color mapping of the true geodesic distance from the center and the *bottom row* shows the difference between the true geodesic distance and the Euclidean distance inside the patch, where darker means larger error

vertices which do not lie on the convex hull back to the hull. This achieves better results than if we parameterize the circle, by conforming more naturally to the patch shape (Fig. 4b). This solution is useful if we require a convex map (e.g., for mean shift, see Sect. 4). However, using this method still introduces some errors in the preservation of geodesic distances.

The most accurate method in term of geodesic distance preservation involves computing the convex hull as above, but instead of pulling the vertices back to the hull, add "filler" triangles and vertices to complete the shape to a convex shape much like in [7]. This will create a convex patch where the original boundary vertices are positioned in the true geodesic distance. Hence, by keeping the natural shape of the patch and adding "filler" vertices we add more flexibility to the patch, and the resulting parametrization preserves geodesic distances much better than the other two solutions (Fig. 4c).

Now that we have a convex boundary of the patch we calculate the parameterizations of the triangles inside the patch by using simple mean value coordinates [4]. In this method each vertex is expressed as a convex combination of its neighbors. This improves over Floater's earlier work [3] utilizing the mean value theorem for Harmonic functions. In practice, the method takes into account both the angles around the vertices and the edge lengths. This method produced good results, although other parameterizations techniques may be applied at this stage. Post processing techniques might also be used to relax the parameterizations and spread the geodesic distance error more evenly across the patch.

Figure 4 shows the different alternatives for parameterizing the patch. We measure the error using the difference between geodesic distance from the center vertex (as calculated by the front marching algorithm) and the Euclidean distance from the center vertex (distance calculated between the vertices on the 2D parametrization).

3 Implementation

3.1 Timing

The computation time for parametrization is dependent on the size of the patch (radius of geodesic neighborhood), the complexity of the mesh (number of seams, etc.) and more. However, as can be seen in the following table, in most cases the whole process takes a few second to compute. Most of the computation time is taken up by the iterative solution which finds the coordinates of vertices inside the patch. This code was not optimized and furthermore, this part is similar in any parametrization method. This means that the added computation for creating the correct geodesic boundary is very small.

Mesh	# Vertices in patch	Type of parametrization	Number of seams	Computation time (s)
Dino-pet	500	Circle	4	3.2
Dino-pet	500	Filled	4	4.4
Dino-pet	650	Circle	4	4.1
Dino-pet	650	Filled	4	5.4
Dino-pet	800	Circle	4	7.0
Dino-pet	800	Filled	4	9.6
Horse	100	Circle	0	1.4
Horse	100	Filled	0	2.0
Horse	1,000	Circle	0	6.7
Horse	1,000	Filled	0	7.5
Eight	800	Circle	1	4.6
Eight	800	Filled	1	5.7
Eight	1,300	Circle	1	9.0
Eight	1,300	Filled	1	9.9

3.2 Artificial Seams

In some cases there is a need to add seams even if the patch is homeomorphic to a disk. Such an example can be seen in Fig. 5. The center vertex is chosen close to the hoof, where the triangles are very large. In contrast, the triangles along the leg are smaller, and the leg itself is long and thin. When the patch is mapped to a circle, the boundary triangles are stretched and the triangles near the center are condensed. This creates a large error in terms of geodesic distance (Fig. 5d). The solution is to find a special vertex while building the geodesic neighborhood, and add a seam to the patch (Fig. 5e).

3.3 Map Storage

Performing geodesic parametrization process on all (or a selected number of) vertices in a two-manifold mesh, creates a collection of local geodesic maps which can be used for a variety of purposes. Nevertheless, since the amount of overlap in these maps can be quite large, there is a need to analyze the storage of these maps carefully, balancing efficiency and speed with storage size and memory requirements.

The first option is to store each local map as a regular mesh, with connectivity and geometry information, and a mapping from the map vertices to the original mesh vertices. This method is efficient, but costly in terms of memory requirement. These requirements depend on the size of the original mesh and the size of the geodesic radius for the maps. If the size of the original mesh is $O(n)$ the required memory for all local maps is roughly $O(n^2)$.

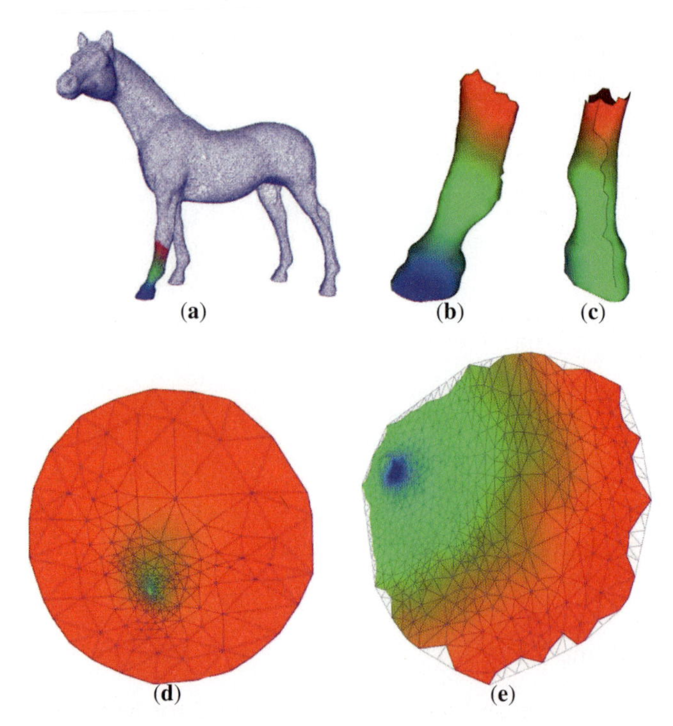

Fig. 5. Adding a seam for better parametrization: compare the mapping of geodesic distance without a seam (**d**), and with a seam (**e**)

Other options alow us to reconstruct the patch using different information. For example, we can save for each local map:

1. The indexes of the vertices in the original mesh which appear in the patch
2. The distance from the center vertex to each vertex in the patch (from the front marching algorithm)
3. The set of half edges which represent the seam tree
4. The location of cach vertex in the patch (original vertices which were split have more than one location)

This storage method saves the space of the connectivity information for the patch, but forces some reconstruction every time we need a map. During reconstruction we must create a mesh object for the patch, use the connectivity information from the original mesh, cut the patch along the seam tree, and then and assign the correct coordinates to each vertex.

In fact, most of the storage space is composed of the geometry information of the patch, i.e., the vertices coordinates. Hence, the last storage method is similar to the previous one without geometry (4). This creates very small memory footprint as it consists only of indices and half edge lists, with no geometry or connectivity. Nevertheless, it forces us to parameterize the actual patch anew every time we need a map resulting in a relatively high processing times to rebuild the patch.

4 Sample Application

There are several possible applications for local geodesic maps such as re-meshing, local texture mapping and more. In this section we concentrate on a specific application where preserving the local geodesic distances from the center is a key constraint.

The mean-shift algorithm is a clustering or filtering algorithm which is widely used on images and video. One step in the mean shift procedure moves a point in feature-space to its neighborhood average point. This shifting is continued until convergence (i.e., until the point is located at the average of its neighborhood). The point of convergence is called the *mode point*, as it is a local maximum in a Parzen window density estimation function. The main operation in this clustering process is therefore a weighted averaging of the neighborhood of a point in a high-dimensional space of "features." The weights for averaging are relative to the distance from the center point. More details can be found in [1].

Recent results [9, 10] have adapted the mean-shift method to work on volumetric as well as manifold meshes (Fig. 6). Since the averaging is dependent on the distance from the center point, measured on the mesh, there is a need to use geodesic distances between points on the mesh for averaging. Nevertheless, unlike images or volumes, where the averaging neighborhood is convex, on manifold meshes weighting points on the mesh can easily result in a point laying outside the mesh. Contrarily to the Laplacian operator, there is a need to constrain the movement of the mean-shift on manifold meshes to remain on the mesh. Furthermore, due to the fact that usually the mesh sampling is not uniform, there is a need to weigh the contribution of each point in the neighborhood by the area it represents.

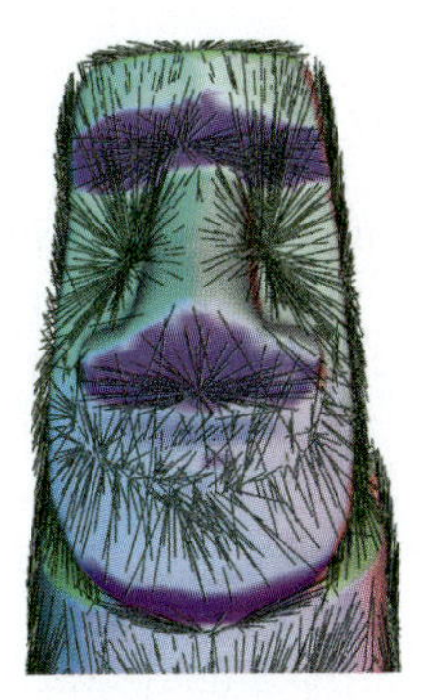

Fig. 6. An example of using the mean-shift algorithm for clustering areas on a manifold mesh. The color signifies the feature value which is the normal direction. This creates a 6D feature space (along with the XYZ coordinates). The lines connect each vertex with its mode in the mean shift process

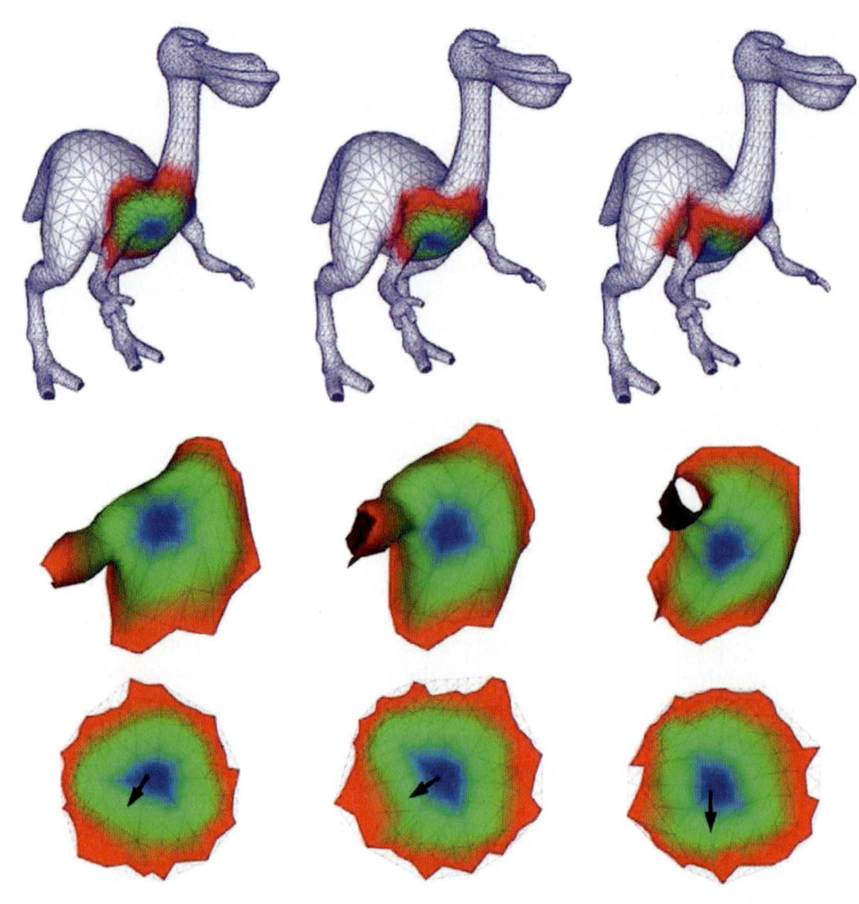

Fig. 7. The mean-shift algorithm on a manifold mesh. Each step the geodesic neighborhood is changed (*top*) and a new patch is parametrized (*bottom*)

Using the local geodesic maps for averaging the geodesic neighborhood of a point solves both these problems for mean-shift on manifolds. Using the convex map, the weighted average of the neighborhood will always fall inside the map, and consequently can be mapped back to the manifold mesh. Furthermore, the map itself preserves distances from the center point as required by the algorithm (see Fig. 7).

5 Conclusion and Future Work

We have presented a new type of parametrization of manifold meshes which we call *local-geodesic*. This type of parametrization is targeted at preserving geodesic distances from one central point to other points in its geodesic neighborhood. This results in a type of local perspective map similar to the point of view of an Ant living on the manifold.

As an example for successful usage of this type of parametrization we presented the mean-shift process of points on a manifold mesh. In future we would like to pursue other applications and uses for local geodesic parametrization. We are also working on methods to constrain the center point to stay exactly in the middle of the patch and to distribute the error between geodesic and Euclidean distance more evenly over the patch.

References

1. D. Comaniciu and P. Meer. Mean shift: A robust approach towards feature space analysis. *IEEE Transactions on Pattern Analysis and Machine Intelligence*, 24:603–619, May 2002.
2. J. Erickson and S. Har-Peled. Optimally cutting a surface into a disk. In *Proceedings of the 18th Annual ACM Symposium on Computational Geometry*, pages 244–253, 2002.
3. M. Floater. Parametrization and smooth approximation of surface triangulations. *Computer Aided Geometric Design*, 14:231–250, 1995.
4. M. Floater. Mean value coordinates. *Computer Aided Geometric Design*, 20:19–27, 2003.
5. M. Floater and K. Hormann. Surface parameterization: a tutorial and survey. In N.A. Dodgson, M.S. Floater, and M.A. Sabin (eds.) *Advances on Multiresolution in Geometric Modelling*. Springer, Heidelberg, 2005, 157–186.
6. R. Kimmel and J.A. Sethian. Computing geodesic paths on manifolds. volume 95, pages 8431–8435, July 1998.
7. Y. Lee, H.S. Kim, and S. Lee. Mesh parameterization with a virtual boundary. In *Computers and Graphics (Special Issue of the 3rd Israel-Korea Binational Conf. on Geometric Modeling and Computer Graphics)*, volume 26, pages 677–686, 2002.
8. B. Levy and J.-L. Mallet. Non-distorted texture mapping for sheared triangulated meshes. In *Proceedings of the 25th annual conference on Computer graphics and interactive techniques*, pages 343–352, 1998.
9. A. Shamir. Feature-space analysis of unstructured meshes. In *Proceedings IEEE Visualization 2003*, pages 185–192, Seattle, Washington, 2003.
10. A. Shamir, L. Shapira, and D. Cohen-Or. Mesh analysis using geodesic mean shift. *The visual Computer*, 22:1–10, 2006.
11. A. Sheffer. Spanning tree seams for reducing parameterization distortion of triangulated surfaces. In *Proceedings of the International Conference on Shape Modeling and Applications 2002 (SMI'02)*, pages 61–66, 2002.
12. A. Sheffer and E. de Sturler. Surface parameterization for meshing by triangulation flattening. In *Proceedings of the 9th International Meshing Roundtable*, pages 161–172, 2000.
13. E. W. Weisstein. Manifold. From MathWorld–A Wolfram Web Resource. `http://mathworld.wolfram.com/Manifold.html`, 2004.
14. G. Zigelman, R. Kimmel, and N. Kiryati. Texture mapping using surface flattening via multi-dimensional scaling. *IEEE Transactions on Visualization and Computer Graphics*, 8(2):198–207, April 2002.

Tensor-Fields Visualization Using a Fabric-like Texture Applied to Arbitrary Two-dimensional Surfaces

Ingrid Hotz, Louis Feng, Bernd Hamann, and Kenneth Joy

Institute for Data Analysis and Visualization (IDAV), Department of Computer Science, University of California, Davis, CA 95616, USA
{ihotz, zfeng, bhamann, kijoy}@ucdavis.edu

Summary. We present a visualization method for the exploration of three-dimensional tensor fields. The representation of the tensor field on a one-parameter family of two-dimensional surfaces as stretched, compressed and bent piece of fabric reflects the physical properties of stress and strain tensor fields. The texture parameters as the fiber density and fiber direction are controlled by tensor field. The surface family is defined as a set of isosurfaces extracted from an additional scalar field. This field can be a "connected" scalar field, for example, pressure or a scalar field representing some symmetry or inherent structure of the dataset. The texture generation consists basically of three steps. The first is the transformation of the tensor field into a positive definite metric. In the second step, we generate a spot noise texture as input for the final fabric generation. Shape and density of the spots are controlled by the eigenvalues of the tensor field. This spot image incorporates the entire information defined by the three eigenvalue fields. In the third step we use line integral convolution (LIC) to provide a continuous representation that enhances the visibility of the eigendirections. This method supports an intuitive distinction between positive and negative eigenvalues and supports the additional visualization of a connected scalar field.

1 Introduction

Tensor data play an important role in mathematical, physical, and several technical disciplines. Mathematical, a tensor is a linear function that relates different vectorial quantities. Its high dimensionality makes it very complex and difficult to understand. Since the physical interpretation and significance of its mathematical features are highly application-specific, we focus on symmetric tensor fields of second order that are similar to stress and strain tensor fields. Such fields appear, for example, in geomechanics and solid state physics, which are our major application areas. Here tensors are used, for example, to express the response of a material to forces. In contrast to other types of tensors, like diffusion tensors, these tensor fields are characterized by the property that they have positive and negative eigenvalues. The sign of the eigenvalues indicates regions of expansion and compression, and it is therefore of

T. Möller et al. (eds.), *Mathematical Foundations of Scientific Visualization, Computer Graphics, and Massive Data Exploration*, Mathematics and Visualization,
DOI: 10.1007/978-3-540-49926-8,

special interest. To understand field behavior it is important to express this behavior in an intuitive way.

We extend a methointerpreting these tensor fields as distortions of a flat metric [8]. A deformation of a fabric-like texture leads to a continuous representation of the main features of the tensor field, regions of compression and tension. Due to occlusion this method is basically restricted to two-dimensional slices of a higher-dimensional domain. We now use a similar approach to investigate three-dimensional datasets, visualizing a tensor field on a family of arbitrary two-dimensional surfaces. By defining these surfaces as a one-parameter family of isosurfaces it is possible to represent an additional related scalar field, e.g., pressure. Another way to define the surfaces could be based on the geometry of the problem. The texture on the surfaces is generated by blurring a three-dimensional input texture along the tensor lines of the tensor field, projected onto the surfaces. The result is a fabric-like texture that is dense in regions of compression and sparse in regions of expansion. The input texture consists of three-dimensional "spots" whose sizes and local densities reflect the eigenvalues of the tensor field. It is precomputed using a reaction-diffusion method.

2 Related Work

Even though several visualization techniques exist for tensor fields, they only cover a few specific applications. Many of these methods are extensions of vector field visualization methods, which focus on a technical generalization without providing an intuitive physical interpretation of the resulting images. They often concentrate on the representation of eigendirections and neglect the importance of eigenvalues. Therefore, in many application areas traditional two-dimensional plots are still being used, which represent the interaction of two scalar variables only.

A basic way to represent a tensor field uses "icons." They illustrate the characteristics of a field at selected points. One simple example icon that represents a symmetric tensor is the ellipsoid. The principal axes of the ellipsoid are aligned to the eigendirections, scaled according to the corresponding eigenvalues (see, for example, Kriz et al. [11] or Haber [3]). Ellipsoid-based methods are very common for medical applications to visualize results of diffusion magnetic resonance imaging (MRI). More complex glyphs were constructed by Leeuw et al. [12] showing additional features using flow probes. An improvement of these icon methods using a reaction-diffusion simulation, introduced by Kindlmann et al. [10], generates a pattern that is closely related to ellipsoids. There, the packing of the texture spots is generated automatically by the simulation. A more advanced but still discrete approach uses hyperstreamlines. This approach is strongly related to streamline methods used for vector field visualization. Hyperstreamlines were introduced by Delmarcelle and Hesselink [2] and were adapted in a geomechanical context by Jeremic et al. [9]. Given a point in the field, one eigenvector field is used to generate a vector field streamline. The other two eigendirections and eigenvalues are represented by the cross section along the streamline. This method extracts more information than icons, but still leaves the problem of choosing appropriate seed points to the user. Thus, both methods have limited value for the exploration of complete data sets and are limited to low-resolution visualization due to cluttering.

An adaptation of hyperstreamlines to diffusion tensors of MRI data was used by Zhukov and Barr [25] with the goal of tracing anatomical fibers. Their method is based on the assumption that there exist one large and two small eigenvalues inside the fibers, and fiber direction corresponds to the dominant eigenvector. An approach that arose in a similar context is an adaptation of direct volume rendering to diffusion tensor fields presented by Kindlmann et al. [20]. After a classification of a field with respect to anisotropy, it is divided into three categories: linear, planar and spherical [21]. This property is then used to define barycentric coordinates of a transfer function over a triangular domain that highlights regions of different anisotropic properties. Approaches like this one are specially designed for the visualization of diffusion tensors that only have positive eigenvalues; thus, they are not appropriate for stress, strain or gradient tensor fields.

To generate a more global view, a widely accepted solution for vector field analysis and visualization is the reduction of the field to its topological structure. Methods based on such an approach generate topologically similar regions that lead to a natural separation of a field. The concept of topological segmentation was also applied to two-dimensional tensor fields [4, 5, 17] and has recently been extended to three-dimensional tensor fields [24]. The topological skeleton consists of field singularities and connecting separatrices. For tensor fields the singularities of interest are those degenerate points where the tensor has multiple eigenvalues. Although the eigenvector field can be reconstructed on the basis of topological structure there is no information about the eigenvalues, and physical interpretation is difficult.

Another class of visualization methods provides a continuous representation, based on textures. The first researchers to use a texture to visualize a tensor field in a medical context were Ou and Hsu [14]. A method similar to LIC adapted to tensor fields was proposed by Zheng and Pang [23]. Here, a white-noise texture is blurred according to the tensor field. In contrast to LIC images, the convolution filter is a two-dimensional area determined by the local tensor field. This visualization is especially powerful for showing the anisotropy of a tensor field with only positive eigenvalues.

A geometrical approach was followed by Hotz et al. [6]. This approach uses a metric interpretation of a tensor field to emphasize the physical meaning of tensors behaving similarly to stress, strain, or gradient tensors. For two-dimensional fields an isometric embedding is used to visualize the resulting metric locally [7]. Using a deformation of a fabric-like texture makes possible a global representation of the metric [8].

3 Mathematical Background and Notation

A tensor is a geometrical entity that generalizes the concept of scalars, vectors, and linear operators in a way that is independent of any chosen coordinate system. It is the mathematical idealization of a geometric or physical quantity that expresses a linearized relation between multidimensional variables. For example, the stress, strain, or elasticity tensors express the response of a material to an applied force.

The tensors we are interested in are tensors of second order defined over three-dimensional Cartesian space. Using a fixed coordinate basis, each tensor can be represented by a 3×3 matrix, given by nine independent scalars: $\mathbf{T} = (t_{ij})$. A tensor $\mathbf{T}$ is called symmetric if $t_{ij} = t_{ji}$ for $i, j = 1, \ldots, n$. It is called antisymmetric if $t_{ij} = -t_{ji}$ for $i, j = 1, \ldots, n$. Every tensor can be decomposed into a symmetric part $\mathbf{S}$ and an antisymmetric part $\mathbf{A}$: $\mathbf{T} = \mathbf{S} + \mathbf{A}$, where $s_{ij} = \frac{1}{2}(t_{ij} + t_{ji})$ and $a_{ij} = \frac{1}{2}(t_{ij} - t_{ji})$. In many applications, the tensor fields are already symmetric by definition. We restrict ourselves to symmetrical tensor fields in this paper.

A symmetric tensor $\mathbf{S}$ is uniquely characterized by its *eigenvalues* λ_1, λ_2, and λ_3 and its corresponding *eigenvectors* w_1, w_2, and w_3, implied by the characteristic equation $\mathbf{S}w_i = \lambda_i w_i$. For symmetric tensors the eigenvalues are always real, and the eigenvectors are mutually orthogonal.

4 Method Overview

To support an intuitive exploration on the entire three-dimensional tensor dataset we define a family of surfaces that move through the volume, controlled by one parameter. The tensor field restricted to these surfaces is represented by a deformed fabric-like texture illustrating the forces on the surface. The texture is stretched or compressed and bent according to the tensor field. Positive eigenvalues, which indicate tension, are illustrated by a texture with low density or a stretched piece of fabric. Correspondingly, negative eigenvalues are represented by a dense texture. Our method consists of three steps, which are described in more detail in the next sections. The three steps are:

1. Interpretation of the tensor field as distortion of a flat metric:
   This step corresponds to a transformation of the tensor field into a metric. The resulting metric reflects the physical meaning of the tensor field.
2. Definition of a family of surfaces:
   To define the surfaces we support two different approaches. The first integrates an additional related scalar field (e.g., pressure or the determinant of the tenors field). The second approach uses some underlying geometry of the numerical simulation to define an artificial family of surfaces.
3. Texture generation:
   Using a fabric-like texture we have enough flexibility to integrate all characteristics of the tensor field. The direction of the fibers reflect the eigenvector fields, their density, thickness, and length represents the eigenvalues. We compute the texture using LIC. The specific definition of the input noise image determines the density of the fabric. The texture is computed in two steps:
   (a) Generation of the three-dimensional spot input image, e.g., using a reaction-diffusion approach.

(b) Blurring the input texture according to the two eigenvector fields of the tensor field projected onto the surfaces to generate two fiber textures.
(c) Overlay of the two fiber textures on the surfaces (as defined in step 2).

5 Metric Definition

For the fabric texture generation we need a tensor field with positive eigenvalues. An interpretation of the original tensor field as a distortion of a flat metric supports an intuitive mapping onto the space of positive definite tensor fields without loosing the information carried by the sign of the eigenvalues. We summarize the basic steps for this mapping, for more details we refer to [8].

Considering a stress or strain tensor field or the symmetrical part of the gradient tensor field of a vector field, positive eigenvalues lead to a separation of particles or expansion of a probe. Eigenvalues equal to zero imply no change in distance, and negative eigenvalues indicate convergence of particles or compression of the probe. For a tensor field $\mathbf{T}$ defined on a domain D this behavior can be expressed by a time-dependent metric $\mathbf{g}$ of the underlying parameter space D with components g_{ik}:

$$\mathrm{d}s^2(t) = \sum_{ik} \underbrace{(a\delta_{ik} + s_{ik} \cdot t)}_{= g_{ik}} \, \mathrm{d}x_i \, \mathrm{d}x_k, \tag{1}$$

where δ_{ik} is the Kronecker-delta symbol. The constant a plays the role of a unit length, and t is a time variable that can be used as a scaling factor. In our implementation, we use a more general mapping that supports more flexibility in the visualization, but still represents the same properties. We use a transformation based on three steps:

1. Diagonalization of the tensor field:

$$\mathbf{T} \mapsto \mathbf{T}' = U \cdot \mathbf{T} \cdot U^T = \mathrm{diag}\,(\lambda_1, \lambda_2, \lambda_3)\,, \tag{2}$$

   where U is the diagonalization matrix.
2. Transformation and scaling of the eigenvalues, to define the metric $\mathbf{g}'$ according to the eigenvector basis:

$$\mathbf{T}' \mapsto \mathbf{g}' = \mathrm{diag}\,(F(\lambda_1), F(\lambda_2), F(\lambda_3))\,, \tag{3}$$

   where $F : [-\lambda_{max}, \lambda_{max}] \rightarrow I\!R^+$ is a positive monotone function, with $\lambda_{max} = max\{|\lambda_i(P)|;\ P \in D, i = 1, 2, 3\}$.
3. Definition of the metric $\mathbf{g}$ in the original coordinate system by inverting the diagonalization step defined by (2):

$$\mathbf{g} = U^T \cdot \mathbf{g}' \cdot U. \tag{4}$$

If the mapping F is linear, these three steps can be combined into one step, and F can be applied to the tensor components, independently of the chosen basis. Since the visual perception of texture attributes is nonlinear, a linear approach is not always a good choice. For the definition of the function F there are a variety of possibilities. The only restriction we apply to F is monotony to guarantee a one to one

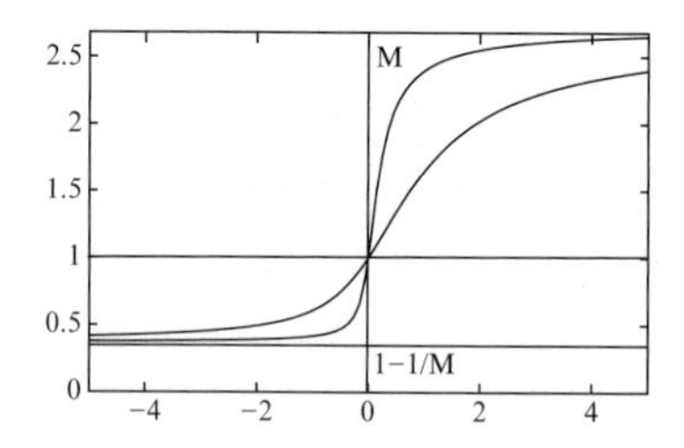

Fig. 1. Example of a nonsymmetric transformation function F for two different slopes at the origin

mapping. A class of functions that have proved to produce good results is given by $F : [-\lambda_{max}, \lambda_{max}] \rightarrow [\frac{a}{M}, aM]$, where

$$F(-\lambda) = a^2/F(\lambda). \tag{5}$$

The constant a defines the unit, aM the maximum, and $\frac{a}{M}$ the minimum value for F with $M > 1$ and $a > 0$. Functions with this property can be constructed using an anti-symmetric function f as exponent:

$$F(\lambda) = a \cdot exp(\sigma \cdot f(\lambda)), \text{ where } f(-\lambda) = -f(\lambda). \tag{6}$$

An example for such a function with $a = 1$ $F(\lambda; c, \sigma) = exp(\sigma \arctan(c \cdot \lambda))$ is shown in Fig. 1. The constant c determines the slope of the function at the origin.

6 Surface Definition and Tensor Projection

To explore the entire three-dimensional volume D we define a family of surfaces $\{S_q, q \in I\}$ with $S_q \subset D$, for all $q \in I$ where I is an interval representing the parameter space with

$$\bigcup_{q \in I} S_q = D. \tag{7}$$

These surfaces can be given explicitly based, for example, on a geometric property or symmetry inherent to the data. A simple possibility is to move planes, cylinders or parts of spheres through the volume. Another option it to define the surfaces implicitly using isosurfaces of an additional scalar field S:

$$S(x, y, z) = q, \tag{8}$$

where q is an isovalue between a minimal and maximum value.

To compute the textures for the surfaces we project the tensor field onto the surfaces (Fig. 2). If we use isosurfaces, the unit surface normals are given by the gradient of the scalar field S:

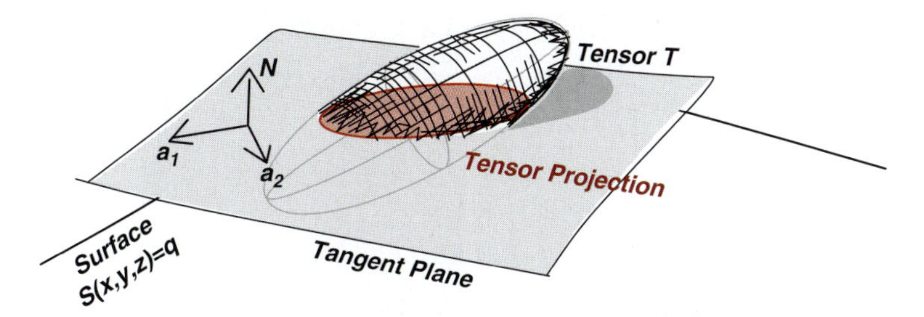

Fig. 2. Projection of the tensor onto the surface S

$$N = \begin{pmatrix} n_x \\ n_y \\ n_z \end{pmatrix} = \frac{\text{grad}S}{|\text{grad}S|} = \frac{1}{\sqrt{S_x^2 + S_y^2 + S_z^2}} \begin{pmatrix} S_x \\ S_y \\ S_z \end{pmatrix}, \tag{9}$$

where n_x, n_y, n_z are the components of the normal, and $S_x = \frac{\partial S}{\partial x}$, $S_y = \frac{\partial S}{\partial y}$, and $S_z = \frac{\partial S}{\partial z}$ the partial derivatives of the scalar function S. The projection $\mathbf{T}'$ of the tensor $\mathbf{T}$ onto the surface defined by N is given by

$$\mathbf{T}' = P \cdot \mathbf{T} \cdot P^T. \tag{10}$$

The projection tensor P is

$$P = \mathbf{1} - NN^T = \begin{pmatrix} (1 - n_x^2) & -n_x n_y & -n_x n_z \\ -n_x n_y & (1 - n_y^2) & -n_y n_z \\ -n_x n_z & -n_y n_z & (1 - n_z^2) \end{pmatrix}, \tag{11}$$

where $\mathbf{1}$ is the unit tensor. It is symmetric ($P^T = P$). The projected tensor $\mathbf{T}'$ has one eigenvector in direction of the surface normal N with eigenvalue zero and two orthogonal eigenvectors, lying in the tangent plane. The eigenvectors of the projected tensor $\mathbf{T}'$ are in general not eigenvectors of the original tensor $\mathbf{T}$.

7 Texture Generation

To visualize the properties of the resulting metric we use a texture that resembles a piece of fabric. The texture is stretched or compressed and bent according to the metric. To generate the texture we use LIC, a method originally developed for vector field visualization [1, 16]. LIC blurs a noise image along the vector field or integral curves. Blurring results in a high correlation of the pixels along field lines, whereas perpendicular to them almost no correlation appears. The resulting image leads to a very effective depiction of flow direction everywhere, even in a dense vector field. We compute a LIC image for both eigendirection fields on the surface. Finally, we overlay these images, choosing everywhere the pixel value with the larger intensity to obtain the fabric-like texture, see Fig. 3.

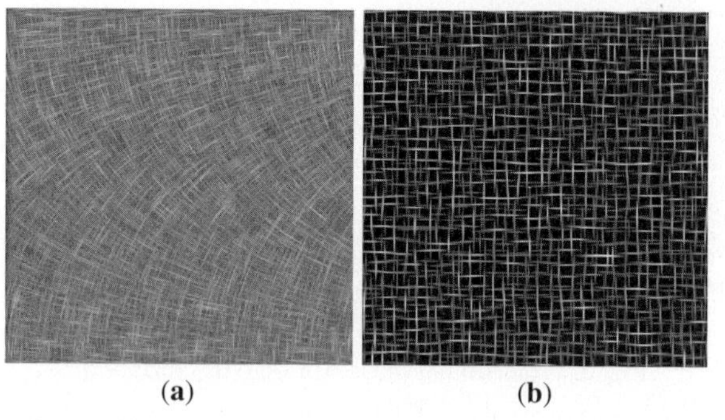

Fig. 3. Overlay of two LIC images to illustrate two direction fields, without integrating the eigenvalues; constant input image and constant convolution length. (**a**) White noise image; (**b**) sparse noise image

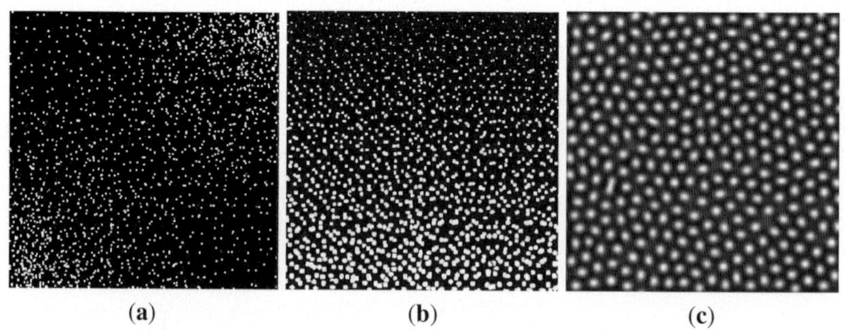

Fig. 4. Different input images. (**a**) Spot noise image with changing density; (**b**) spot noise image with changing spot size; (**c**) spot noise image generated with isotropic reaction

7.1 Input Texture Definition

Besides the direction field we need for each LIC image a specific noise input. The parameters of this input image determine the properties of the texture. The standard white noise input is the input that supports the highest resolution, but it is not flexible enough to represent a stretched or compressed structure. For this reason, we use sparse input images with lower density and larger spot size even if we obtain a lower resolution. A regular, homogeneous input spot image results in a piece of fabric with constant density and fiber width. An impression of a stretching or compressing can be achieved by changing density, width, and length of the fibers. Figures 4 and 5 show examples of different input images and their impact on fiber structure.

7.2 Fiber Density and Fiber Width: Forces Orthogonal to the Fibers

Stretching and compressing forces orthogonal to the fiber direction are directly related to the eigenvalues of the orthogonal eigenvector field. The change of the fiber density and fiber width can be controlled by the density and spot-size of the input

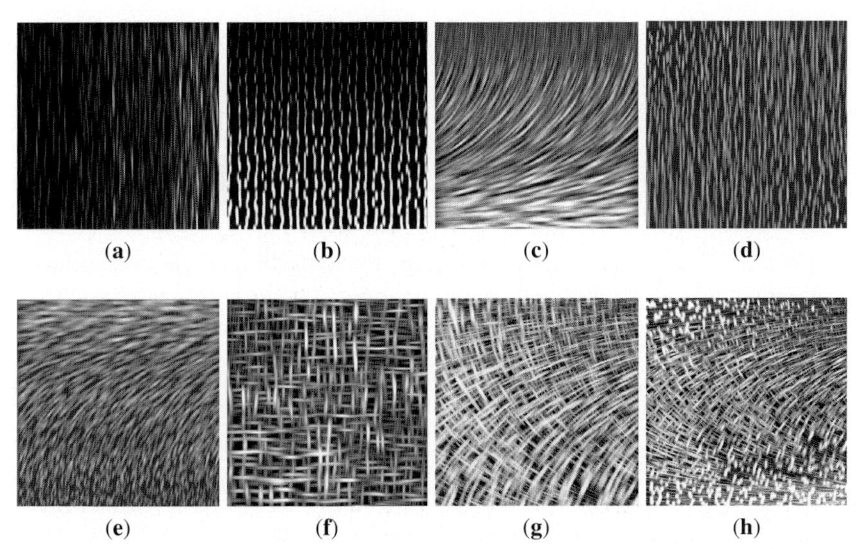

Fig. 5. Effect of changing image parameters; one eigenvector field of a simple synthetic tensor field. In (**a**)–(**c**), only the input image is changed corresponding to the eigenvalues of the orthogonal eigenvector field; (**a**) change of density; (**b**) change of spot size; (**c**) change of density and spot size. Image (**d**) illustrates the effect of changing the convolution length, where the parameters of the input noise image are constant. Image (**e**) shows a combination of the three parameters (density, spot size, and convolution length). Images (**f**)–(**h**) show a combination of both eigenvector fields. In images (**g**) and (**h**), only density and spot size are changed; (**i**) shows a combination of the three parameters

spot texture. For each direction field $\mathbf{w}_i$, $i = 1, 2$, on the surface, we define a specific density d_i and spot size r_i, $i = 1, 2$:

$$\begin{array}{l} d_i(x, y, z) = d_0 \cdot \frac{1}{F(\lambda_j)} \\ r_i(x, y, z) = r_0 \cdot F(\lambda_j) \end{array}, \text{ with } j = \begin{cases} 2 \text{ if } i = 1 \\ 1 \text{ if } i = 2 \end{cases}, \tag{12}$$

where the function F is defined by (3), λ_1 and λ_2 are the two eigenvalues of the projected tensor field, d_0 and r_0 define the "unit density," and the "unit-radius" of the circular spots, respectively. The density and spot size are spatially varying and depend on the definition of the surfaces. Since the change of the texture parameters in direction of the integration is hardly noticeable we can combine the two textures by using ellipses instead of circles and a direction-dependent density. The resulting texture still depends on the definition of the surface. This means that every time we define a new set of surfaces we have to recompute all input textures.

An alternative construction of the spot textures on the surfaces is the generation of a three-dimensional input image, where the spots are replaced by ellipsoids whose principal axes and radii $r_i = r_0 F(\lambda_i)$ are defined by the metric $\mathbf{g}$ given by (4). The texture on the surface then results from an intersection of the surface with the three-dimensional texture. The direction-dependent radius can be expressed by the tensor $\mathbf{r}$, i.e.,

$$\mathbf{r} = r_0 \cdot \mathbf{g} = r_0 \cdot U^T \cdot \begin{pmatrix} F(\lambda_1) & 0 & 0 \\ 0 & F(\lambda_2) & 0 \\ 0 & 0 & F(\lambda_3) \end{pmatrix} \cdot U. \tag{13}$$

The direction-dependent density is defined by the tensor $\mathbf{d}$ with the same principal directions but inverse eigenvalues, i.e.,

$$\mathbf{d} = d_0 \cdot U^T \cdot \begin{pmatrix} 1/F(\lambda_1) & 0 & 0 \\ 0 & 1/F(\lambda_2) & 0 \\ 0 & 0 & 1/F(\lambda_3) \end{pmatrix} \cdot U. \tag{14}$$

The parameters r_0 and d_0 define a unit radius and unit density. Using this approach the input texture only has to be computed once, independently of the surface definition, and this operation can be done in a preprocessing step.

7.3 Fiber Length: Forces Along the Fibers

Stretching and compressing forces parallel to the fibers changes length. This effect can most easily be controlled by the filter length used for the convolution, and it is merely influenced by the parameters of the input spot texture. We define for each direction field a spatially varying convolution length

$$l_i = l_0 \cdot F(\lambda_i), \tag{15}$$

where l_0 is a constant defining the unit length.

7.4 Color and Color Intensity

In addition to these three "structural" parameters, color intensity can be used to enhance the impression of compression and stretching. We use red for compression and green for tension. We apply a continuous color map ranging from red for the smallest negative eigenvalues, white for zero eigenvalues, and to green for positive eigenvalues.

7.5 Input Texture Computation

To synthesize the input texture an algorithm is needed that places ellipsoids with varying radii and direction-dependent density in the three-dimensional domain. "Reaction diffusion" is a method that generates a texture with the desired properties automatically. Using a reaction-diffusion approach, a large variety of patterns arising from local nonlinear interactions of two chemicals can be generated. This mechanism was first discussed by Alan Turing [18] to describe the biological process of morphogenesis in biological cells. The basic assumption is that two concurrently operating processes define the biological patterns. The two processes are diffusion, which transports chemicals form points of higher concentration to points of lower concentration, and reaction, which produces or destroys a chemical. Reaction diffusion has been used in many different computer graphics applications, including, for example, the generation of natural texture patterns [13, 19, 22] and for vector field visualization [15]. An application very similar to ours was described by Kindlmann et al. [10], who used reaction diffusion to visualize diffusion tensor fields. We provide an overview of the method in the following. For more detail we refer to the referenced papers [10, 13, 15, 19, 22].

Reaction Diffusion

Reaction diffusion of two chemicals can be described by a set of two nonlinear partial differential equations describing the concentrations of the chemicals c_1 and c_2 as a function of time. The change of the concentration is determined by two terms representing the diffusion, and the reaction of the chemicals:

$$\frac{\partial c_l}{\partial t} = \underbrace{d_l \nabla^2 c_l}_{\text{diffusion}} + \underbrace{R_l(c_1, c_2)}_{\substack{\text{reaction} \\ +\text{dissipation}}} \;, \qquad l = 1, 2. \tag{16}$$

The constants d_l are the diffusion rates for the chemicals, and $\nabla^2 c_l$ is the Laplacian of the concentrations. The functions R_l, $l = 1, 2$, control the reaction of the two chemicals determining the resulting “Turing patterns.” We choose a set of functions proposed by Turing and used by Kindlmann et al. [10] for the visualization of diffusion tensor fields:

$$R_1 = k(16 - c_1 c_2) \tag{17}$$

$$R_2 = k(c_1 c_2 - c_2 - 12 + \beta) \tag{18}$$

Here, k is the reaction rate relative to the diffusion. The value β is the decay rate of c_2; it is a random value in a small interval around zero. This value is the source of slight irregularities in the concentrations which lead to the pattern formation. As initial condition for the concentrations we chose $c_1 = c_2 = 4$ everywhere. Further, we apply periodic boundary conditions.

Equation (16) describes a uniform diffusion rate in all directions at all positions. It generates a texture with spherical bubbles of constant size and density. A generalization, resulting in an ellipsoid pattern, uses anisotropic spatial varying diffusion. This is achieved by replacing the scalar diffusion rate d by a diffusion tensor matrix D.

$$\frac{\partial c}{\partial t} = \nabla(D \nabla c) + R. \tag{19}$$

The radii of the ellipsoids are proportional to the square root of the eigenvalues of D, oriented parallel to the principal directions of D. To generate ellipsoids with the aspect ratio radii as defined in (13) we use the diffusion tensor

$$D = r_0^2 \cdot U^T \cdot \begin{pmatrix} F^2(\lambda_1) & 0 & 0 \\ 0 & F^2(\lambda_2) & 0 \\ 0 & 0 & F^2(\lambda_3) \end{pmatrix} \cdot U. \tag{20}$$

The size of the spots can be adjusted by changing the parameter k in (17). If we assume that the spatial change of the Diffusion tensor occurs on a much lower scale than the change of the chemical concentration, we can treat D as a constant in (19).

Discretization

In our implementation, we represent the concentration c as a three-dimensional array C of discrete samples $C_{i,j,k}$. The Laplacian is approximated using finite differences

$$\frac{\partial^2 c}{\partial x^2} \simeq \frac{C_{i+1,j,k} + C_{i-1,j,k} - 2C_{i,j,k}}{h^2} \tag{21}$$

and defined similarly for the other dimensions.

In three dimensions, the Laplacian can be expressed as a convolution of the concentration array with a $3 \times 3 \times 3$ mask. Together with the anisotropic diffusion tensor D the convolution mask M has the following form for an uniform and equidistant grid:

$$\begin{aligned} z = k+1: &\quad \begin{pmatrix} 0 & D_{yz}/2 & 0 \\ -D_{xz}/2 & D_{zz} & D_{xz}/2 \\ 0 & -D_{yz}/2 & 0 \end{pmatrix}, \\ z = k: &\quad \begin{pmatrix} -D_{xy}/2 & D_{yy} & D_{xy}/2 \\ D_{xx} & -2\,\mathrm{trace}(D) & D_{xx} \\ D_{xy}/2 & D_{yy}/2 & -D_{xy}/2 \end{pmatrix}, \text{ and} \\ z = k-1: &\quad \begin{pmatrix} 0 & -D_{yz}/2 & 0 \\ D_{xz}/2 & D_{zz} & -D_{xz}/2 \\ 0 & D_{yz}/2 & 0 \end{pmatrix}. \end{aligned} \tag{22}$$

Convolution with the mask M is computation of the weighted sum of the concentration values of the neighboring voxels, see Fig. 6:

$$(M * C)_{i,j,k} = \sum_{\substack{x=i-1,i,i+1 \\ y=j-1,j,j+1 \\ z=k-1,k,k+1}} M_{xyz}\, C_{x,y,z}. \tag{23}$$

For the simulation of the reaction-diffusion process we use forward Euler integration of the finite-difference equations. The discretized reaction diffusion equation is given as

$$C_{t+\Delta t} = C_t + \Delta t\, (M * C + R(C)). \tag{24}$$

Here, the index t specifies the concentration at some time t, and Δt is the time-step used in the integration.

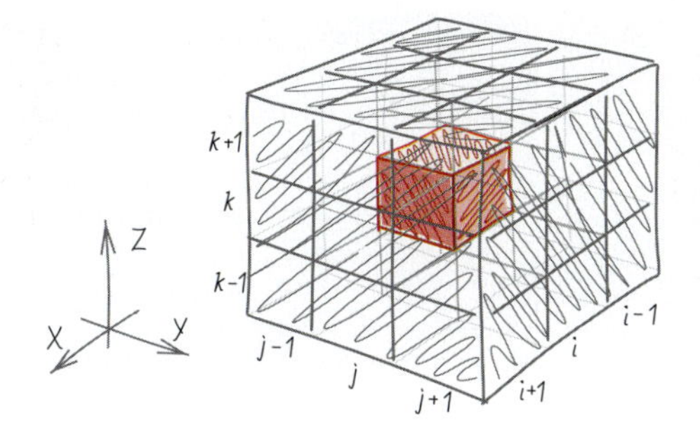

Fig. 6. The 27 voxels involved in the convolution operation with mask M

7.6 Convolution

The next step concerns the creation of two fiber textures using line convolution. The noise input is the ellipsoid image generated with reaction diffusion. The direction field for the integration is defined by the eigenvectors of the tensor field projected onto the set of surfaces. For integration we use a Runge–Kutta method of fourth order. The LIC image is computed using "Fast-LIC" as proposed by Hege et al. [16]. The convolution length depends on the eigenvalues of the tensor field. It is defined by (15).

7.7 Visualization

We show the texture on one surface that is moved through the volume, defined by an isovalue or other parameter. We use two approaches to represent the surfaces:

- Volume rendering based on a transfer function illustrating the volume only in an ϵ-neighborhood of the chosen isovalue
- Explicit extraction and visualization of the surfaces

8 Results and Conclusions

We have used two data sets of stress fields being the results of applying different load combinations to a solid block. These datasets are well-understood and therefore appropriate to evaluate our method. The tensor field resulting from the numerical simulation is continuous inside each cell, but not on cell boundaries. In Figs. 7 and 8, we show two-dimensional planar slices through a one-point and two-point load dataset. These images were generated using a specific input noise texture for each eigenvector field. The input textures consist of circular spots of varying size, density and for Fig. 8 also color. Figure 9 shows results for the same dataset when using only one input texture generated with reaction diffusion. These images provide also a good representation of the tensor field, but the features are not as clearly visible when compared with the corresponding results based on the spot noise input. Figure 10 shows

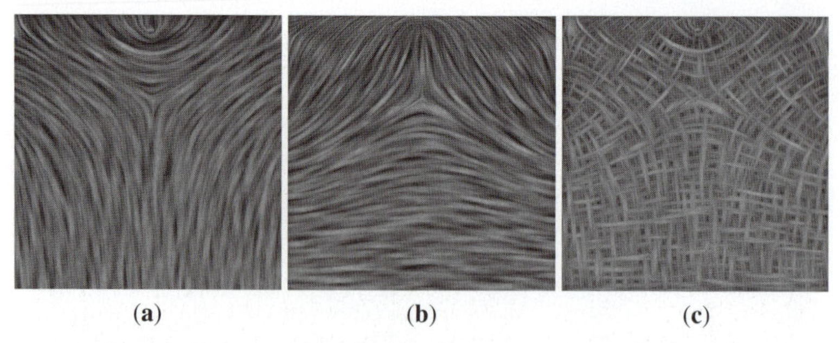

Fig. 7. yz-plane slice of single top-load data set, where a force is applied in z-direction. Images (**a**) and (**b**) illustrate the two eigenvector fields separately; in (**c**) they are overlaid. Spot size and density are changed according to eigenvalues

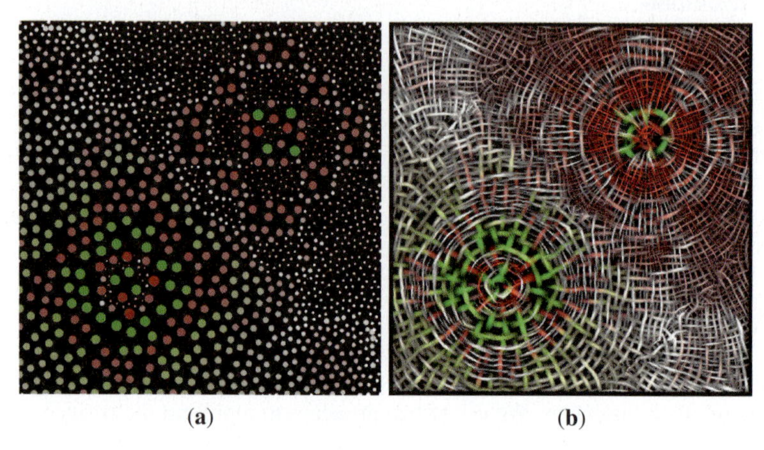

Fig. 8. xz-plane slice of a two-force dataset. (**a**) represents one of the two spot noise input textures for the final image. (**b**) The *lower left circle* corresponds to the pushing and the right to the pulling force

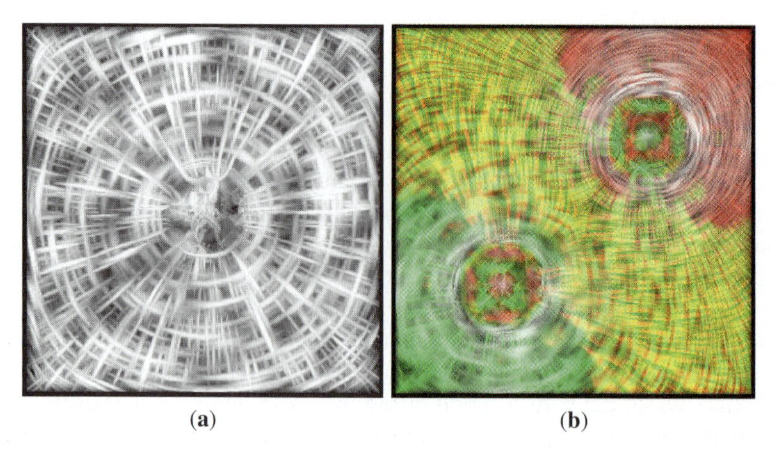

Fig. 9. xz-plane slice of (**a**) a one-force and (**b**) a two-force dataset using three-dimensional spot noise input generated with diffusion reaction

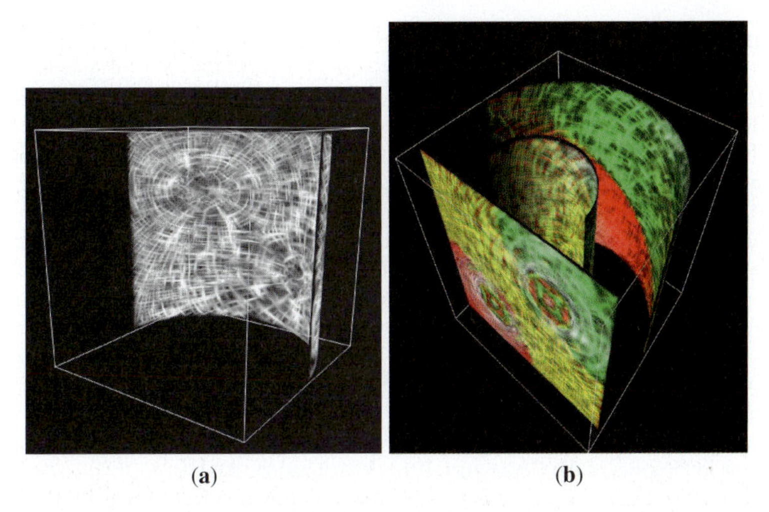

Fig. 10. Representation of several curved surfaces to explore the three-dimensional domain for the two-point load dataset

examples of textures on curved surfaces for the two-point load dataset. All images support a good visual segmentation of regions of compression and expansion.

The interpretation of a tensor field as a distortion of a flat metric can be used to produce visualizations based on the physical effect of the tensor field. Using a three-dimensional input texture simplifies texture generation for arbitrary surfaces substantially. Having a precomputed input texture, the three-dimensional domain can be examined easily using different sets of surfaces. Reaction diffusion is capable of generating input textures with the desired properties with automatic placement and packing of spots. There are also disadvantages when using reaction diffusion. The generated features are not as "crisp" as they are when compared to spot noise input. The determination of appropriate parameters in the reaction-diffusion equation is not intuitive, and its numerical computation is very time-consuming. A more efficient algorithm for reaction-diffusion simulation was proposed by Witkin and Kass [22]. We plan to investigate other ways to generate three-dimensional input textures, ways that are less expensive and produce crisp features.

Acknowledgments

This work was supported by the National Science Foundation under contracts ACI 9624034 (CAREER Award) and ACI 0222909, through the Large Scientific and Software Data Set Visualization (LSSDSV) program under contract ACI 9982251, and a large Information Technology Research (ITR) grant. We gratefully acknowledge the support of the W.M. Keck Foundation provided to the UC Davis Center for Active Visualization in the Earth Sciences (CAVES) We thank the members of the Visualization and Computer Graphics Research Group at the Institute for Data Analysis and Visualization (IDAV) at the University of California, Davis.

References

1. Brian Cabral and Leith Leedom. Imaging vector fields using line integral convolution. In: SIGGRAPH 1993 Conference Proceedings, 1993, pages 263–272.
2. Thierry Delmarcelle and Lambertus Hesselink. Visualization of second order tensor fields and matrix data. In: Proceedings of the Visualization 1992 Conference, 1992, pages 316–323.
3. Robert B. Haber. Visualization techniques for engineering mechanics. Computing Systems in Engineering, 1(1), 1990, pages 37–50.
4. Lambertus Hesselink, Thierry Delmarcelle and James L. Helman . Topology of second-order tensor fields. Computers in Physics, 9(3), 1995, pages 304–311.
5. Lambertus Hesselink, Yuval Levy and Yingmei Lavin. The topology of symmetric, second-order 3D tensor fields. IEEE Transactions on Visualization and Computer Graphics, 3(1), 1997, pages 1–11.
6. Ingrid Hotz. Visualizing second order symmetric tensor fields by metric surfaces. In: Work in Progress of the Conference IEEE Visualization, 2001.
7. Ingrid Hotz. Isometric embedding by surface reconstruction from distances. In: Proceedings of the IEEE Visualization 2002 Conference, 2002, pages 251–258.
8. Ingrid Hotz, Louis Feng, Hans Hagen, Bernd Hamann, Boris Jeremic and Kenneth Joy. Physically based methods for tensor field visualization. In: Proceedings of the IEEE Visualization 2004 Conference, 2004, pages 123–130.
9. B. Jeremic, Gerik Scheuermann and Jan Frey. Tensor visualization in computational geomechanics. International Journal for Numerical and Analytical Methods in Geomechanics, 26, 2002, pages 925–944.
10. Gordon Kindlmann, David Weinstein and David Hart. Strategies for direct volume rendering of diffusion tensor fields. IEEE Transactions on Visualization and Computer Graphics, 6(2), 2000, pages 124–138.
11. Ron D. Kriz, Edward H. Glaessgen and J.D. MacRae. Visualization blackboard: visualizing gradients in composite design and fabrication. IEEE Computer Graphics and Applications, 15, 1995, pages 10–13.
12. W. C. de Leeuw and J. J. van Wijk. A probe for local flow field visualization. In: Proceedings of the Visualization 1993 Conference, 1993, pages 39–45.
13. James D. Murray. Mathematical biology I. An introduction, 3rd edition in 2 volumes. Springer, Berlin, 2002, ISBN: 0-387-95223-3.
14. J. Ou and E. Hsu. Generalized line integral convolution rendering of diffusion tensor fields. In: Proceedings of the 9th Scientific Meeting and Exhibition of the International Society for Magnetic Resonance in Medicine (ISMRM), 2001, page 790.
15. Allen R. Sanderson and Chris R. Johnson and Robert M. Kirby. Display of vector fields using a reaction-diffusion model. In: Proceedings of the IEEE Visualization 2004, pages 115–122, 2004.
16. Detlev Stalling and Hans-Christian Hege. Fast and resolution independent line integral convolution. In: SIGGRAPH 1995 Conference Proceedings, 1995, pages 149–256.
17. Xavier Tricoche, Gerik Scheuermann and Hans Hagen. Tensor topology tracking: a visualization method for time-dependent 2D symmetric tensor fields. In: A. Chalmers and T.-M. Rhyne, editors, EG 2001 Proceedings vol. 20(3) of Computer Graphics Forum. 2001, pages 461–470.
18. Alan Turing. The chemical basis of morphogenesis. Philosophical Transactions of the Royal Society London B, 237, 1952, pages 37–72.
19. Greg Turk. Generating textures on arbitrary surfaces using reaction diffusion. In: proceedings SIGGRAPH 1991, volume 25, pages 289–298. Addison Wesley, 1991.

20. David M. Weinstein, Gordon L. Kindlmann and Eric C. Lundberg. Tensorlines: advection-diffusion based propagation through diffusion tensor fields. In: proceedings IEEE Visualization 1999 conference,1999, pages 249–254.
21. Carl-Fredrik Westin, Sharon Peled, Hakon Gudbjartsson, Ron Kikinis and Ferenc A. Jolesz. Geometrical diffusion measures for MRI from tensor basis analysis. In: ISMRM 1997, 1997, page 1742.
22. Andrew Witkin and Michael Kass. Reaction-diffusion textures. In: Proceedings SIGGRAPH 1991, vol. 25, pages 299–308. Addison Wesley, 1991.
23. Xiaoqiang Zheng and Alex Pang. HyperLIC. In: Proceedings of the IEEE Visualization 2003 Conference, 2003, pages 249–256.
24. Xiaoqiang Zheng and Alex Pang. Topological lines in 3D tensor fields. In: Proceedings of the IEEE Visualization 2004 Conference, 2004.
25. Leonid Zhukov and Alan H. Barr. Oriented tensor reconstruction: tracing neural pathways from diffusion tensor MRI. In: Proceedings of the IEEE Visualization 2002 Conference, 2002, pages 387–394.

Flow Visualization via Partial Differential Equations

Tobias Preusser[1], Martin Rumpf[2], and Alex Telea[3]

[1] Center for Complex Systems and Visualization, University of Bremen, Universitätsallee 29, 28359 Bremen, Germany
tp@cevis.uni-bremen.de

[2] Institute for Numerical Simulation, Bonn University, Nussallee 15, 53115 Bonn, Germany
martin.rumpf@ins.uni-duisburg.de

[3] Department of Mathematics and Computer Science, Eindhoven University, Den Dolech 2, Eindhoven, Netherlands
alext@win.tue.nl

Summary. The visualization of stationary and time-dependent flow is an important and challenging topic in scientific visualization. Its aim is to represent transport phenomena governed by vector fields in an intuitively understandable way. In this paper, we review the use of methods based on partial differential equations (PDEs) to post-process flow datasets for the purpose of visualization. This connects flow visualization with image processing and mathematical multi-scale models. We introduce the concepts of flow operators and scale-space and explain their use in modeling post processing methods for flow data. Based on this framework, we present several classes of PDE-based visualization methods: anisotropic linear diffusion for stationary flow; transport and diffusion for non-stationary flow; continuous clustering based on phase-separation; and an algebraic clustering of a matrix-encoded flow operator. We illustrate the presented classes of methods with results obtained from concrete flow applications, using datasets in 2D, flows on curved surfaces, and volumetric 3D fields.

1 Introduction

A great variety of different approaches for the visualization of vector field data has been presented in the past. The methodology ranges from simple discrete arrow plots applied to steady two-dimensional vector fields to advanced hardware-accelerated volumetric techniques for visualizing multivariate data for three-dimensional, unsteady flow problems and multi-scale feature detection and tracking techniques for complex time-dependent CFD problems.

The recent increase of the number of flow visualization techniques has been driven by two main factors. On one hand, the exponential growth in size of datasets produced by CFD simulations requires flow visualization methods to be able to display more data in shorter time. On the other hand, specific application fields, ranging from weather simulation, meteorology, and ground water flow, to automotive, aerodynamics, and machine design, have each their own particular requirements and questions to be answered regarding flow datasets. As a central and generally accepted

T. Möller et al. (eds.), *Mathematical Foundations of Scientific Visualization, Computer Graphics, and Massive Data Exploration*, Mathematics and Visualization,
DOI: 10.1007/978-3-540-49926-8,

high-level goal, flow visualization should provide intuitively better receptible methods which give overall as well as detailed views on the flow patterns and behavior.

Given the above, several classifications of flow visualization methods have been recently proposed from different points of view. In their State of the Art report, Laramee et al. [20] have classified flow visualization into direct, dense texture-based, geometric, and feature-based methods, following a model of the flow data in discrete samples, continuous (dense) scalar fields, geometric integral primitives, and application-specific feature-based representations. A second overview of flow visualization methods is given by Weiskopf and Erlebacher [45]. Here, three classifications of flow visualization methods are proposed: based on the *visual primitive* used (points, curves, or features); based on the *density* of the produced image (sparse vs. dense, texture-based); and finally based on the data *structure* (2D, 2.5D, and 3D methods) and *discretization* (on various grid types). A more recent report of Laramee et al. [21] presents a comparison of major visualization techniques evaluated from the point of view of a specific application – the understanding of swirl and tumble flow data. Here, visualizations are classified into texture-based methods, clustering approaches, analyses of the vector field topology, and feature-tracking approaches.

In this paper, we give an overview of flow visualization methods based on partial differential equations (PDEs) [9,12,13]. These methods use a particular model of the flow data, as follows. The flow domain is seen as a subset of the continuous $\mathbb{R}^2$ or $\mathbb{R}^3$ space. The visualization process is described now in terms of a continuous, physical process, such as diffusion, advection, or phase separation. The particular type of PDE and its boundary conditions are used as instruments to model different visualization questions, such as: Which are the laminar, transient, and turbulent regions? How does the material density vary in time in the dataset? How does the flow look like on small spatial scales (flow details) as opposed to a global, coarse scale (flow overview)? Once the type and parameters of the proper PDE are established, the flow domain is discretized, usually by a finite-element or finite-difference scheme, and the underlying PDE is solved with appropriate solvers for the resulting equation or system of equations. Finally, the solution is displayed, thereby answering the initial set of visualization questions that target the flow data.

The above leads us to an outline of several characteristics of PDE-based flow visualization methods. For this, we shall use the terminology employed by the classifications presented in [20,21,45]. First, PDE-based flow visualizations are *dense* methods, by definition. Second, they work in all dimensions where flow visualization is of interest, i.e. 2D, 2.5D (curved surfaces embedded in 3D), and 3D. Third, they are applicable to both steady and unsteady (time-dependent) flow datasets. In terms of actual visual representation, PDE-based methods are naturally closely related to texture-based methods. Although the results of PDE-based methods can be displayed using also other techniques, such as slice planes, streamlines [12], and isosurfaces [9], their inherent continuous, dense nature makes them natural candidates for using 2D and 3D texture-based display techniques. In this sense, PDE-based methods can be seen as a front-end, that translate a given direct flow representation (vector field plus additional scalar quantities such as pressure or concentration) to a second dense, usually scalar, representation, which is then visualized by a texture-driven back-end. On a high level, this translation, encoded in the PDE, enriches the data with application and question-specific semantics, in order to emphasize the specific aspects of the flow the user is interested in. Conversely, many texture-based visualization methods implemented using (programmable) graphics hardware have at their core a model based on an advection ordinary differential equation (ODE), as described further in Sect. 2. Finally, in terms of data discretization (grid type), all PDE-based methods can essentially be used on any grid, given a suitable underlying finite element discretization implemented on that grid (cf. Sect. 8).

From another point of view, PDE-based methods share many common aspects with multi-scale flow visualization methods. Overall, the main goal of such methods is to provide a multi-scale representation of the flow field, such that users can subsequently navigate between detailed, low-level views of the flow and global, overview pictures thereof. Several multi-scale methods exist in flow visualization [20, 45]. Clustering methods [14, 35] group similar flow dataset points together based on a task or application-dependent similarity measure, or correlation. Energy minimization techniques can be used to produce streamline visualizations at several levels of detail, represented by different streamline spatial densities [18, 19, 38]. PDE-based methods bring several powerful tools for defining the multi-scale, based on scale space theory (cf. Sect. 3), and are able to accommodate several scale notion definitions, ranging from continuous to discrete.

Given this close connection between PDE-based and texture-based flow visualization, we give first an overview of the texture-based methods in Sect. 2 followed by a brief introduction to the basic multi-scale methods in image processing, which motivate the approaches to be discussed here. Next, in Sect. 3 we review the connections to scale-space methodology from image processing. Together with the differential operator defined in Sect. 4, this leads to the presentation of the anisotropic diffusion method in Sect. 5. The extension of this model towards time-dependent flow fields is discussed in Sect. 6. In the remaining, we review clustering methods based on PDE techniques and we start with the model based on anisotropic phase separation in Sect. 7. A discussion of hierarchical multi-scale clustering using algebraic multigrid (AMG) methods follows in Sect. 9. Section 10 closes this article drawing conclusions.

2 Review of Texture Based Flow Visualization

Texture-based flow visualization is a notion generally used for those methods that output a full spatial coverage of the flow field to be described, in the region of interest chosen by the user. By full coverage, we mean that the discreteness of the output data is limited to the one inherent to the image, or texture primitives used in the visualization. Given this dense representation of the flow field, the texture image will mostly encode some continuous color, luminance, or transparency variation that conveys insight into the flow data. Often, the above continuous signal is naturally generated by physically-based methods. Texture-based methods differ, thus, mainly in how, and what, they generate and encode in the output texture. We outline below the most important classes of texture-based methods, and refer for an in-depth overview to [20].

The spot noise method proposed by van Wijk [18] pioneered the use of textures in flow visualization. Elliptic splats of intensity based on a noise function and which are oriented along the field are blended together on 2D or 3D surface domains. The original first order approximation of the flow was improved by de Leeuw and van Wijk in [8] by using higher order polynomial deformations of the spots in areas of significant vorticity. By animating the intensity, spots appear to move along flow streamlines. Several applications of spot noise were presented in the context of smog prediction and turbulent flows [6], non-continuous flow visualization [22], and flow topology [7].

Line Integral Convolution (LIC) methods represent the second major class of texture-based visualizations. Introduced by Cabral and Leedom [2], LIC integrates the fundamental ODE describing streamlines forward and backward in time at every discrete domain point. White noise is convolved, using a Gaussian kernel, along these particle paths. The resulting value gives the intensity of the starting pixel. The resulting texture exhibits strong correlations along streamlines and weak correlation across, giving the perception of streamline-like filaments of varying intensity. Essentially, LIC is equivalent to a diffusion process along the vector field. Hege and Stalling [17] increased the performance of LIC by reusing portions of the convolution integral already computed on points along a given streamline. Forssell [11] proposed a similar method on surfaces, whereas Max et al. [24] approach flow visualization by texturing iso-surfaces. UFLIC [31] extended LIC to unsteady flows. Interrante and Grosch [16] generalized line integral convolution to 3D in terms of volume rendering of line filaments. Multivariate flow fields are visualized with LIC using a color mapping technique called color weaving [39]. OLIC [43] and its fast version FROLIC [29] add up- and downstream cues to the basic LIC by varying the intensity along the streamline. Finally, we mention 3D LIC, which uses texture-based volume rendering to compute and display LIC visualizations for 3D flow fields [27]. Again, the above is just a short overview of a wealth of existing LIC techniques. For a more detailed overview, see [20].

Yet another class of texture-based methods arc the ones based on texture advection. Here, the visualization primitive is directly supported by the graphics unit or GPU. Consequently, the term texture in these methods often refers to the graphics hardware term. GPU-based methods are classified based on the primitive they advect, or warp (pixel or polygon) and the advection direction (forward or backward). Max and Becker [23] presented one of the first texture-advection methods using triangles. Image-based flow visualization (IBFV) proposes an injection of noise (stored as textures), advecting it by warping a polygon mesh, and blending the result for smooth visualization, with applications in 2D [41], curved surfaces [42], and 3D volumes [34]. Lagrangian–Eulerian Advection (LEA) is another such model, where particle positions are advected individually (Lagrangian step) and the color texture is updated in-place (Eulerian step) [17, 47]. Recently, the above (and other) frameworks, were united in UFAC (Unsteady flow advection-convolution), using an implementation based on programmable GPUs [46]. Interestingly, the emergence of the "framework" for GPU-based methods as a collection of tightly-woven conceptual, modeling, and implementational aspects seems to be driven by the large importance of the implementational aspect in the whole process, in contrast to, e.g. LIC methods.

Especially for 3D velocity fields, particle tracing is a very popular tool. However, even relatively many seed particles released by the user can hardly cope with the complexity of 3D vector fields. Zöckler et al. [33] use pseudo randomly distributed, illuminated and transparent streamlines to give a denser and more receptible representation, which shows the overall structure and enhances important details.

Most notably every subclass of texture-based method seems to produce visualizations that carry an easily recognizable visual signature. For example, it is easy to tell spot-noise from LIC; the various IBFV and LEA methods have also a distinct visual appearance, probably due to the specific noise functions used; illuminated streamlines are also a class apart; reaction-diffusion methods create regular repetitive patterns which noise-injection methods cannot replicate. We believe that a perceptual classification of texture-based flow visualizations would bring valuable insight in the effectiveness (and limitations) of such methods and lead to a better understanding of flow data, although we are not aware of any such classification.

3 A Brief Introduction to Scale Space Methods in Image Processing

Textures used in various flow visualization approaches can be regarded as images and thus the type of flow post processing above discussed can be considered as image processing. In the last two decades powerful PDE based image processing methods for several fundamental tasks in imaging such as segmentation and de-noising have been introduced. In particular so called *scale space methods* introduce a natural scale of image representations. Most of the methods in flow post processing lack such a perspective of multiple scales. They have in common that the generation of a coarser scale requires a re-computation. For instance, if we ask for a finer or coarser scale of the line integral convolution patterns, the computation has to be restarted with a coarser initial image intensity. In case of spot noise larger spots have to be selected and their stretching along the field has to be increased. To motivate our PDE based approach, let us briefly review scale space methods based on anisotropic nonlinear diffusion in imaging.

Discrete diffusion type methods have been known for a long time. Perona and Malik [26] have introduced a continuous diffusion model which allows the denoising of images together with the enhancement of edges. Alvarez et al. [1] have established a rigorous axiomatic theory of diffusive scale space methods. The recovery of lower dimensional structures in images is analyzed by Weickert [44], who introduced an anisotropic nonlinear diffusion method where the diffusion matrix depends on the so called structure tensor of the image.

In PDE-based scale-space methods of image processing we consider a function $u : \mathbb{R}_0^+ \times \Omega \to \mathbb{R}$ which solves the parabolic problem

$$\begin{aligned} \partial_t u - \mathcal{A}[u] &= f(u) \quad \text{in } \mathbb{R}^+ \times \Omega\,, \\ u(0,\cdot) &= u_0 \qquad \text{on } \Omega\,, \end{aligned} \tag{1}$$

for given initial density $u_0 : \Omega \to [0,1]$. Here, the differential operator $\mathcal{A}[\cdot]$ is defined by

$$\mathcal{A}[u] := \operatorname{div}\left(a(\nabla u_\epsilon)\nabla u\right)$$

and we prescribe Neumann boundary conditions $a(\nabla u_\epsilon)\nabla u \cdot \nu = 0$, where "$\cdot$" denotes the scalar product in $\mathbb{R}^+$ and $a : \mathbb{R}^n \to \mathbb{R}$ is the diffusion coefficient that controls the amount of spatial diffusion (blurring) in a given direction in $\mathbb{R}^n$. For the sake of robustness and well-posedness, a pre-smoothed version of the current density $u_\epsilon = \chi_\epsilon * u$ is used. In our setting we interpret the density as an image intensity, a scalar grey scale or a vector valued color. Thus, the solution family $u(\cdot)$ can be regarded as a family of images $\{u(t)\}_{t\in\mathbb{R}_0^+}$, where the time t serves as a scale parameter. Let us remark that by the trivial choice $a = 1$ and $f(u) = 0$ we obtain the standard linear heat equation with its isotropic smoothing and coarsening effect.

In image de-noising, u_0 is a given noisy initial image and the goal is to remove this noise while keeping the important content of the given image. Thus, the diffusion is supposed to be controlled by the gradient of the image intensity. Large gradients mark edges in the image, which should be enhanced, whereas small gradients indicate areas of approximately equal intensity. For that purpose we prescribe a diffusion coefficient

$$a = g(\|\nabla u_\epsilon\|)$$

where $g : \mathbb{R}_0^+ \to \mathbb{R}^+$ is a monotone decreasing function with $\lim_{d\to\infty} g(d) = 0$ and $g(0) = \delta \in \mathbb{R}^+$, e. g. $g(d) = \frac{\delta}{1+\|d\|^2}$. A suitable choice for the pre-smoothing is Gaussian filtering or the convolution with the heat equation kernel. That is, we define $u_\epsilon = \tilde{u}(t = \epsilon^2/2)$ where $\tilde{u}$ is the solution of the heat equation with initial data u. Then ϵ is the variance of the corresponding Gaussian filter. The function f may serve as a penalty which forces the scale of images to stay close to the initial image, e. g. choosing $f(u) = \gamma(u_0 - u)$ where γ is a positive constant. Figure 1 gives an example of image smoothing and edge enhancement by nonlinear diffusion.

Fig. 1. The image on the *left* is successively smoothed by nonlinear diffusion. With increasing scale more and more fine-scale details vanish while the significant content is retained and enhanced

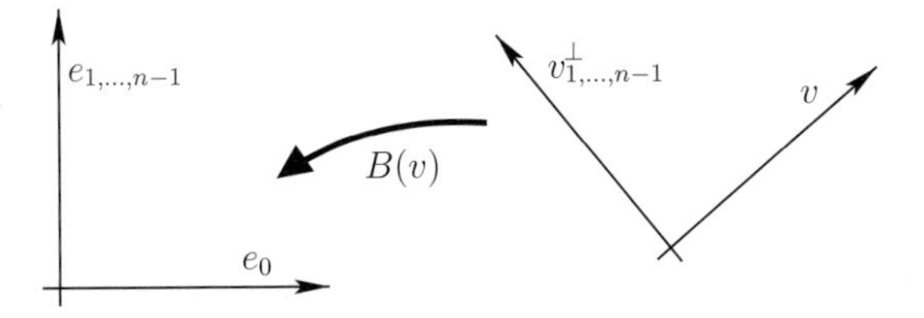

Fig. 2. The coordinate transformation $B(v)$

4 A Flow Aligned Differential Operator

Above, we have modeled an edge aligned operator $\mathcal{A}[\cdot]$, which enabled the feature sensitive de-noising of images. For the subsequent use, let us now define a streamline-aligned differential operator for flow fields. For a given vector field $v : \Omega \to \mathbb{R}^n$ we model linear diffusion in the direction of the vector field and a Perona–Malik type diffusion orthogonal to the field. Let us suppose that v is continuous and $v \neq 0$ on Ω. Then there exists a family of continuous orthogonal mappings $B(v) : \Omega \to SO(n)$ such that $B(v)v = \|v\| e_0$, where $\{e_i\}_{i=0,\dots,n-1}$ is the standard base in $\mathbb{R}^n$ (cf. Fig. 2).

We consider a diffusion tensor $a(v, \nabla u_\epsilon)$ which we define as

$$a(v,d) = B(v)^T \begin{pmatrix} \alpha(\|v\|) & \\ & g(\|d\|)\mathrm{Id}_{n-1} \end{pmatrix} B(v) \,,$$

where $\alpha : \mathbb{R}^+ \to \mathbb{R}^+$ controls the linear diffusion in vector field direction, i. e. along streamlines, and the above introduced edge enhancing diffusion coefficient $g(\cdot)$ acts in the orthogonal directions (Id_{n-1} is the identity matrix in dimension $n-1$).

We may either choose a linear function α or, in case of, e.g. a velocity field which spatially varies over several orders of magnitude, we select a monotone function α (cf. Fig. 2) with $\alpha(0) > 0$ and $\lim_{s\to\infty} \alpha(s) = \alpha_{\max}$.

The differential operator based on this diffusion tensor is finally given by

$$\mathcal{A}[v,u] := \mathrm{div}\,(a(v, \nabla u_\epsilon)\nabla u) \ . \tag{2}$$

It encodes a strong coupling along the velocity field and in case of steep gradients in u a weak coupling in directions perpendicular to the field.

5 Anisotropic Diffusion for Stationary Flow

We shall now make use of the differential operator defined in Sect. 4 to define a diffusion process, which generates texture patterns aligned to a flow field. These patterns will grow upstream and downstream, whereas the edges tangential to them are successively enhanced. Still there is some diffusion perpendicular to the field which supplies us for evolving time with a scale of progressively coarser representation of the flow field.

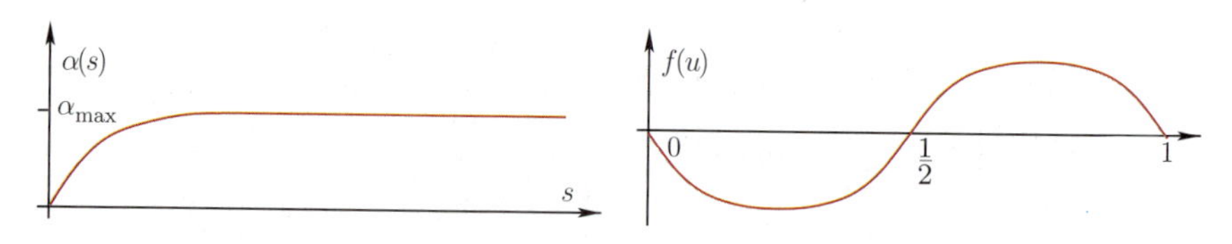

Fig. 3. *Left:* Graph of the velocity dependent linear diffusion coefficient $\alpha(\cdot)$. *Right:* Graph of the scalar contrast enhancing right-hand side $f(\cdot)$

In general it does not make sense to consider a specific initial image for such a diffusion process. As initial data u_0 we thus choose some random noise of an appropriate frequency range. If we run the evolution (1) for vanishing right-hand side f the image contrast will unfortunately decrease due to the diffusion along streamlines. The asymptotic limit will turn out to be an averaged grey value. Therefore, we strengthen the image contrast during the evolution, selecting an appropriate continuous function $f : [0, 1] \to \mathbb{R}^+$ (cf. Fig. 3) with

(F1) $f(0) = f(1) = 0$,
(F2) $f < 0$ on $(0, 0.5)$, and $f > 0$ on $(0.5, 1)$.

Neglecting the diffusive term of the evolution at a first glance we realize that this right-hand side pushes values below the average value 0.5 towards the zero and accordingly values above 0.5 towards 1. A more detailed analysis of the contrast enhancement including the diffusive term is discussed in Sect. 6.1. However, well-known maximum principles ensure that the interval of grey values $[0, 1]$ is not enlarged running the nonlinear diffusion. Here the property (F1) is of great importance.

Finally, we end up with the method of nonlinear anisotropic diffusion to visualize complex vector fields. Thereby we solve the nonlinear parabolic problem

$$\partial_t u - A[v, u] = f(u) \tag{3}$$

starting from some random initial data $u(0, \cdot) = u_0$ and obtain a scale of images representing the vector field in an intuitive way (cf. Fig. 4).

5.1 Enhancing the Resulting Texture

If we ask for point-wise asymptotic limits of the evolution, we expect an almost everywhere convergence to $u(\infty, \cdot) \in \{0, 1\}$ due to the choice of the contrast enhancing function f. Analytically, 0.5 is a third, but unstable fixed point of the dynamics, which, thus, will not turn out to be locally dominant numerically.

The space of asymptotic limits significantly influences the richness of the developing vector field aligned structures. We may ask how to further enrich the pattern which is settled by anisotropic diffusion. This turns out to be possible by increasing the set of asymptotic states. We no longer restrict the considerations to a scalar density u but consider a vector valued $u : \Omega \to [0, 1]^2$ and a corresponding system of parabolic equations. The coupling is given by the nonlinear diffusion coefficient $g(\cdot)$ which now depends on the norm $\|\nabla u\|$ of the Jacobian of the vector

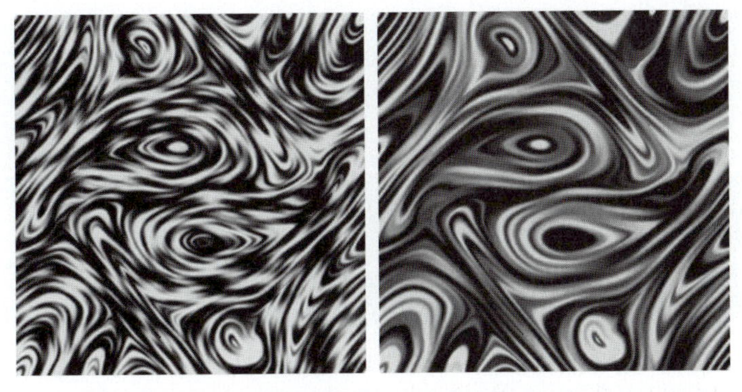

Fig. 4. A vector field from a 2D magneto-hydrodynamics simulation (MHD) is visualized by nonlinear diffusion. A discrete white noise is considered as initial data. We run the evolution on the *left* for a small and on the *right* for a large constant diffusion coefficient α

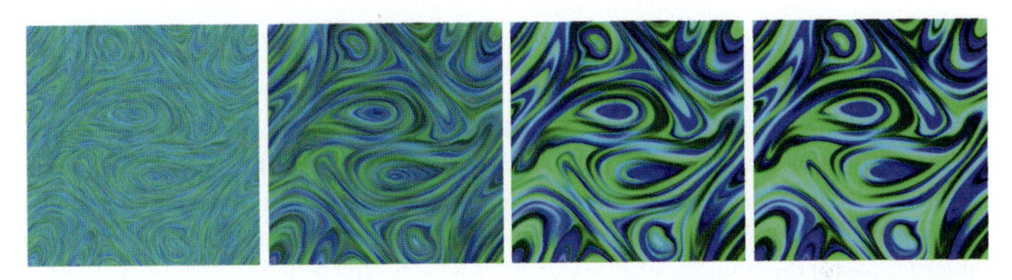

Fig. 5. Different snapshots from the multi-scale based on anisotropic diffusion are depicted for a 2D MHD simulation vector field (cf. Fig. 4). Here we consider a two-dimensional diffusion problem and interpret the resulting density as a color in a *blue/green* color space

valued density ∇u and the right-hand side $f(u)$. We define $f(u) = h(\|u\|)u$ with $h(s) = \tilde{f}(s)/s$ for $s \neq 0$, where $\tilde{f}$ is the old right-hand side from the scalar case, and $h(0) = 0$. Furthermore we select an initial density which is now a discrete white-noise with values in $B_1(0) \cap [0, 1]^2$. Thus, the contrast enhancing now pushes the point wise vector density u either to the 0 or to some value on the sphere sector $S^1 \cap [0, 1]^2$. Again a straightforward application of the maximum principle ensures $u(t, x) \in B_1(0) \cap [0, 1]^2$ for all t and $x \in \Omega$.

Figure 5 shows an example for the application of the vector valued anisotropic diffusion method applied to a 2D flow field from a MHD simulation. The field shows the dynamics of an electrically conducting fluid. The performed clustering outlines the so-called magnetic domains (thick clusters), i.e. domain regions where the fluid circulates, and current sheets (thin, closed clusters bordering the magnetic domains), which contain most of the current. Furthermore, Fig. 6 shows results of this method applied to several time steps of a convective flow field. An incompressible Bénard convection is simulated in a rectangular box with heating from below and cooling from above. The formation of convection rolls will lead to an exchange of temperature. We recognize that the presented method is able to nicely depict the global structure of the flow field, including its saddle points, vortices, and stagnation points on the boundary.

Fig. 6. Convective patterns in a 2D flow field are displayed and emphasized by the method of anisotropic nonlinear diffusion. The images show the velocity field of the flow at different time steps. Thereby the resulting alignment is with respect to streamlines of this time dependent flow

5.2 3D Flow Fields

The anisotropic nonlinear diffusion operator (2) has been formulated for arbitrary space dimension. It results in a scale of vector field aligned patterns which we then have to visualize. In 2D this has already been done in a straightforward manner in the above figures. In 3D we have somehow to break up the texture-volume and open up the view to inner regions. Otherwise we must confine ourselves with some pattern close to the boundary representing solely the shear flow.

Here we can benefit from the vector valued diffusion. Since for $m = 2$ the non-trivial asymptotic limits are in mean equally distributed on $S^1 \cap [0, 1]^2$, we can we reduce the image-content and focus on a ball shaped neighborhood $B_\delta(\omega)$ of a certain point $\omega \in S^1 \cap [0, 1]^2$. Now we can either use a volume rendering to visualize this type of sub-volumes or look at iso-surfaces of the function

$$\sigma(x) = \|u(x) - \omega\|^2 .$$

Then the parameter δ^2 allows us to depict the boundary of the pre-image of $B_\delta(\omega)$ with respect to the mapping u (cf. Fig. 7).

5.3 Flow Fields on 2D Surfaces

So far we considered vector fields on domains which are subsets of the 2D or 3D Euclidean space. It is straight-forward to extend the methodology to tangential flow fields on surfaces, such as weather-map wind-fields over the earth, flow fields on stream-surfaces, or vector fields from differential geometry. We have to replace the Euclidean gradient ∇ and the divergence operator div by their geometric counterparts $\nabla_{\mathcal{M}}$ and $\mathrm{div}_{\mathcal{M}}$ respectively. Here the index $\mathcal{M}$ indicates that we are working with the tangential gradient and divergence on the surface or manifold $\mathcal{M}$. Proceeding as in Sect. 4 the differential operator describing the given flow field is given by

$$\mathcal{A}[u] := \mathrm{div}_{\mathcal{M}}(a(\nabla_{\mathcal{M}} u_\epsilon)\nabla_{\mathcal{M}} u)$$

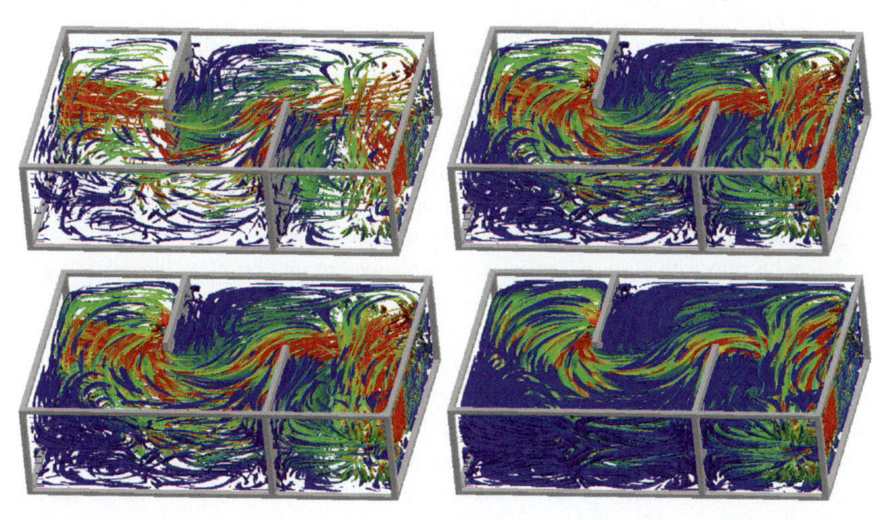

Fig. 7. The incompressible flow in a water basin with two interior walls and an inlet (on the *left* of the box) and an outlet (on the *right* of the box) is visualized by the anisotropic non-linear diffusion method. Iso-surfaces show the pre-image of $\partial B_\delta(\omega)$ under the vector valued mapping u for some point ω on the sphere sector. From top left to bottom right the radius δ is successively increased. A color ramp *blue–green–red* indicates an increasing absolute value of the velocity. The diffusion is applied to initial data which is a relatively coarse grain random noise

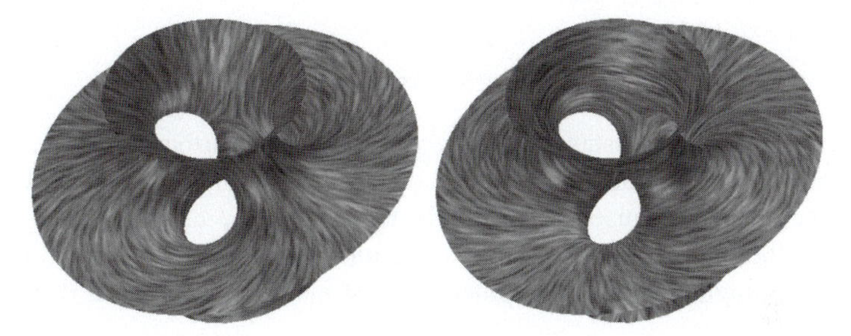

Fig. 8. The principal directions of curvature of a minimal surface are visualized using the anisotropic diffusion equation on surfaces

for C^2 functions u on the manifold $\mathcal{M}$. As an illustration Fig. 8 shows the visualization of the principal directions of curvature on a minimal surface.

5.4 Flow Segmentation

The above applications already show the capability of the anisotropic nonlinear diffusion method to outline the flow structure not only locally. In particular for larger evolution times in the diffusion process the topological skeleton of a vector field becomes clearly visible. We will now investigate a possible flow segmentation by means of the anisotropic diffusion. Let us restrict to the two-dimensional case of an incompressible flow with vanishing velocity v at the domain boundary $\partial\Omega$. Then topological regions are separated by homoclinic, respectively heteroclinic orbits connecting critical points in the interior of the domain and stagnation points on the boundary. Critical points, by definition points with vanishing velocity $v = 0$, may either be saddle points or vortices. Furthermore we assume critical points to be non-degenerate, i. e. ∇v is regular. Saddle points are characterized by two real eigenvalues of ∇v with opposite sign, whereas at vortices we obtain complex conjugate eigenvalues with vanishing real part. Stagnation points on $\partial\Omega$ are similar to saddles. For details we refer to [15].

In each topological region there is a family of periodic orbits close to the heteroclinic, respectively homoclinic orbit. This observation gives reason for the following segmentation algorithm. At first, we search for critical points in Ω and stagnation points on $\partial\Omega$. We calculate the directions which separate the different topological regions. In case of saddle points these are the eigenvectors of ∇v. Next, we successively place an initial spot in each of the sectors and perform an appropriate field aligned anisotropic diffusion.

Let us suppose that a single sector is spanned by vectors $\{w_+, w_-\}$ where the sign $\pm$ indicates incoming and outgoing direction. The method described by (3) would lead to a closed pattern along one of the above closed orbits for time t large enough. To fill out the interior region we modify the diffusion by selecting an orientation for a "one sided" diffusion (cf. Fig. 9). That is, we select a unique normal $v^\perp$ to v and consider the diffusion matrix

$$a(v, \nabla u_\epsilon) = B(v)^T \begin{pmatrix} \alpha & \\ & G((\nabla u_\epsilon \cdot v^\perp)_+) \end{pmatrix} B(v)\,,$$

where α is a positive constant and $(s)_+ := \max\{s, 0\}$. Furthermore we consider a non negative, concave function $f : \mathbb{R}_0^+ \to \mathbb{R}_0^+$ with $f(0),\ f(1) = 0$ as a source term in the diffusion equation. If the orientation of $\{w_+, w_-\}$ coincides with that of $\{v, v^\perp\}$, then linear diffusion in the direction towards the interior will fill up the complete topological region. A segmentation of multiple topological regions at the

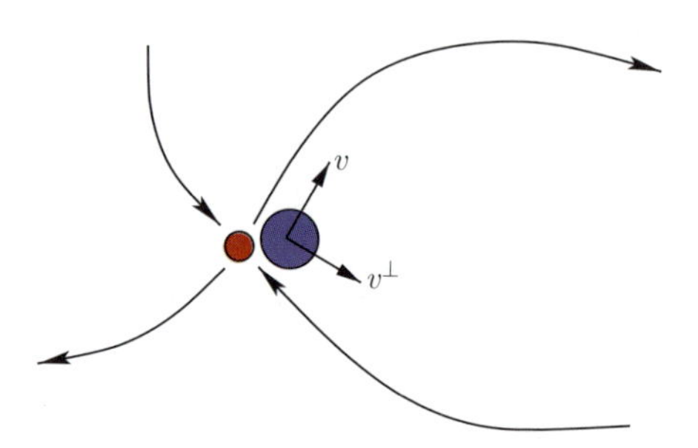

Fig. 9. A sketch of the four sectors at a critical point (indicated by *red disk*), the initial spot (*blue disk*) for the diffusion calculation and the oriented system $\{v, v^\perp\}$

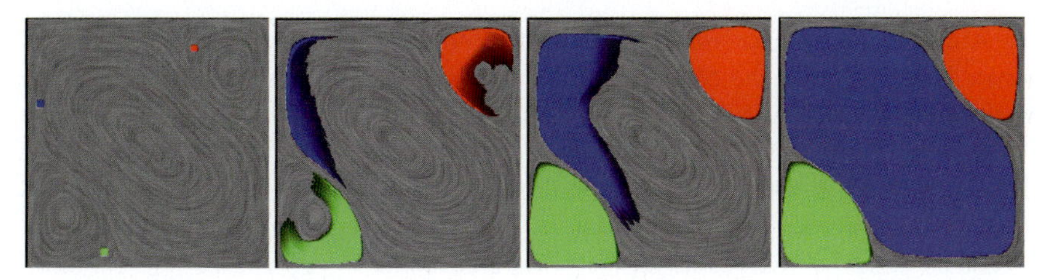

Fig. 10. Several time-steps from the nonlinear diffusion segmentation applied to a velocity field from a Bénard convection are shown. We have placed the seed-points as close as possible in terms of the grid size in the sectors spanned by the eigenvalues of the Jacobian ∇v of the velocity. Only to emphasize the evolution process a single grey-scale image from the diffusion calculation is underlying the sequence of segmentation time steps

same time is possible, if we carefully select the sectors where we release initial spots. Figure 10 shows different time steps of the segmentation applied to a convective incompressible flow.

6 Transport and Diffusion for Non-stationary Flow Fields

So far, the above anisotropic diffusion method generates streamline type patterns, which are aligned to trajectories of the vector field for a fixed given time. I. e. for a time-dependent vector field $v : \mathbb{R}^+ \times \Omega \to \mathbb{R}^d$ on a computational domain $\Omega \subset \mathbb{R}^d$ and $d = 2, 3$, we have been considering integral lines $\{x(s) \mid s \in \mathbb{R}\}$ with $\frac{d}{ds}x(s) = v(t, x(s))$ for a fixed time t. Thus, the method intuitively visualizes the vector field freezed at time t but offers only very limited insight in the actual transport process governed by the underlying time-dependent flow field.

To ensure that our visualization actually displays this process we have to consider the true transport problem and its particle lines. Hence we take into account the particle motion obeying the equation $\frac{d}{dt}x(t) = v(t, x(t))$ and the induced transport of a given density $u(t, x)$. The fact that such a purely advected density u stays constant along particle trajectories leads to the conservation law

$$\frac{D}{dt}u := \frac{d}{dt}u(t, x(t)) = \frac{\partial}{\partial t}u + \nabla u \cdot v = 0$$

which means a vanishing of the material derivative.

In addition to this conservation law, we have to incorporate a mechanism for the generation, the growth and enhancement of flow aligned patterns. Here we pick up the previous model (3) and consider a simultaneous anisotropic nonlinear diffusion process with linear diffusion along the particle line and sharpening in the perpendicular direction. Let us emphasize here that this diffusion process acts in forward *and* backward direction of the particle line. Thus, a careful control of the parameters is indispensable to avoid an artificial propagation in downwind direction with the accompanying visual impression of a wrong velocity. In the next section we will discuss in detail a suitable balance of parameters.

Altogether our basic *transport diffusion* model for time-dependent vector fields looks as follows: On the computational domain $\Omega \subset \mathbb{R}^d$ we consider for a given vector field $v : \mathbb{R}^+ \times \Omega \to \mathbb{R}^d$ the boundary and initial value problem:

$$\begin{aligned} \partial_t u + \nabla u \cdot v - \mathcal{A}[v, u] &= f(u) && \text{in } \mathbb{R}^+ \times \Omega, \\ (A \nabla u) \cdot \nu &= 0 && \text{on } \mathbb{R}^+ \times \partial\Omega, \\ u(0, \cdot) &= u_0 && \text{in } \Omega, \end{aligned}$$

where $\mathcal{A}[v, u]$ is the diffusion tensor (2) already known from the anisotropic diffusion for steady flow fields. The initial data u_0 is again assumed to be a white noise of appropriate frequency. Still the role of the right-hand side f is to ensure contrast enhancement. Consequently we apply functions f which fulfill the properties (F1), (F2) mentioned previously.

The new model generates and stretches patterns along the flow field and transports them simultaneously. The resulting motion texture is characterized by a dense coverage of the domain with streakline type patterns, which do not have a fixed injection point but move in time with the fluid (cf. Fig. 11). The method is applicable in any dimension, in particular on 3D domains and 2D surfaces as for the static flow case before, although we have not performed such computations yet.

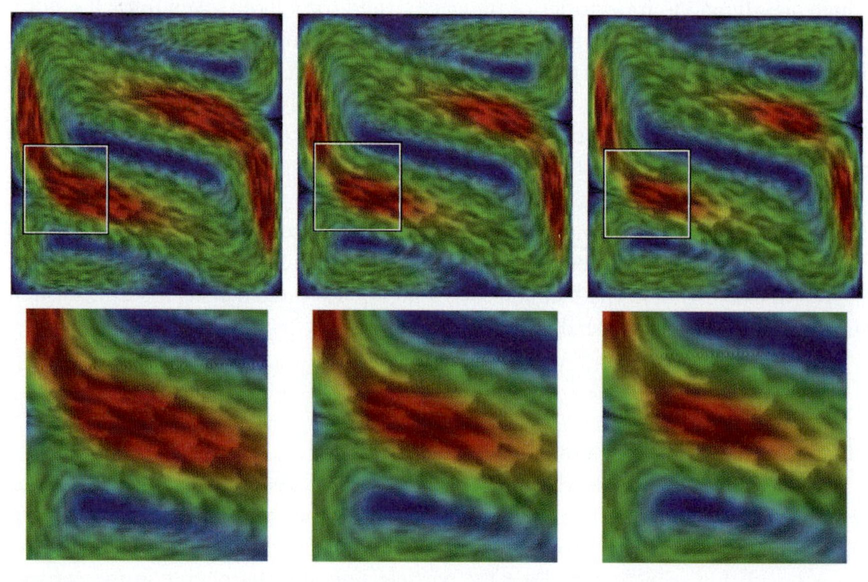

Fig. 11. Three successive time-steps of the transport diffusion process generating directed patterns of a Bénard convection (cf. Sect. 6.1). The additional coloring indicates the speed of the flow field. *Red colors* indicate high velocity, whereas *blue colors* indicate low velocity. To emphasize the transport of patterns we have magnified the marked sections of the images in the *lower row*

6.1 Balancing Parameters

In general, transport and diffusion are contrary processes. Our goal in mind – the generation and transport of patterns which simultaneously diffuse along the flows – there has to be a careful weighting of the parameters that steer the transport and the diffusion respectively. Otherwise the diffusion may overrun the transport, resulting in a process that is rather diffusion than transport with some pattern generating diffusion.

Let us suppose the temporal resolution of the given vector-field data is of size τ. It is well known that the solution of the heat equation at a time t corresponds to the convolution of the initial data with a Gaussian kernel of variance $\sqrt{2t}$. Since the diffusion tensor $a(v)$ invokes linear diffusion with a coefficient $\alpha(\|v(x)\|)$ in the direction of the velocity $v(x)$ for every $x \in \Omega$, we consider the corresponding variance

$$\mathcal{D}(\alpha(x)) := \sqrt{2\tau\alpha(x)}$$

to be a measure for the diffusion within the transport diffusion process for the time τ. Of course a measure for the corresponding expected transport distance is

$$\mathcal{T}(x) := \tau \, \|v(x)\| \, .$$

Typically $\mathcal{T}(\cdot)$ is more or less fixed, since τ is in general prescribed by the underlying CFD data. Thus, we would like to adjust α locally such that $\mathcal{D}$ is balanced with $\mathcal{T}$. To this end we introduce a balancing parameter $\eta \in \mathbb{R}^+$ and consider the balancing condition

$$\mathcal{D}(\alpha(x)) = \eta\mathcal{T}(x).$$

Roughly speaking we then have the following relations:

$\eta \ll 1$	Transport dominates the model,
$\eta = 1$	Transport $\approx$ Diffusion,
$\eta \gg 1$	Diffusion dominates the model.

Hence, choosing $\eta < 1$ fixed, and solving the balance condition for $\alpha(x)$, we get a suitable diffusion coefficient

$$\alpha(\|v\|)(x) = \frac{\eta^2 \, \|v(x)\|^2 \, \tau}{2}$$

as a function on the domain Ω which instead of the one defined in Sect. 4 is inserted into the diffusion tensor $a(v, \nabla u_\epsilon)$ of our transport diffusion model.

Let us furthermore study the amplification of certain frequencies of the initial image due to the right-hand side of our model. Our focus will be on the influence of the shape of f on the contrast enhancing property of the model. To this end let us consider a much simpler setting of a high frequency initial data given by

$$u_0(x) = \frac{1}{2}\left[\sin\left(\frac{x}{\epsilon}\right) + 1\right]$$

and restrict ourselves to a simple diffusion equation along a (one-dimensional) streamline, which is given by

$$\partial_t u - \alpha \Delta u = f(u) \qquad \text{in } [0, 1].$$

We consider the linearization of f around $\frac{1}{2}$

$$\tilde{f}(u) = \gamma \left(u - \frac{1}{2} \right),$$

where γ is the slope of the original f at 1/2. Now let us take into account the ansatz $u(t) = b(t) \left(u_0(x) - \frac{1}{2} \right) + \frac{1}{2}$ for the evolution of a one-dimensional image-density. Inserting the ansatz into the linear diffusion-equation we obtain

$$\frac{1}{2} \left[b' + \left(\frac{\alpha}{\epsilon^2} - \gamma \right) b \right] \sin \left(\frac{x}{\epsilon} \right) = 0$$

and so

$$b(t) = \exp \left[\left(\gamma - \frac{\alpha}{\epsilon^2} \right) t \right].$$

This means that frequencies above $\sqrt{\alpha / \gamma}$ are damped, whereas frequencies below this threshold are amplified. Given an upper threshold $1/\epsilon$ for the frequencies which we want to amplify, we choose

$$\gamma = \frac{\alpha}{\epsilon^2}.$$

Finally, we construct our nonlinear right-hand side $f(\cdot)$ in such a way that the slope at $1/2$ equals γ.

6.2 A Blending Strategy for Long-Term Animation

With the anisotropic diffusion model for steady flow fields we have generated a whole scale of representations. Here, the scale was identified with the time t of the evolution process. But as proposed in the last section, the scale parameter is now coupled to the actual transport process in our transport diffusion model. In particular for long-time visualization purposes this coupling leads to unsatisfactory results. Because due to the nature of our model, we are unable to freeze the scale and solely consider the evolution of suitable patterns at that specific scale in time, which would be the optimum process.

The solution we propose here is a compromise based on the blending of different results from the transport diffusion evolution started at successive time-points. First, we select a suitable interval for the scale parameter $[s_0, s_1]$ with $s_1 > s_0 > 0$ around our preferred multi-scale resolution for the resulting images. Based on a smooth blending function $\psi : \mathbb{R} \to [0, 1]$ having support in $(-1, 1)$ and such that

$$\psi(t) = \psi(-t) = 1 - \psi(1 - t),$$
$$\psi(0) = 1,$$

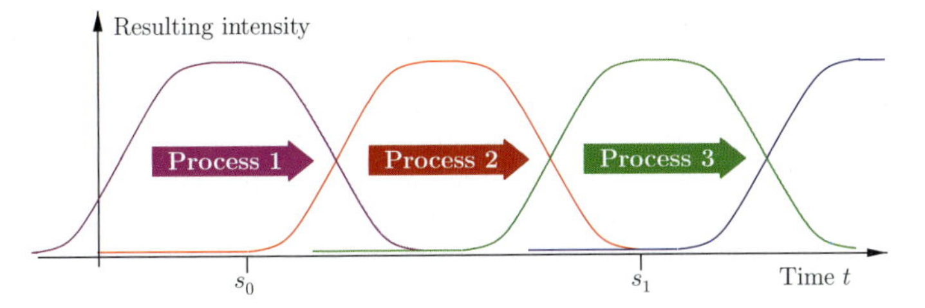

Fig. 12. The weighting factors in the blending operation together with the overlapping scale/time intervals of the considered transport diffusion processes are shown in a diagram over time

we can construct a partition of unity $\{\psi_i\}$ on the real line $\mathbb{R}$. That is, we define $\psi_i(t) = \psi(\frac{2t-(i+1)(s_0+s_1)}{s_1-s_0})$. Now, for all $i = 0, 1, \ldots$ we separately solve the above transport diffusion problem for different starting times $t_i = i\,(\frac{s_1-s_0}{2})$ always considering some white noise of a fixed frequency range as initial data and denoting the resulting solution by u_i. For negative time we suppose a suitable extrapolation of the velocity field to be given. Finally, applying blending of at least two different solutions we compute

$$u(t,x) = \sum_i \psi_i(t) u_i(t,x)\,.$$

This intensity function is well defined for arbitrary times and characterized by the initially prescribed scale parameter interval.

We use this construction for an animation of the flow over a certain time interval (cf. Fig. 12 for a graph of the blending functions). Such an animation involves all solutions u_i for which the blending function ψ_i has a non vanishing overlap with the given time interval. Other constructions of a partition of unity and corresponding blending functions are near at hand and especially multiple overlaps can be considered which requires the blending of more than two intensity functions at the same time. We emphasize that the application of this blending technique does not introduce any inaccuracy, because for any time t the resulting image $u(t,x)$ consists of images $u_i(t,x)$ showing streaklines at time t and at slightly varying scale.

7 Continuous Clustering via Anisotropic Phase Separation

So far, we have discussed the generation of flow aligned multi scale textures. Let us now look the hierarchical clustering of flow data, ranging from small clusters showing strong local coherence of the flow to large global cluster sets gathering large flow patterns. As before, we will discuss this task in the framework of continuous models. Before we detail the application of such a model for the actual flow clustering let us review the underling physical model for the coarsening of structures in metal alloys, which goes back to Cahn and Hilliard [5].

7.1 The Cahn–Hilliard Model

The Cahn–Hilliard model was introduced to describe phase separation and coarsening in binary alloys. Phase separation occurs when a uniform mixture of the alloy is quenched below a certain critical temperature underneath which the uniform mixture becomes unstable. As a result a micro-structure of two spatially separated phases with different concentrations develops. In later stages of the evolution on a much slower time scale than that of the initial phase separation the structures become coarser: either by merging of particles or by growing of bigger particles at the cost of smaller ones. This coarsening can be understood as a clustering, where the system mainly tries to decrease the surface energy of the particles which leads to coarser and coarser structures during the evolution. In the basic Cahn–Hilliard model this surface energy is isotropic. There are no preferred directions of the interfaces. Hence the particles tend to be ball shaped (cf. Fig. 13).

In the following paragraph we briefly outline the basic concept of the Cahn–Hilliard model. For more details we refer to the review papers by Elliott [10] and Novick-Cohen [25]. The model is based on a Ginzburg–Landau free energy which is a functional in terms of the concentration difference u of the two material components. The Ginzburg–Landau free energy E is defined to be

$$E(u) := \int_\Omega \left\{ \Psi(u) + \frac{\gamma}{2}|\nabla u|^2 \right\},$$

where Ω is a bounded domain. The first term $\Psi(u)$ is the chemical energy density and typically has a double-well form. In this paper we take

$$\Psi(u) = \frac{1}{4}(u^2 - \beta^2)^2$$

with a constant $\beta \in (0, 1]$ (cf. Fig. 14). We note that the system is locally in one of the two phases if the value of u is close to one of the two minima $\pm\beta$ of Ψ.

Now, the diffusion equation for the concentration u is given by

$$\frac{\partial u}{\partial t} - \Delta w = 0$$

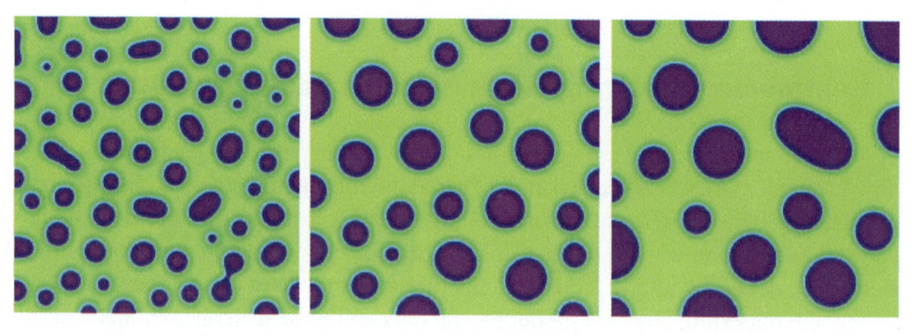

Fig. 13. Three time-steps of the original Cahn–Hilliard phase separation

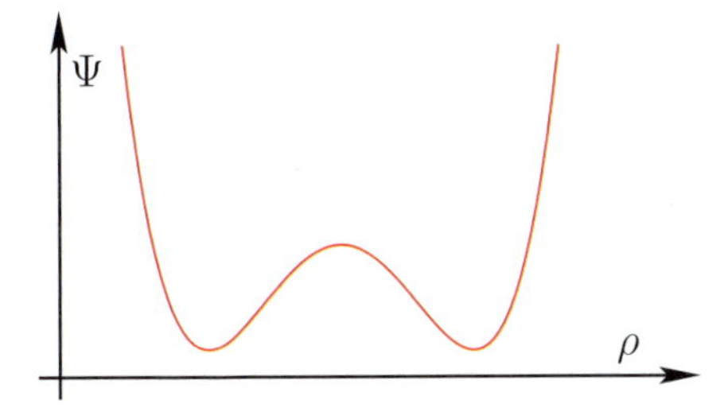

Fig. 14. Chemical energy as function of concentration

on $\mathbb{R}^+ \times \Omega$, where w is the local chemical potential difference, which is given as the variational derivative of E with respect to u

$$w = \frac{\delta E}{\delta u} = -\gamma \Delta u + \Psi'(u).$$

As boundary and initial conditions we request $\partial_\nu w = \partial_\nu u = 0$, where ν is the outer normal on $\partial\Omega$, and $u(0,\cdot) = u_0(\cdot)$ for some initial concentration distribution u_0.

Starting with a random perturbation of a constant state $\bar{u}_0$, which has a values in the unstable concave part of Ψ, we observe the following: In the beginning the chemical energy decreases rapidly whereas the gradient energy increases. This is due to the fact that during phase separation u attains values which are at large portions of the domain close to the minima of the chemical energy Ψ. Since regions of different phase are separated by transition zones with large gradients of u, the gradient energy increases during phase separation. In the second stage of the evolution – the actual clustering – when the structures become coarser, the total amount of transition zones decreases. Correspondingly the amount of gradient energy becomes smaller again.

7.2 Anisotropic Interface Energy

Let us now turn to the clustering model for flow data. We introduce a cluster mapping $u : \mathbb{R}_0^+ \times \Omega \to \mathbb{R}$ which will be the solution of an appropriate evolution problem. Thereby, time will again serve as the scale parameter leading from fine cluster granularity to successively coarser clusters. For fixed scale t our definition of the set of clusters $\mathcal{C}(t)$ is

$$\mathcal{C}(t) = \{x \mid u(t,x) \geq 0\}.$$

This set splits up into the connected components of $\mathcal{C}(t)$

$$\mathcal{C}(t) = \bigcup_i \mathcal{C}_i(t).$$

The evolution problem steering the clusters via the quantity u should satisfy the following properties:

- The number of clusters generically decreases in time.
- The shape of the cluster components strongly corresponds to correlations in the data field.
- The volume fraction covered by $C(t)$ is approximately constant in t, i. e. $\frac{|C(t)|}{|\Omega|} \approx \Theta$ for $\Theta \in (0, 1)$.

These conditions motivate us to pick up the physical Cahn–Hilliard model with the double-well separation potential $\Psi(u)$, a separation energy $E_s = \int_\Omega e_s(u)$ and energy density $e_s(u) = \Psi(u)$. Among all u with $\int_\Omega u = \bar{u}_0 = const.$ the energy E_s attains its minimum if u has the values $\pm\beta$ only. This leads to a binary decomposition of the domain into two parts, where one part corresponds to $\{x \mid u(x) = \beta\}$. However this set can have many connected components and may even be very unstructured. Furthermore there is no mechanism which enforces a successive coarsening and thus a true multi-scale of clusters.

We remedy this behavior by introducing a term which penalizes the occurrence of many disconnected cluster components with high interfacial area. To this end we choose a gradient energy $E_\partial = \int_\Omega e_\partial$ with local energy density e_∂ that penalizes rapid spatial variations of u. In order to have flexibility to choose an anisotropic and inhomogeneous gradient energy, an appropriate definition of an interfacial energy density is given by

$$e_\partial(\nabla u) = \frac{\gamma}{2} a \nabla u \cdot \nabla u,$$

where γ is a scaling coefficient and $a \in \mathbb{R}^{n \times n}$ is some symmetric positive definite matrix that may depend on the space variable and other quantities involved.

In the following we will refer to the set $\partial\{x \mid u(x) = 0\}$ as the interface. The orientation of the interface can be described by its normal which, in the case that $\nabla u \neq 0$, is given by the normalized gradient

$$\nu = \frac{\nabla u}{\|\nabla u\|}.$$

For $a = \mathrm{Id}$ all gradients of u and hence, all interfaces are penalized equally independent of their orientation. But with respect to our clustering intention we consider an anisotropic energy density which strongly depends on the orientation of the local interface and thereby on the direction of ∇u.

For a given static vector-field $v : \Omega \to \mathbb{R}^n$ a natural clustering should emphasize the coherence along the induced streamlines. Thus, interfaces aligned across streamlines have to be penalized significantly by the gradient energy whereas interfaces oriented along streamlines are tolerated. We choose the diffusion tensor similar to the ones used above (cf. Sect. 4)

$$a(v) := B(v)^T \begin{pmatrix} 1 & 0 \\ 0 & \alpha(\|v\|)\mathrm{Id}_{n-1} \end{pmatrix} B(v)$$

Since interfaces that cross streamlines shall have larger energy we choose a positive function α with $\alpha \leq 1$.

Altogether we have defined the energy

$$E(u) := \int_{\Omega} \left\{ \Psi(u) + \frac{\gamma}{2} a \nabla u \cdot \nabla u \right\}$$

and proceed as for the basic Cahn–Hilliard model. We define the first variation of the energy and arrive at the potential

$$w = \Psi'(u) - \gamma \mathcal{A}[v, u],$$

where $\mathcal{A} = \mathrm{div}(a(v)\nabla u)$ is defined as in the previous sections.

Let us continue as before and assume that the evolution of the cluster mapping u is governed by diffusion where the corresponding flux linearly depends on the negative gradient of the first variation of energy. Thus, we choose $\partial_t u - \Delta w = 0$ as above and end up with the following fourth order differential equation:

$$\partial_t u - \Delta \left(\Psi'(u) - \gamma \mathcal{A}[v, u] \right) = 0$$

with boundary conditions $\partial_\nu u = \partial_\nu w = 0$ and prescribed initial data $u(0, \cdot) = u_0(\cdot)$. After an initial short period of phase separation it is mainly the interfacial energy contribution which is successively reduced.

As in the texture generation approaches it does not make sense to consider certain initial data, if no a priori information on the clustering is known. Consequently we choose a constant value $\bar{u}_0$ plus some random perturbations as initial data u_0. The constant $\bar{u}_0$ depends on the volume fraction Θ of the domain which shall be covered by the clusters, i.e. by the sets $\{x \mid u(t, x) \geq 0\}$. Therefore, we choose $\bar{u}_0 = \Theta\beta - (1 - \Theta)\beta$.

During the evolution very rapidly cluster patterns will grow without any prescribed location and orientation. This is in order to decrease $E_s = \int_\Omega \Psi(u)$ which forces the solution to obtain values close to $\pm\beta$ in most of the domain Ω. After this starting phase the clusters orient themselves in an anisotropic way to decrease the amount of the anisotropic gradient energy E_∂. In addition they become coarser and coarser due to the fact that smaller particles shrink and larger ones grow. In particular one observes that a large particle being surrounded by smaller ones grows to the expense of the smaller ones. This implies that as time evolves locally only the main features of the clusters will be kept (Fig. 15).

Finally we obtain a scale $u(t, \cdot)$ of cluster mappings and induced cluster sets $\mathcal{C}(t)$. They represent a successively coarser representation of simulation data and continuously enhances coherences in the underlying simulation dataset, where the cluster set $\mathcal{C}(t)$ will cover a volume of approximate size $\Theta|\Omega|$.

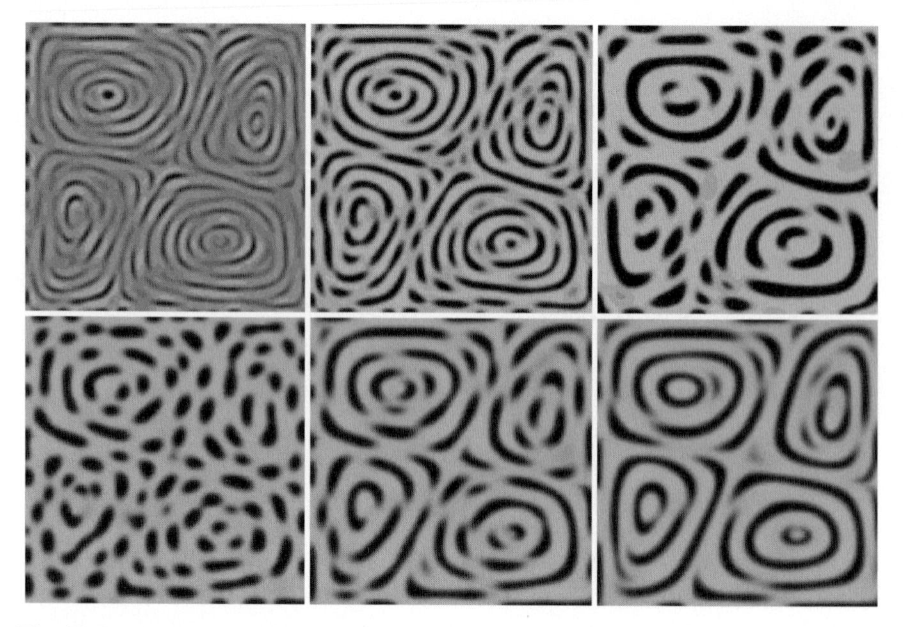

Fig. 15. *Top:* Successive stages of the continuous clustering of a Bénard convective flow field. *Bottom:* Effect of increasing anisotropy α. The computations are based on a grid of resolution 257^2

8 Remarks on the Finite Element Implementation

So far, we have not yet discussed discretization in space and time of the above introduced continuous time-dependent partial differential equations. Hence, we deal with the variational formulation of the different PDE problems introduced above. We propose to consider a finite element discretization in space and a semi-implicit backward Euler scheme in time.

The semi-implicit temporal discretization means that the nonlinear diffusion tensor $\mathcal{A}$ and the nonlinear term on the right-hand side $f(u)$ as well as the derivatives of the non convex potential ψ are evaluated at the old time-steps.

For the spatial discretization we can restrict the numerical considerations to regular hexahedral grids in 2D and 3D. On these grids we have bilinear, respectively trilinear finite element spaces. However below we will use triangular elements as well. In any case we can base numerical integration on the lumped-masses product $(\cdot,\cdot)^h$ [36] for the L^2 product and a midpoint quadrature rule for the bilinear form $(A\nabla\cdot,\nabla\cdot)$.

Finally, in each step of the discrete evolution we have to solve a single system of linear equations for the vector of nodal values for the density function u and the chemical potential w, respectively. In case we need pre-smoothed data, we consider a single discrete, implicit time-step for the computation of the heat equation with the density being smoothed as initial data.

9 Clustering Based on Hierarchical Decomposition of a Differential Operator

In the previous section we have discussed a continuous physical model for the clustering of flow fields. Instead of involving methods adapted from continuum mechanics, we might ask for a direct hierarchical decomposition of the differential operators $\mathcal{A}$ from Sect. 4, which represent diffusion processes strongly aligned to the flow field.

9.1 Review of Algebraic Multigrid

The idea we develop in this section uses algebraic multigrid (AMG) methodology to decompose the corresponding discrete operator. AMG methods were first introduced in the early 1980s [2, 3, 28] for the solution of discrete linear systems $AU = F$ of equations coming from the discretization of linear differential equations $\mathcal{A}[u] = f$ on domains Ω with suitable boundary conditions. We refer to [37] for a detailed introduction. Thereby U is supposed to be a finite element approximation of the continuous solution u and A the finite element stiffness matrix corresponding to $\mathcal{A}$. Finally, F is the corresponding discrete right-hand side.

The development of AMG was led by the idea to mimic classical (geometric) multigrid methods for applications where a hierarchy of nested meshes is either not available at all, or cannot reflect particular properties such as strength of diffusion of the discretized operator appropriately on coarse grid levels.

Consequently one has to work with the matrix A and its algebraic structure. The general procedure is sketched in Fig. 16. AMG tries to coarsen this matrix independently from any underlying fine grid discretization, where n is the number of degrees of freedom. It computes a sequence of prolongation matrices P^l which encodes how coarse scale (l) basis functions are combined using the basis functions on the finer scale ($l-1$). This induces a sequence of corresponding matrices A^l, defined by the so-called Galerkin projection $A^l := R^l A^{l-1} P^l$, where the restriction R^l is given

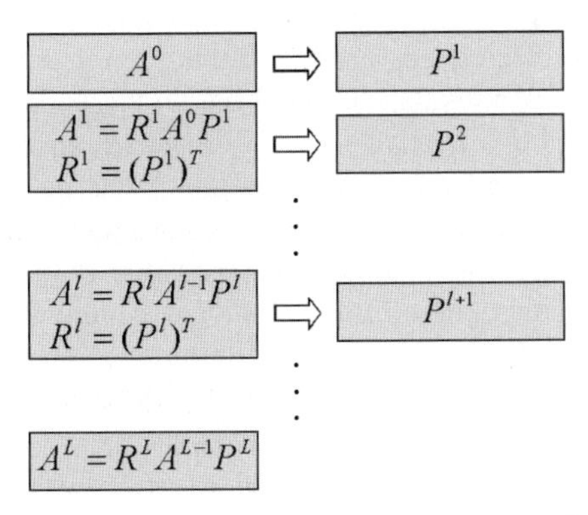

Fig. 16. General AMG construction. From the fine scale matrix A^0 input, AMG computes prolongations P^l, restrictions R^l, and coarse scale matrices A^l on successively coarser scales $l = 1, \ldots, L$

as the transpose of the prolongation P^l ($R^l := (P^l)^T$). The prolongation matrices $\{P^l\}_{l=1,\dots,L}$ are computed using information from the matrix A^{l-1} on the previous level $l-1$ only. The sequence of prolongation matrices allows for the construction of a problem-dependent basis $\{\Psi^{l,i}\}$. One constructs a coarser basis $\{\Psi^{l,i}\}$ which captures the appropriate features relevant for the approximation of the corresponding continuous problem.

9.2 AMG for Flow Field Clustering

The theory and design of efficient AMG tools is rather involved. However, we emphasize that our flowing clustering requires just basic AMG capabilities. We perform no specific tuning of the AMG for flow clustering. Let us apply the method to the concrete discrete finite element matrix A of the differential operator $\mathcal{A}$. This stiffness-matrix A can be regarded as a description of the structure of the flow field v, because as in the operator $\mathcal{A}$ the flow alignment is encoded in this matrix. Indeed, the matrix simultaneously represents dominant flow patterns as well as successively finer, more detailed flow structures.

With the AMG we find a tool which is able to represent flow patterns in a hierarchical multi-scale fashion. AMG delivers a set of descriptions of the flow-induced coupling in terms of matrices A^l for $l = 0, \dots, L$, ranging from detailed ($A^0 = A$) to very coarse (A^L).

Let us illustrate how AMG works using two simple examples. Consider the flow fields $v_1(x) = (-1, 1)$ and $v_2(x) = (1, 1)$ on the square domain $\Omega = [-1; 1]^2 \subset \mathbb{R}^2$. We can define a simple diffusion tensor

$$a(v) = B^T \begin{pmatrix} \|v\| + \epsilon & 0 \\ 0 & \epsilon \end{pmatrix} B := B^T \begin{pmatrix} \sqrt{2} + 0.001 & 0 \\ 0 & 0.001 \end{pmatrix} B,$$

where B is a rotation of ∓ 45 degrees and we chose a small diffusion value $\epsilon = 0.001$ orthogonal to the direction of the vector field v. We need a non-zero diffusion value in this direction too. Indeed, the diffusion value couples neighbor domain points (or element nodes, in the discretized version). Having a non-zero diffusion across the field v to be clustered ensures that we shall obtain thick clusters on coarse scales, as described later. Having a relatively large diffusion along the field v ensures that these clusters will be much larger in the direction of the field than across, which is the desired result. The choice of the ϵ value is not important as long as it stays a few orders of magnitude smaller than the average value of $\|v\|$. We then consider the corresponding differential operator $\mathcal{A}[u] = \mathrm{div}(a\nabla u)$ and apply the AMG method to the matrix which results from the discretization of $\mathcal{A}$ on a regular triangulation. Figure 17 shows the coupling strengths encoded in the matrices A^l for the first three finest levels $l = 0, 1, 2$, for the fields $v_1 = (1, 1)$ and $v_2 = (-1, 1)$, using a blue-to-red colormap. For the same fields, Fig. 18 shows selected basis functions on the four coarsest decomposition levels.

For the actual flow field clustering application we consider the differential operator $\mathcal{A}[u] = \mathrm{div}(a(v)\nabla u)$ where the diffusion tensor is

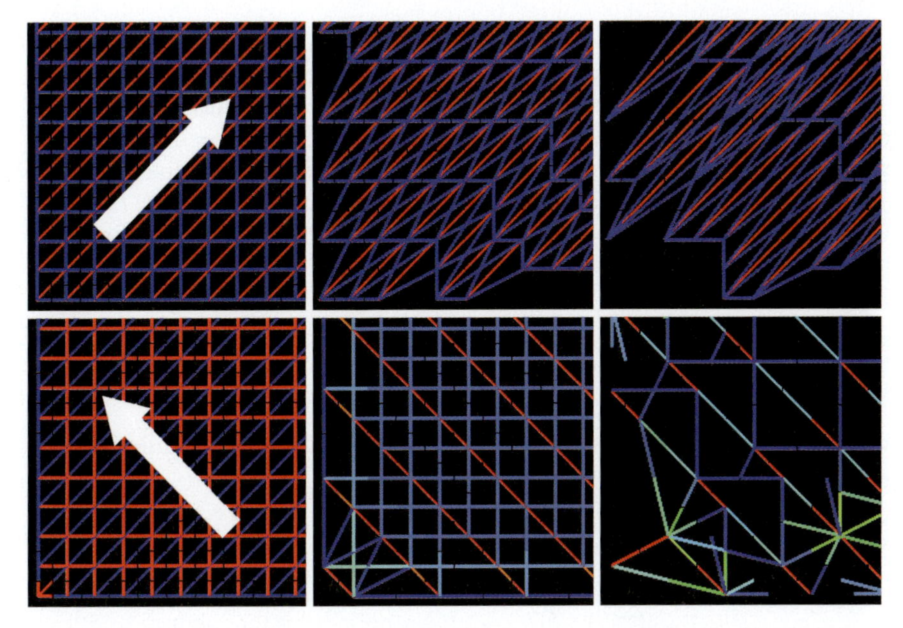

Fig. 17. Color-coded coupling strength (zoomed in) on the computational grid. Three finest levels (*left to right*) are shown for the fields $v_1 = (-1, 1)$ (*bottom row*) and $v_2 = (1, 1)$ (*top row*). The *white arrows* show the field direction

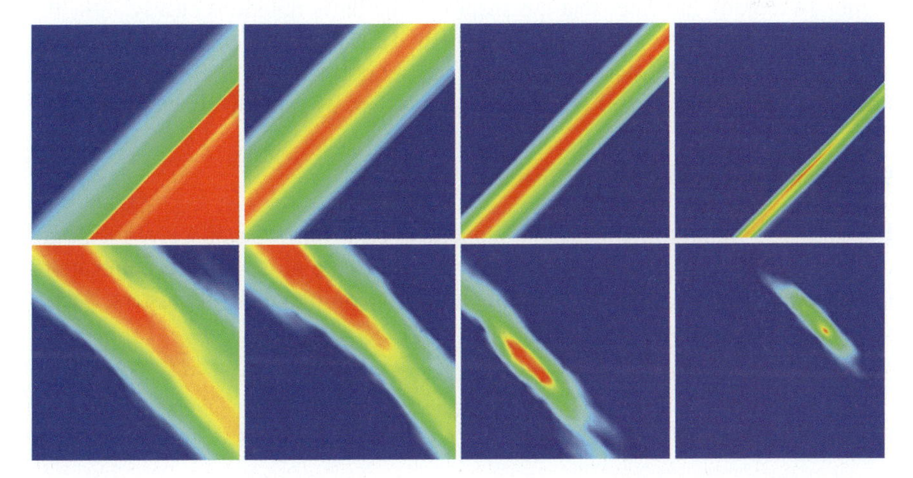

Fig. 18. For the two vector fields $v_1 = (-1, 1)$ (*bottom row*) and $v_2 = (1, 1)$ (*top row*) basis functions on the four coarsest levels are shown. Obviously the basis functions are clearly aligned to the flow field (cf. Fig. 17)

$$a(v) := B(v)^T \begin{pmatrix} \alpha(\|v\|) & 0 \\ 0 & \mathrm{Id}_{n-1} \end{pmatrix} B(v)$$

and $B(v)$ is the same rotation as above.

When we apply the AMG algorithm to the matrix $A \in \mathbb{R}^{n,n}$ corresponding to the above differential operator, we obtain a sequence of prolongation matrices $P^l \in \mathbb{R}^{n_{l-1},n_l}$ as output, where n_l for $l = 0, \ldots, L$ are the decreasing numbers of remaining unknowns and $n_0 = n$. The entries in each column $i = 1, \ldots, n_l$ of P^l give the coefficients of the linear combination of the finer basis functions $\Phi^{l-1,j}$ for $j = 1, \ldots, n_{l-1}$ corresponding to the coarser basis function $\Phi^{l,i}$ on level l. In other words, each matrix A^l delivered by the AMG, starting with the initial, finest one $A^0 = A$ down to the coarsest one A^L, approximates the fine grid operator using the (matrix-dependent) basis $\{\Phi^{l,i}\}_{i=1,\ldots,n_l}$:

$$A^l_{ij} = A\overline{\Phi^{l,i}} \cdot \overline{\Phi^{l,j}} = \int_\Omega a(v)\nabla\Phi^{l,i} \cdot \nabla\Phi^{l,j} ,$$

where $\overline{\Phi^{l,i}}$ is the nodal vector corresponding to the function $\Phi^{l,i}$, i. e. denoting the initial basis functions with Θ^j we have $\Phi^{l,i} = \sum_{j=1,\ldots,n} (\overline{\Phi^{l,i}})_j \Theta^j$.

Hence, the following simple recursive recipe can be used to calculate the multi-scale of basis functions $\Phi^{l,i}$

$$\Phi^{l,i} := \sum_{j=1,\ldots,n_{l-1}} P^l_{ji} \Phi^{l-1,j} \quad \forall i = 1, \ldots, n;\ l = 1, \ldots, L$$

$$\Phi^{0,i} := \Theta^i \quad \forall i_1, \ldots, n$$

Figure 18 already indicates that the shapes of the basis functions clearly show the strength of the local coupling. The AMG method clusters vertices along a streamline already on a fine scale, since they are strongly coupled. Vertices not aligned to the flow are clustered on coarser scales, since their coupling is relatively weaker.

9.3 From Basis Functions to Clusters

But as usual with finite elements, the supports of basis functions on a given scale are overlapping. Therefore, we need to derive a multi-scale of domain decompositions from the set of basis functions to partition the domain into disjoint clusters. Such a domain decomposition

$$\mathcal{D}^l := \{\mathcal{D}^{l,i}\}_{i=1,\ldots,n_l}$$

can easily be defined for every $l = 0, \ldots, L$ as follows:

$$\mathcal{D}^{l,i} := \{x \in \Omega \mid \Psi^{l,i}(x) \geq \Psi^{l,j}(x)\, \forall j = 1, \ldots, n_l\}$$

In other words, a domain $\mathcal{D}^{l,i}$ on level l is the set of points where the basis $\Phi^{l,i}$ is dominant on that level.

Now, several observations can be made:

- The domains on different scales need not be *strictly* spatially nested – the supports of the shape functions are, but the decomposition arising from the maximum property is not. However, the domains are clearly aligned to the flow field.

- All domains on a given level l have comparable sizes and the average domain size is reduced by a factor, roughly equal to 2, from level l to level $l+1$. These properties are inherited from the bottom-up coarsening scheme used by the AMG method.
- The clustering of the field $v_1 = (-1, 1)$ (Fig. 17 top row) is perfectly aligned with the field (cf. the basis functions in Fig. 18 top row). However, the clustering of the field $v_2 = (1, 1)$, although very similar, is less regular (Figs. 17 and 18 bottom row). This is the unavoidable impact of the underlying operator discretization (which is here a mesh containing triangles). Since v_2 is perpendicular to the initial mesh edges, this is the worst-case scenario. However, even in this case, the constructed domains are still very much aligned with the field.
- The supports of the basis functions, respectively the induced domains on a given level, do not have *exactly* the same size (area), since AMG cannot evaluate (integrate) the *mass* of the basis functions. Indeed it does not employ any *geometric* nodal information, but only a matrix of coupling strengths. However these restrictions cause no practical problems for visualizing real-world datasets.

Finally we can show the color-coded domains and in addition velocity-colored curved arrow icons (cf. [12, 35]). For every domain $\mathcal{D}^{l,i}$, we draw one such icon, using a streamline seeded at the point where the corresponding basis is maximum. Figure 19 shows the decomposition of a magneto-hydrodynamics (MHD) flow dataset.

9.4 Clustering 3D Flow Fields

Our method works identically for 3D (volumetric) vector fields. The only difference is the use of tetrahedral, instead of triangular, meshes. However, direct visualization of a color-coded domain decomposition, as in the 2D case, is not effective due to the volumetric occlusion. Hence, we use a few post-processing steps. For every domain $\mathcal{D}^{l,i}$ on a given level l, we construct a closed triangle mesh that bounds $\mathcal{D}^{l,i}$. Next, we smooth these meshes using, e.g. a Laplacian filter or a windowed sinc filter [30]. As a result, the meshes become slightly smaller, which allows us to better separate them visually. Next, we implement an interactive navigation scheme in which domains $\mathcal{D}^{l,i}$ can be made half or completely transparent by a mouse click. Users can interactively "carve" into the flow volume to, e.g. remove uninteresting areas and bringing inner flow structures into sight, see Fig. 20. Alternatively, we can visualize the flow at a given level of detail using the same colored arrow glyphs as in the 2D case. Figure 20 shows the first three coarsest decomposition levels of a 3D helix flow and of a 3D laminar flow with $v = (1, 1, 1)$ respectively. We use the interactive technique sketched above to remove the outer domains and to expose the more interesting inner flow structure. The remaining smoothed domains are shown in the top row of Fig. 20 for the helix flow and in the bottom trow of Fig. 20 for the laminar flow (compare the latter with the 2D field in Fig. 17). In the center row of Fig. 20 the same domains as in the top row are shown, but this time half transparent and equipped with an arrow icon.

Fig. 19. For the vector field from a magneto-hydrodynamics simulation (MHD) shown in Figs. 4 and 5 the hierarchical decomposition is shown. *From top to bottom* the clusters on the five coarsest levels are indicated with a color-coding (*left*). The origins of the cluster-corresponding basis functions serve as the starting point for the integration of trajectories (*right*)

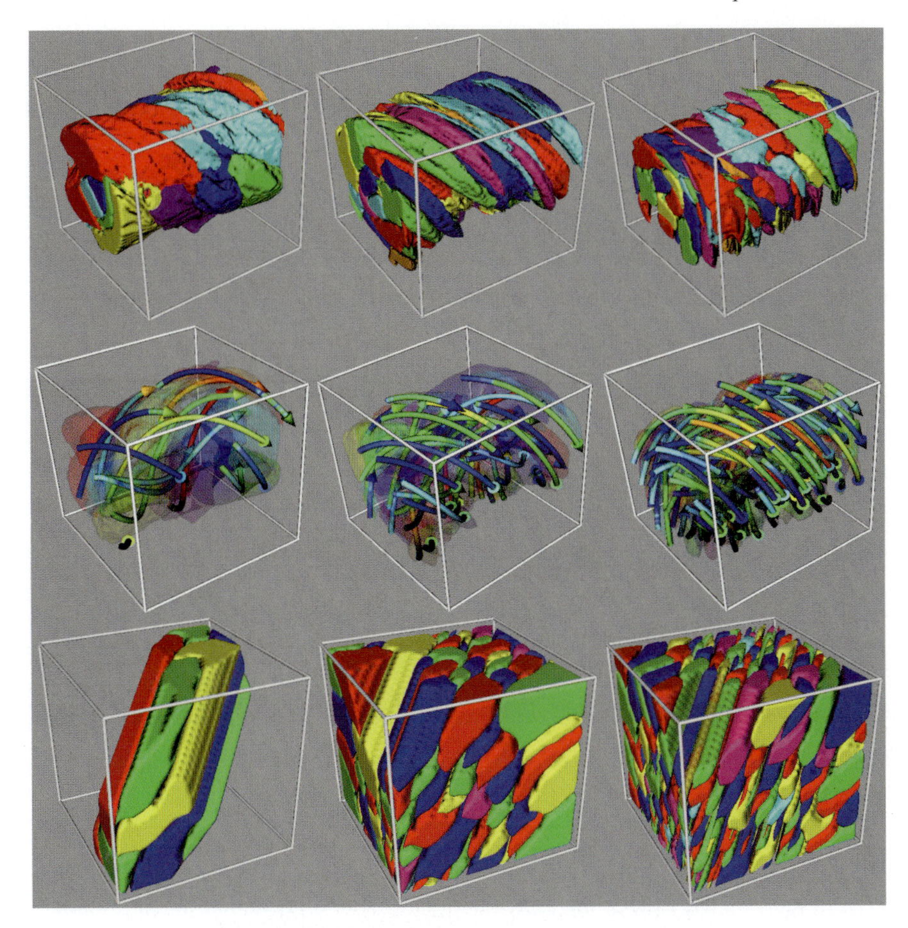

Fig. 20. Helix flow, selected domains (*top row*), half-transparent domains with *arrow icons* (*middle row*). Diagonal flow, selected domains (*bottom row*)

Finally, we consider the incompressible flow in a water basin with two interior walls, an inlet and an outlet – the same dataset as shown in Fig. 7. Figure 21 shows several multi-scale levels, visualized with curved arrow icons. These images show that our method scheme works in 3D just as well as in 2D.

9.5 Clustering Vector Fields on 2D Surfaces

For the clustering approach we considered vector fields on Euclidean domains so far. Since we have seen in Sect. 5.3 that we can extend to differential operator towards surfaces, we can apply the same generalization to the AMG clustering as well. Again we replace the Euclidian gradient and divergence operators by their geometric counter parts and apply the AMG to a discretization of

$$\mathcal{A}[u] := -\mathrm{div}_{\mathcal{M}}(A\nabla_{\mathcal{M}} u).$$

Fig. 21. For the water basin dataset (cf. Fig. 7) we show the three coarsest levels of the hierarchical decomposition

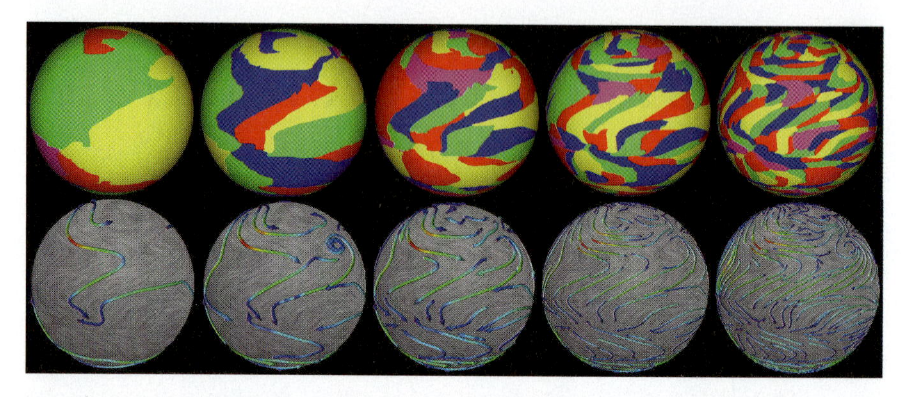

Fig. 22. Climate dataset decomposition, five coarsest levels (*left to right*). Domains (*top row*) and flow texture overlaid with *curved arrow icons* (*bottom row*)

The finite element discretization is now completely analogous to the above Euclidean case. In fact, we use exactly the same code for all our applications. We approximate the surface $\mathcal{M}$ by a triangulation $\mathcal{M}_h$ and compute in the same way as on flat domains the stiffness matrix A corresponding to the operator $\mathcal{A}$.

As an illustration, we show the multi-scale decomposition of the average wind stress field on the surface of the Earth in Fig. 22 (the dataset is taken from [42]). The flow texture in the bottom row was produced with the IBFV method described in [42].

10 Conclusions

We have presented an overview of flow field visualization methods using partial differential equations (PDEs). These methods cover a broad area between pure simulation of physical processes based on such equations, and pure post-processing of such simulation data obtained by other techniques. PDE-based visualization methods have a number of strong advantages. First, they are dense methods that produce visualizations where every point of the considered spatial domain is represented. This naturally associates them with, and brings them close to, texture-based visualization methods. Second, PDE-based visualizations can naturally target any spatial dimension, e.g. from 2D to time-dependent 3D datasets, in a uniform modeling and computational framework. Third, one can use existing, well-proven and well understood numerical techniques to solve the underlying PDE discretizations, yielding an overall robust approach to data visualization. Fourth, many such methods have a natural support of the multiscale notion, being able to capture and represent flow data details at different spatial scales. Often, time serves as the parameter controlling the scale. However, probably the most attractive aspect of PDE-based visualizations is their ability to model a wide range of phenomena and data enhancement operations, ranging from simple filtering to sophisticated multiscale feature-preserving decomposition techniques, in a well-founded mathematical way, that leads to novel and insightful visualizations.

References

1. L. Alvarez, F. Guichard, P.-L. Lions, and J.-M. Morel. Axioms and fundamental equations of image processing. *Arch. Ration. Mech. Anal.*, 123 (3):199–257, 1993.
2. A. Brandt. Algebraic multigrid theory: the symmetric case. In *Preliminary Procs. of the Intl. Multigrid Conf.*, Copper Mountain, Colorado, April 1983.
3. A. Brandt, S. F. McCormick, and J. W. Ruge. Algebraic multigrid for sparse matrix equations. In D. J. Evans, editor, *Sparsity and Its Applications*. Cambridge University Press, Cambridge, 1984.
4. B. Cabral and L. Leedom. Imaging vector fields using line integral convolution. In J. T. Kajiya, editor, *Computer Graphics (SIGGRAPH '93 Proceedings)*, volume 27, pages 263–272, Aug. 1993.
5. J. Cahn and J. Hilliard. Free energy of a non-uniform system I. Interfacial free energy. *J. Chem. Phys.*, 28:258–267, 1958.
6. W. de Leeuw. Divide and conquer spot noise. In *Proceedings of Supercomputing 97 (CD-ROM)*. ACM SIGARCH and IEEE, 1997.
7. W. de Leeuw and R. van Liere. Multi-level topology for flow visualization. *Comput. Graph.*, 24(3):325–331, 2000.
8. W. C. de Leeuw and J. J. van Wijk. Enhanced spot noise for vector field visualization. In *Proceedings Visualization '95*, 1995.
9. U. Diewald, T. Preusser, and M. Rumpf. Anisotropic diffusion in vector field visualization on euclidean domains and surfaces. *IEEE Trans. Vis. Comput. Graph.*, 6(2):139–149, 2000.
10. C. M. Elliott. The Cahn–Hilliard model for the kinetics of phase separation. *Num. Math.*, pages 35–73, 1988.
11. L. Forssell. Visualizing flow over curvilinear grid surfaces using line integral convolution. In *Proceedings IEEE Visualization '94*, pages 240–246, 1994.
12. H. Garcke, T. Preusser, M. Rumpf, A. Telea, U. Weikard, and J. J. van Wijk. A phase field model for continuous clustering on vector fields. *IEEE TVCG*, 7(3):230–241, 2000.
13. M. Griebel, A. Schweitzer, T. Preusser, M. Rumpf, and A. Telea. Flow field clustering via algebraic multigrid. In *Proceedings of IEEE Visualization 2004*, 2004.
14. B. Heckel, G. Weber, B. Hamann, and K. I. Joy. Construction of vector field hierarchies. *Proceedings IEEE Visualization '99*, pages 19–25, 1999.

15. J. L. Helman and L. Hesselink. Visualizing Vector Field Topology in Fluid Flows. *IEEE Comput. Graph. Appl.*, 11(3):36–46, 1991.
16. V. Interrante and C. Grosch. Stragegies for effectively visualizing 3D flow with volume LIC. In *Proceedings IEEE Visualization '97*, pages 285–292, 1997.
17. B. Jobard, G. Erlebacher, and Y. M. Hussaini. Lagrangian-eulerian advection for unsteady flow visualization. In *Proceedings of IEEE Visualization '01*, San Diego, CA, October 2001.
18. B. Jobard and W. Lefer. Creating evenly-spaced streamlines of arbitrary density. In W. Lefer, M. Grave (eds.), *Visualization in Scientific Computing '97*, 1997.
19. B. Jobard and W. Lefer. The motion map: Efficient computation of steady flow animations. In *Proceedings IEEE Visualization '97*, pages 323–328, 1997.
20. R. S. Laramee, H. Hauser, H. Doleisch, B. Vrolijk, F. H. Post, and D. Weiskopf. The state of the art in flow visualization: Dense and texture-based techniques. *Comput. Graph. Forum*, 23(2):203–221, 2004.
21. R. S. Laramee, D. Weiskopf, J. Schneider, and H. Hauser. Investigating swirl and tumble flow with a comparison of visualization techniques. In *Proceedings of IEEE Visualization '04*, pages 51–58, 2004.
22. H. Löffelmann, T. Kuνcera, and E. Gröller. Visualizing Poincaré maps together with the underlying flow. In *Mathematical Visualization*, pages 315–328. Springer, Berlin, 1998.
23. N. Max and B. Becker. Flow visualization using moving textures. In *Proceedings of the ICASE/LaRC Symposium on Visualizing Time-Varying Data*, pages 77–87, 1995.
24. N. Max, R. Crawfis, and C. Grant. Visualizing 3D Velocity Fields Near Contour Surface. In *Proceedings IEEE Visualization '94*, pages 248–254, 1994.
25. A. Novick-Cohen. The Cahn–Hilliard equation: mathematical and modelling perspectives. *Adv. Math. Sci. Appl.*, 8:965–985, 1998.
26. P. Perona and J. Malik. Scale space and edge detection using anisotropic diffusion. *IEEE Trans. Pattern Anal. Mach. Intell.*, 12:629–630, 1990.
27. C. Rezk-Salama, P. Hastreiter, C. Teitzel, and T. Ertl. Interactive exploration of volume line integral convolution based on 3D texture mapping. In *Proceedings IEEE Visualization '99*, pages 233–240, 1999.
28. J. W. Ruge and K. Stüben. Efficient Solution of Finite Difference and Finite Element Equations by Algebraic Multigrid. In D. J. Paddon and H. Holstein, editors, *Multigrid Methods for Integral and Differntial Equations*, The Institute of Mathematics and its Applications Conference Series. Clarendon, Oxford, 1985.
29. R. Wegenkittl and E. Gröller. Fast oriented line integral convolution for vector field visualization via the internet. In *Proceedings IEEE Visualization '97*, pages 309–316, 1997.
30. W. Schroeder, K. Martin, and B. Lorensen. *The Visualization Toolkit: An Object-Oriented Approach to 3D Graphics*. Prentice Hall, New Jersey, 1995.
31. H. W. Shen and D. L. Kao. UFLIC: a line integral convolution algorithm for visualizing unsteady flows. In *Proceedings IEEE Visualization '97*, pages 317–323, 1997.
32. D. Stalling and H.-C. Hege. Fast and resolution independent line integral convolution. In *SIGGRAPH 95 Conference Proceedings*, pages 249–256. ACM SIGGRAPH, Addison-Wesley, New York, 1995.
33. D. Stalling, M. Zöckler, and H.-C. Hege. Fast display of illuminated field lines. *IEEE Trans. Vis. Comput. Graph.*, 3(2), 1997. ISSN 1077-2626.
34. A. Telea and J. J. van Wijk. 3D IBFV: Hardware-accelerated 3d flow visualization. *Proceedings IEEE Visualization '03*, pages 223–240, 2003.
35. A. C. Telea and J. J. van Wijk. Simplified representation of vector fields. *Proceedings IEEE Visualization '99*, pages 35–42, 1999.

36. V. Thomée. *Galerkin – Finite Element Methods for Parabolic Problems*. Springer, Berlin, 1984.
37. U. Trottenberg, C. W. Osterlee, and A. Schüller. *Multigrid*, Appendix A: An Introduction to Algebraic Multigrid by K. Stüben, pages 413–532. Academic, San Diego, 2001.
38. G. Turk and D. Banks. Image-guided streamline placement. In *Computer Graphics (SIGGRAPH '96 Proceedings)*, 1996.
39. T. Urness, V. Interrante, I. Marusic, E. Longmire, and B. Ganapathisubramani. Effectively visualizing multi-valued flow data using color and texture. In *Proceedings IEEE Visualization '03*, pages 115–122, 2003.
40. J. J. van Wijk. Spot noise-texture synthesis for data visualization. In T. W. Sederberg, editor, *Computer Graphics (SIGGRAPH '91 Proceedings)*, volume 25, pages 309–318, July 1991.
41. J. J. van Wijk. Image based flow visualization. *Computer Graphics (Proc. SIGGRAPH '02), ACM, pp. 263–279*, 2001.
42. J. J. van Wijk. Image based flow visualization for curved surfaces. *Proceedings IEEE Visualization '03*, pages 123–130, 2003.
43. R. Wegenkittl, E. Gröller, and W. Purgathofer. Animating flow fields: Rendering of oriented line integral convolution. In *CA '97: Proceedings of the Computer Animation*, pages 15–21, 1997.
44. J. Weickert. *Anisotropic diffusion in image processing*. Teubner, Stuttgart, 1998.
45. D. Weiskopf and G. Erlebacher. Flow visualization overview. In *Handbook of Visualization*, pages 261–278. Elsevier, Amsterdam, 2005.
46. D. Weiskopf, G. Erlebacher, and T. Ertl. A texture-based framework for spacetime-coherent visualization of time-dependent vector fields. In *Proceedings IEEE Visualization '03*, pages 107–114, 2003.
47. D. Weiskopf, G. Erlebacher, M. Hopf, and T. Ertl. Hardware-accelerated lagrangian-eulerian texture advection for 2D flow visualizations. In *Proceedings of the Vision Modeling and Visualization Conference 2002 (VMV-02)*, pages 439–446, 2002.

Iterative Twofold Line Integral Convolution for Texture-Based Vector Field Visualization

Daniel Weiskopf

Institute of Visualization and Interactive Systems, University of Stuttgart
weiskopf@vis.uni-stuttgart.de

Summary. Iterative twofold convolution is proposed as an efficient high-quality two-stage filtering method for dense texture-based vector field visualization. The first stage employs a compact filter, evaluated via Lagrangian particle tracing. This stage facilitates a flexible design of filters and is a means of avoiding numerical diffusion. The second stage uses semi-Lagrangian texture advection with iterative alpha blending to efficiently implement a large-scale exponential filter. A discussion of frequency-space properties and adequate sampling rates shows that this order of convolution operations permits large integration step sizes without loss of quality. Twofold convolution can be used for steady and unsteady vector fields, dye and noise advection, as well as vector fields on flat manifolds or curved surfaces. The proposed approach is prepared for an efficient GPU implementation to achieve interactive visualizations.

1 Introduction

Vector field visualization plays an important role in computer graphics and in various scientific and engineering disciplines alike. For example, the analysis of computational fluid dynamics (CFD) simulations in the aerospace and automotive industries relies on effective visual representations. Another field of application is the visualization of surface shape by emphasizing principal curvature vector fields [6].

This paper focuses on texture-based vector field representations that densely cover the domain with particle traces. The fundamental idea is to introduce high correlation along particle traces by filtering an input noise texture along these traces, which can, e.g., be achieved by Line Integral Convolution (LIC) [2]. This correlation along lines is needed for the human observer to be able to recognize them.

Filtering is an important operation in various applications. Texture filtering [8] is a typical example in the field of computer graphics. Filtering techniques, in general, have to address the following issues: (1) quality and properties of the filter, (2) accuracy of the computation, and (3) efficiency of the computation. This paper covers all three issues in the specific context of convolution along curves.

The starting point for the discussion is iterative alpha blending, which is often used for interactive vector field visualization [10, 19]. This kind of filtering method just operates on data for the current time step and, therefore, is very fast and

T. Möller et al. (eds.), *Mathematical Foundations of Scientific Visualization, Computer Graphics, and Massive Data Exploration*, Mathematics and Visualization,
DOI: 10.1007/978-3-540-49926-8,

memory-friendly, which is particularly important for time-dependent vector fields. However, iterative alpha blending is affected by rather bad quality due to the restriction to an exponential filter kernel (see Appendix A) and numerical diffusion from semi-Lagrangian advection [21]. In contrast, direct convolution with completely Lagrangian particle tracing [22] is not subject to these problems but is much more costly.

The objective of this paper is to combine the performance benefits of iterative alpha blending with the high quality and flexibility of Lagrangian particle tracing. This goal is achieved by introducing iterative twofold convolution, which consists of two convolution stages. The first stage applies a user-specified compact filter kernel, based on Lagrangian particle tracing. The subsequent iterative alpha blending works on the prefiltered texture and is responsible for a large-scale exponential filter.

After briefly reviewing previous work in Sect. 2, a continuous description of the filter process is presented, including a discussion of filters in frequency space (Sect. 3). Section 4 focuses on a discrete numerical solution and pays special attention to sampling aspects in order to avoid aliasing artifacts. It is shown that the idea of twofold convolution can be used for steady and unsteady vector fields, dye and noise advection, as well as vector fields on flat manifolds or curved surfaces (Sect. 5). A detailed discussion of performance behavior, memory footprint, and obtained visualization quality follows in Sect. 6. Section 7 demonstrates that the proposed approach is prepared for an efficient GPU implementation to achieve interactive visualizations. The paper closes with a short conclusion and an outlook on possible future work.

2 Previous Work

A large body of research has been published on noise-based and dense vector field visualization. Comprehensive presentations can be found in the review articles by Sanna et al. [15], Laramee et al. [11], and Erlebacher et al. [4]. Spot noise [18] and Line Integral Convolution (LIC) [2] are early texture-synthesis techniques for dense flow representations and serve as the basis for many subsequent papers that provide a variety of extensions and improvements to these original methods. Many recent techniques are based on the closely related concept of texture advection, of which the basic idea is to represent a dense collection of particles in a texture and transport this texture along the vector field [13]. Lagrangian–Eulerian Advection (LEA) [10] employs a Lagrangian integration of particle positions and an Eulerian advection of particle colors. Image Based Flow Visualization (IBFV) [19] is a variant of 2D texture advection in which a second texture is blended into the advected texture at each time step. While LEA and IBFV employ iterative alpha blending to compute the line integral convolution with an exponential filter, a permanent recomputation of the complete integral is used within a spacetime framework for texture-based vector field visualization [22].

Jobard et al. [10] combine the LEA method with LIC postfiltering to remove aliasing artifacts. Similarly, Okada and Lane [14] apply LIC twice in the form of "double LIC." Both articles contain the same basic idea as this paper – the subsequent application of two convolution steps. However, they focus on other aspects of vector field visualization and, therefore, do not discuss the mathematical background, the quality issues of filter design, and efficiency improvements, as it is done in this paper. In another related paper, Hege and Stalling [9] express the convolution integral as a linear combination of repeated integrals to derive a general Fast LIC algorithm for piecewise polynomial filter kernels.

The performance and quality of filtering has been extensively investigated in the context of texture filtering; see, e.g., the survey by Heckbert [8]. Particularly interesting is the efficient computation of convolution with large filter kernels, without the time-consuming transformation to and from Fourier space. Affine filters, for example, can be realized by repeated integration [7], which generalizes the idea of summed-area tables [3]. A related method implements large linear filters by adding up the translated outputs of sum-box filters [16]. Unfortunately, most optimization strategies for 2D texture filtering cannot be directly applied to convolution along curved lines because crucial underlying assumptions cannot be met.

3 Continuous Twofold Convolution Along Straightened Lines

Line integral convolution [2, 17] evaluates the integral

$$D(\mathbf{r}) = \int_{s_0-L_s}^{s_0+L_e} \tilde{k}(s - s_0) T(\boldsymbol{\sigma}(s))\, \mathrm{d}s \quad , \tag{1}$$

to compute a gray-scale value in the visualization image D at position $\mathbf{r}$. Vectors are generally denoted by boldface letters and, typically, exist in 2D or 3D Cartesian space. The filter kernel is described by $\tilde{k}(s)$ and has support $[-L_s, L_e]$. The streamline is parameterized by arc length s and yields the position $\boldsymbol{\sigma}(s_0) = \mathbf{r}$. A streamline is computed from a given vector field $\mathbf{u}$ by solving the ordinary differential equation

$$\frac{\mathrm{d}\boldsymbol{\sigma}(\tau)}{\mathrm{d}\tau} = \mathbf{u}(\boldsymbol{\sigma}(\tau)) \quad , \tag{2}$$

with the initial condition $\boldsymbol{\sigma}(\tau_0) = \mathbf{r}$. In general, τ does not provide an arc-length parametrization. However, τ can be transformed to an arc-length parameter s by reparametrization. Arc-length parametrization can be directly obtained from the alternative definition of a streamline according to

$$\frac{\mathrm{d}\boldsymbol{\sigma}(s)}{\mathrm{d}s} = \frac{\mathbf{u}(\boldsymbol{\sigma}(s))}{|\mathbf{u}(\boldsymbol{\sigma}(s))|} \quad , \tag{2a}$$

with the initial value $\boldsymbol{\sigma}(s_0) = \mathbf{r}$. Here, the streamline is traced in a normalized vector field.

For the time being, the original definition of line integral convolution is reformulated to facilitate a direct application of Fourier and convolution theory. A continuous vector field can be straightened in a neighborhood of $\mathbf{r}$ if there is no critical point at $\mathbf{r}$ [5]. The idea of straightening is adopted for single streamlines that are treated individually and independently. It is assumed that the straightening process results in a continuous deformation of the original streamline so that $\boldsymbol{\sigma}(s)$ becomes a straight line along the x axis, still parameterized by arc length. When this straightening is feasible for the complete domain of the line integral, the original equation (1) can be rewritten as

$$d(x) = \int_{x-L_s}^{x+L_e} \tilde{k}(\xi - x) t(\xi) \, \mathrm{d}\xi \quad .$$

The lower-case letters d and t indicate the analogs of the respective upper-case terms from (1) after straightening. By replacing

$$k(x) = \tilde{k}(-x) \quad ,$$

and thanks to the finite support of $k(x)$, the LIC computation is obtained as 1D convolution:

$$d(x) = [t * k](x) = \int_{-\infty}^{\infty} k(x - \xi) t(\xi) \, \mathrm{d}\xi \quad . \tag{3}$$

Twofold convolution applies a second convolution operation to an already filtered image. Since convolution is associative, twofold convolution with kernels k and κ,

$$d = (t * k) * \kappa = t * (k * \kappa) \quad ,$$

can be regarded as a single convolution of the input texture t with the combined kernel $\hat{k} = (k * \kappa)$.

In this paper, the second-stage convolution is always based on iterative alpha blending and, therefore, the filter kernel κ is fixed to an exponential function. As shown in Appendix A and [4], the first-order approximation of the normalized exponential filter $\beta \exp(-\beta x)$ corresponds to an iterative alpha blending with $\alpha = \beta \Delta x$, where Δx is the discretization step length along the x axis. The exponential filter is defined for all positive values x and can be described by the Fourier transform pair

$$f_{\exp}(x) = e^{-2\pi \nu_0 x} H(x) \quad \longleftrightarrow \quad F_{\exp}(\nu) = \frac{1}{2\pi} \frac{\nu_0}{\nu^2 + \nu_0^2} + \frac{-\mathrm{i}}{2\pi} \frac{\nu}{\nu^2 + \nu_0^2} \quad , \tag{4}$$

with $2\pi \nu_0 = \beta$, and without normalization. $H(x)$ is the Heaviside function. Appendix B derives the Fourier transform of the exponential filter. In general, upper-case letters denote a function in frequency space and lower-case letters in the time domain. The complex modulus of $F_{\exp}(\nu)$ – its power spectrum – is

$$|F_{\exp}(\nu)| = \frac{1}{2\pi} \frac{1}{\sqrt{\nu^2 + \nu_0^2}} \quad . \tag{5}$$

Figure 1 illustrates modulus, real, and imaginary parts of the Fourier transform of the exponential function. Note that the real part of $F_{\exp}(\nu)$ is a Lorentz function.

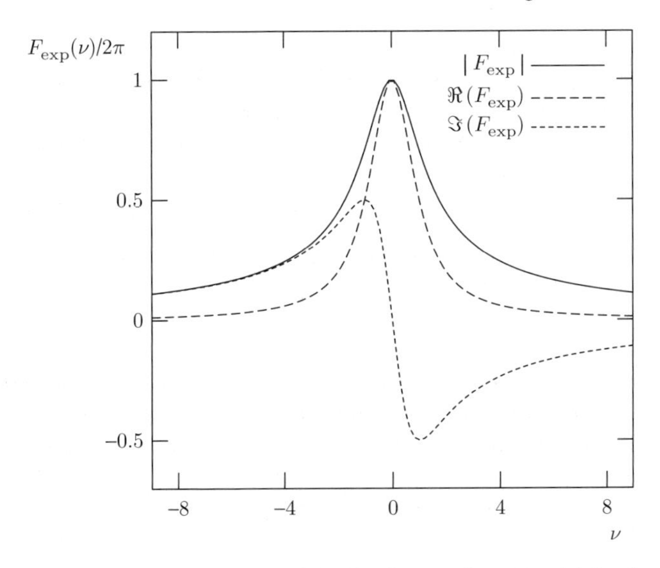

Fig. 1. Modulus, real, and imaginary parts of the Fourier transform $F_{exp}(\nu)$ for the exponential filter $f_{exp}(x) = \exp(-2\pi \nu_0 x) H(x)$ with $\nu_0 = 1$

The frequency domain is adequate to assess the quality of a filter kernel in the context of dense vector field visualization. The fundamental idea behind any dense representation is to introduce high correlation along streamlines and no correlation perpendicular to streamlines. In this way, information about different streamlines is optimally encoded – in the form of missing correlation between these lines. In contrast, correlation along lines is needed for the human observer to be able to recognize these curves.

The convolution equation (3) does not change the frequency spectrum perpendicular to the x axis, i.e., the spectrum of the input texture remains invariant for any direction that is perpendicular to the x direction. Therefore, the input texture should contain high frequencies to achieve uncorrelated neighboring streamlines. The input texture is usually generated as band-limited noise to avoid aliasing artifacts in a discrete setting (see Sect. 4). The convolution process targets a removal of high frequencies – it serves as a low-pass filter along streamlines. The convolution theorem relates convolution in the time domain to multiplication in frequency space,

$$f(x) * g(x) \quad \longleftrightarrow \quad F(\nu)\, G(\nu) \quad .$$

Although the input texture t is band-limited, it usually contains a rather high maximum frequency. A final low-frequency image can only be obtained with a filter kernel that has a sufficiently fast fall-off in frequency space.

Twofold convolution provides flexibility in choosing the filter function k and the combined kernel $\hat{k}$. This is important because the exponential kernel κ has a rather slow fall-off characteristic: Its power spectrum, (5), diminishes inversely with frequency only to the first power. Therefore, methods that only employ an exponential filter retain much high-frequency noise along streamlines. An important benefit of twofold convolution is that the additional kernel k can be chosen to design an overall filter with a better fall-off behavior. The Bartlett (or triangular) window, which has finite support in the time domain, is a typical example for such a filter kernel. The Fourier transform pair of the normalized Bartlett function of total width $2w$ is

$$f_{\mathrm{bartlett}}(x) = \begin{cases} \frac{w-|x|}{w^2} & \text{if } |x| \le w \\ 0 & \text{if } |x| > w \end{cases} \quad \longleftrightarrow \quad F_{\mathrm{bartlett}}(\nu) = \mathrm{sinc}^2(w\nu) \quad ,$$

with $\mathrm{sinc}(x) = \sin(\pi x)/(\pi x)$ [1].

The power spectrum of the Bartlett window attenuates inversely with frequency to the second power. Therefore, the Fourier transform of the combined filter, which is the product of the Fourier transforms of the exponential and Bartlett functions, has an even more pronounced fall-off. Other kernels, such as Hamming or Gaussian windows, could also be applied to reduce high frequencies. All these filters have in common a continuous kernel function. In contrast, the exponential filter has a discontinuous onset at its origin, which contributes much high-frequency energy. A purely iterative computation of the exponential filter, however, does not permit to change this fact.

Actual filter design can be performed in two different ways. The three filter kernels are related to each other according to $\hat{k} = (k * \kappa)$. Only κ is fixed to an exponential function. Either the overall filter $\hat{k}$ can be specified, or the first-stage convolution filter k. From a user-given $\hat{k}$, k is best determined in frequency space, in which convolution corresponds to an easily invertible multiplication.

4 Discretization and Sampling

A vector field visualization algorithm usually discretizes the continuous convolution process to compute images numerically. The discretization typically comprises two elements: (1) the approximation of the convolution integral by a Riemann sum, and (2) an explicit integration scheme for (2) to compute streamlines $\sigma(s)$ as basis for straightening.

We first address the approximation of the integral according to the Riemann sum

$$d(x) \approx \sum_j k(x - \xi_j) t(\xi_j)\, \Delta\xi \quad .$$

Here, a constant sampling rate $\Delta\xi$ is assumed so that the sampling positions are equidistant, i.e., $\xi_j = j\Delta\xi$. The summation involves a sampling of the input texture and, therefore, the sampling theorem has to be applied to ensure an adequate sampling rate. According to the sampling theorem, the sampling frequency has to be at least the Nyquist frequency $\nu_{\mathrm{nyquist}} = 2\nu_{\max}$, where $\nu_{\max}$ is the maximum frequency of the band-limited signal.

Twofold convolution first applies the filter k to the input texture t, and then the filter κ to the result of the first convolution. Therefore, the first convolution step is governed by the Nyquist frequency $2\nu_t$, where ν_t is the maximum frequency of t. The input texture t is typically created in a preprocessing step, based on accurate filtering. Therefore, ν_t should be well known, and the sampling rate for the first convolution can be set to $\Delta\xi = 1/(2\nu_t)$.

The second convolution, implemented as iterative alpha blending, applies the filter κ to the result of the first convolution stage, $\hat{t} = t * k$. The maximum frequency of this input data is given by the minimum of ν_t and ν_k, where ν_k is the maximum frequency of the kernel function k. The kernel k is typically chosen so that ν_k is much smaller than ν_t. In practice, ν_k is often set to the frequency at which the contribution to the Fourier transform K becomes negligible. Accordingly, the sampling distance for the second convolution computation can be set to $\Delta x = 1/(2\nu_k)$, which is larger than for the first convolution step. Twofold convolution allows us to use larger integration steps for iterative alpha blending when compared to previous methods that rely on the exponential filter only. To achieve precise results, the actual blending equation is not based on standard alpha blending, $d(x) = \alpha\hat{t}(x) + (1-\alpha)d(x-\Delta x)$, but on the more accurate expression

$$d(x) = \alpha\hat{t}(x) + e^{-\alpha}d(x - \Delta x) \quad , \text{ with } \alpha = \beta\Delta x \quad , \tag{6}$$

which is derived in Appendix A. The data flow for the complete twofold convolution process is depicted in Fig. 2.

The other numerical issue concerns the explicit integration scheme that solves the ordinary differential equation (2) to determine streamlines and the corresponding straightening. When taking into account the straightening of the actual nD vector field, the term $(x - \Delta x)$ in (6) corresponds to the backward integration of a particle trace for one integration step. The often-used Euler integration just needs the vector field at the current time step (this is true even for time-dependent data, see Sect. 5). The evaluation of $d(x - \Delta x)$ corresponds to accessing the display texture d at the previous time step. Since the display texture is stored in discrete form on a uniform grid (i.e., a texture) and $(x - \Delta x)$ usually does not correspond to grid point positions, the access requires an interpolation scheme. Tensor-product linear interpolations are usually employed for resampling.

The great advantage of this semi-Lagrangian advection scheme is that the display texture d, the input texture $\hat{t}$, and the vector field need to be stored only for the current

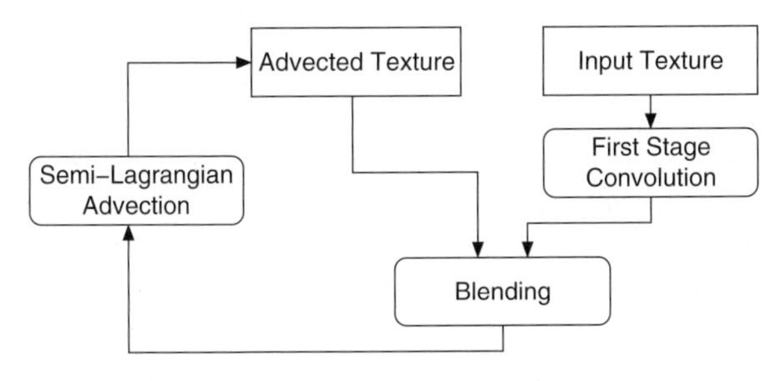

Fig. 2. Dataflow for twofold convolution

integration step. Previous computations can be overridden and data from earlier time steps can be released from memory. A major disadvantage is numerical diffusion, caused by a continuous resampling of the display field [21].

The degree of artificial diffusion increases with the number of resampling operations. Twofold integration significantly reduces this artifact by decreasing the number of integration and resampling steps during iterative alpha blending. Furthermore, direct convolution for the first convolution stage has to evaluate all previous positions in t along the support of the filter kernel k and, therefore, anyway requires a completely Lagrangian integration, which is not affected by numerical diffusion.

Another beneficial feature of twofold integration is the possibility to reuse results of streamline computation. The direct convolution for the first stage covers a finite distance in space – the support of k – in which several finely spaced streamline locations are computed. The second convolution with the exponential filter just needs a single backward particle-tracing step. The extent of this backward integration lies within already computed streamline positions from the first convolution.

The Bartlett window can be used to illustrate this behavior. The Fourier transform of the Bartlett window of total width $2w$ has its first zero crossing at $\nu_{\mathrm{bartlett}} = 1/w$. Assuming ν_{bartlett} as (approximate) maximum frequency, the integration step size for the exponential filter should be $\Delta x = w/2$. This is one half of the maximum distance that has been computed during the Lagrangian integration, which covers a distance w in positive and negative directions.

Other compact filters k have a similar qualitative behavior. In any case, the accurately computed streamlines from the first convolution stage can be directly used for an accurate semi-Lagrangian integration of the second convolution.

5 Extended Scenarios

This section extends the fields of application of twofold convolution beyond the visualization of LIC-like streamlines.

The first extension eliminates the restriction to an arc-length parametrization of streamlines. For example, the magnitude of the vector field can be used as a measure for the velocity of the parametrization in order to visualize different speeds by different lengths of streamline streaks. This extension does not require any modification of the actual convolution methods. Only particle tracing is slightly changed to take into account a varying spacing of streamline positions. A side effect is that the spatial frequency of the input texture t is modified by straightening. The maximum frequency after straightening should be used to determine the sampling distance during convolution in order to avoid aliasing.

This sampling distance can be easily computed for a velocity-based streamline parametrization. A single Euler step of streamline integration covers a length $h\,|\mathbf{u}|$, where $\mathbf{u}$ is the current vector field value and h the Euler step size. Let $u_{\max}$ be the maximum magnitude of the vector field. Then, $hu_{\max}$ is the maximum integration distance, which must not exceed the reciprocal of the Nyquist frequency of the input texture, ν_{nyquist}. Consequently, the Euler step size should be chosen as $h = 1/(\nu_{\text{nyquist}} u_{\text{max}})$ to avoid aliasing. An analogous reasoning is possible for any other particle tracing scheme.

Critical points of a vector field are another aspect that has not been covered so far. Streamline straightening breaks down at critical points because streamline integration cannot cross these points. A possible solution would be to stop the complete integration and convolution process at critical points, but this solution has not yet been implemented.

Another extension includes the visualization of time-dependent vector fields. IBFV [19] and LEA [10] produce streakline-like structures for unsteady flow fields [4]. Analogously, twofold convolution can be readily applied to time-dependent vector fields. The only modification concerns the computation of convolution lines. Here, the time dependency of the vector field has to be taken into account to compute streaklines instead of streamlines. A closely related extension allows for time-dependent input (noise) textures. Their time dependency has to be considered during the first convolution stage by sampling at adequate spatial and temporal locations.

A fundamental problem of streakline-based convolution is the fact that streaklines can cross each other. At a fixed time, a neighboring point B on a streakline that ends at point A may be the ending point for a completely different streakline. For example, a rotating uniform 2D flow $(\cos\tau, \sin\tau)$ (where τ is time) leads to circular streaklines of constant radius. These circular streaklines exist at any point of the domain and, therefore, densely cross each other. In this case, the model of correlation along curves breaks down because neighboring points on curves do not receive a related value from the convolution integral. This problem affects not only twofold convolution, but any streakline-oriented noise-based vector field visualization. A good solution is to reduce the density of the visual representation similarly to Oriented LIC (OLIC) [20]. A sparse input texture leads to sparse and individually distinguishable streaklines after convolution. In this case, the display of a streakline is not "diluted" by neighboring texels. This approach is particularly effective for short streaklines – streaklets – which reduce the probability of mutual crossing.

Dye advection is an example for an extremely sparse vector field representation. Similarly to dye advection in IBFV and LEA, either a very slow fall-off of the exponential filter is chosen or no fade-out is applied at all. Otherwise, there is no change required for dye advection via twofold convolution.

Finally, twofold convolution can be extended to vector fields on curved surfaces. Adopting the idea of image-space advection [12], a 3D vector field is first projected onto the image plane. In the second step, 2D vector field visualization is applied to the projected vector field. This 2D visualization can directly be generated by twofold convolution working on the image plane. However, the use of hybrid image-space/object-space Lagrangian integration is favorable for the first convolution stage because of several quality improvements described in previous work [23].

6 Discussion of Costs and Quality

This section discusses the performance behavior and memory footprint of twofold convolution. Issues of visualization quality are also taken into account.

Performance characteristics are considered first. A constant filter length and a constant sampling rate are assumed for the first convolution stage (i.e., the direct convolution). Therefore, the total number of sampling steps per convolution is fixed to n_{direct}. Then, the total computation time for this part is estimated as

$$n_{\mathrm{direct}} \times (t_{\mathrm{accum}} + t_{\mathrm{ODE}}) \quad ,$$

where t_{ODE} is the time needed for one numerical integration step for the ordinary differential equation (ODE) from (2) and t_{accum} is the time for one accumulation operation for the Riemann sum of the convolution integral. The second part reuses the above ODE computation and just executes one alpha-blending operation. The total time for this part is estimated as t_{blend}.

To roughly compare these performance costs to those for purely iterative alpha blending, the following assumptions are made. As illustrated for the Bartlett window in Sect. 4, a symmetric filter kernel k of total width $2w$ is assumed to result in a sampling distance $w/2$ for the second convolution stage. With the same ODE integration step size as above, we have to perform purely iterative alpha blending $(n_{\mathrm{direct}}/4)$ times to cover the same distance as before. Accordingly, the total time for purely iterative alpha blending would be

$$\frac{1}{4} n_{\mathrm{direct}} \times (t_{\mathrm{blend}} + t_{\mathrm{ODE}}) \quad .$$

Neglecting the difference between t_{blend} and t_{accum}, the estimates for computation times approximately form a 1:4 ratio for purely iterative alpha blending vs. twofold convolution. A comparison with completely Lagrangian schemes is not possible because Lagrangian integration hardly supports exponential filters.

The memory footprint of twofold convolution is determined by the following aspects. Memory arrays in nD (for an nD vector field) are required to store accumulated gray-scale values and intermediate coordinates from Lagrangian ODE integration. A constant and very small number of these arrays is needed. Details are discussed for an actual GPU implementation in Sect. 7. Purely iterative alpha blending has comparable memory requirements. For time-dependent vector fields and input (noise) textures, however, twofold convolution has a higher memory footprint because it needs access to both types of nD data structures for the time span covered by the first convolution stage. These memory costs can be controlled by the size of the filter kernel k.

The last part of this section covers visualization quality. Compared to purely iterative alpha blending, twofold convolution has two important advantages: flexible filter design and reduced numerical diffusion. The exponential filter leads to inappropriate high-frequency noise along streamlines or streaklines. This problem is greatly reduced by smoother filters within twofold convolution.

Analogously to filter design in frequency space, the flexibility in choosing kernel functions can also be useful for filter design in the time domain. This design approach plays an important role in sparse representations that follow the idea of OLIC [20]. Here, the input texture consists of isolated injection points that are transformed into distinguishable streaklines after convolution. The texture input is similar to a collection of δ impulses and, therefore, the intensity profile along a streakline reflects the impulse response of the filter. As this response is described by the kernel function, a filter design in the time domain allows us to directly control the visual representation of streaklines.

The other quality aspect is connected to numerical diffusion by semi-Lagrangian advection. This diffusion is not only along streamlines or streaklines but also perpendicular to them. This artifact removes high frequencies perpendicular to convolution curves and thus reduces the information contents of the visualization. Essentially, the resolution of the vector field representation is decreased and fine details could be glossed over. In contrast, twofold integration significantly reduces numerical diffusion by decreasing the number of resampling steps during iterative alpha blending.

Twofold convolution is related to postfiltering for LEA [10], where a LIC filter for a time-independent "frozen" vector field is applied after an iterative alpha blending based on LEA. As pointed out by Erlebacher et al. [4], LEA generates (inverse) streaklines. Both streaklines and pathlines are different from streamlines for unsteady flow. Twofold convolution computes streaklines in both convolution stages and therefore achieves an appropriate overall convolution along consistent curves. In addition, twofold convolution can reuse ODE integration results for the large exponential filter, which is not possible with LEA postfiltering. Similarly, "double LIC" by Okada and Lane [14] neither reuses ODE results nor exploits a reduced sampling rate for the second convolution computation.

7 Implementation and Results

The GPU implementation of twofold convolution is based on C++ and DirectX 9.0, and was tested on Windows XP machines with ATI Radeon 9800 XT (256 MB), ATI Radeon X800 XT Platinum Edition (256 MB), and NVIDIA GeForce 6800 Ultra (256 MB) GPUs. GPU states and programs (i.e., vertex and pixel shader programs) are configured within effect files. A change of this configuration can be incorporated by modifying the clear-text effect files, without recompiling the C++ code. Shader programs are partly formulated with high-level shading language (HLSL), partly as assembler-level programs. Most elements of the visualization algorithms take place on a texel-by-texel level. This essentially reduces the role of the surrounding C++ program to allocating memory for the required textures and executing the pixel shader programs by drawing single domain-filling quadrilaterals. The shader programs require the functionality of a DirectX 9.0 compliant GPU.

Lagrangian integration in Cartesian 2D space makes use of a previous implementation that already supports arbitrary user-defined filter kernels [22]. Any texture-related data is stored in 2D textures; a time-dependent vector field is represented by a stack of 2D textures. Multi-pass rendering is employed to compute particle tracing and simultaneously accumulate the gray-scale contributions to the convolution integral. Intermediate positions along particle traces are held in a coordinate array, accumulated gray-scale values in a property texture. Both textures are updated according to ping-pong rendering, utilizing render-to-texture functionality. Alternatively, multi-pass rendering can be replaced by a loop in the pixel shader that computes the complete particle trace and the convolution integral in a single render pass on a Shader Model 3 compliant GPU such as an NVIDIA GeForce 6800 Ultra. The second convolution stage accesses a coordinate array with previously computed particle positions and applies alpha blending based on the property texture. The property texture has 16-bit resolution to allow for an adequate accuracy for the Riemann sum of a large number of samples along a convolution curve. A 16-bit texture format is also used to represent coordinate arrays and the vector field, while 8-bit resolution is sufficient for input noise textures.

The implementation of the first convolution stage for vector fields on curved surfaces is adopted from previous work [23]. 32-bit floating-point textures are used to represent coordinates during particle integration, 16-bit textures are used for the accumulated gray-scale values and the flow data alike, and the input noise is held in an 8-bit texture. The second convolution stage works directly on the image plane and is analogous to the second stage for 2D Cartesian vector field visualization.

Figure 3 shows results of GPU-based dense 2D vector field visualization, based on various parameter settings. Histogram equalization is applied to all images. The vector field originates from a numerical simulation of convection flow on a uniform grid. Low-pass filtered white noise (Fig. 3a) is used as input texture. Figure 3b shows the intermediate result of the first stage of twofold convolution. Here, a Bartlett window is applied with sampling at 2×10 positions (the factor 2 indicates that the filter kernel is symmetric). Figure 3c displays the final visualization generated by twofold convolution. In contrast, Figs. 3d–f are produced by purely iterative alpha blending. Figure 3d uses the same small sampling distance as in the first stage of twofold convolution to avoid aliasing artifacts; the alpha blending factor is modified to resemble Fig. 3c. Therefore, Figs. 3c,d should be examined for a faithful comparison between twofold convolution and purely iterative alpha blending. This comparison demonstrates that iterative blending provides rather poor quality and, therefore, supports the theoretical discussion of quality from Sect. 6. Figure 3e uses the same large sampling distance and alpha factor as for the second stage of twofold convolution; severe aliasing artifacts occur due to missing prefiltering. Finally, Fig. 3f employs the same sampling rate as in (b) and (d), along with the alpha value from (e); this leads to very short streaks.

Figure 4 shows results of sparse 2D vector field visualization. The same flow field and visualization methods are used as in Fig. 3. Histogram equalization is applied to all images as well. Only the input texture is different – it contains a rather small number of white blobs. In this way, a visualization is achieved in the tradition of OLIC [20]. Figure 4a displays the input texture. Figure 4b shows the intermediate result of the first stage of twofold convolution (Bartlett window of size 2×10). Figure 4c displays the result of twofold convolution, while Figs. 4d–f are produced

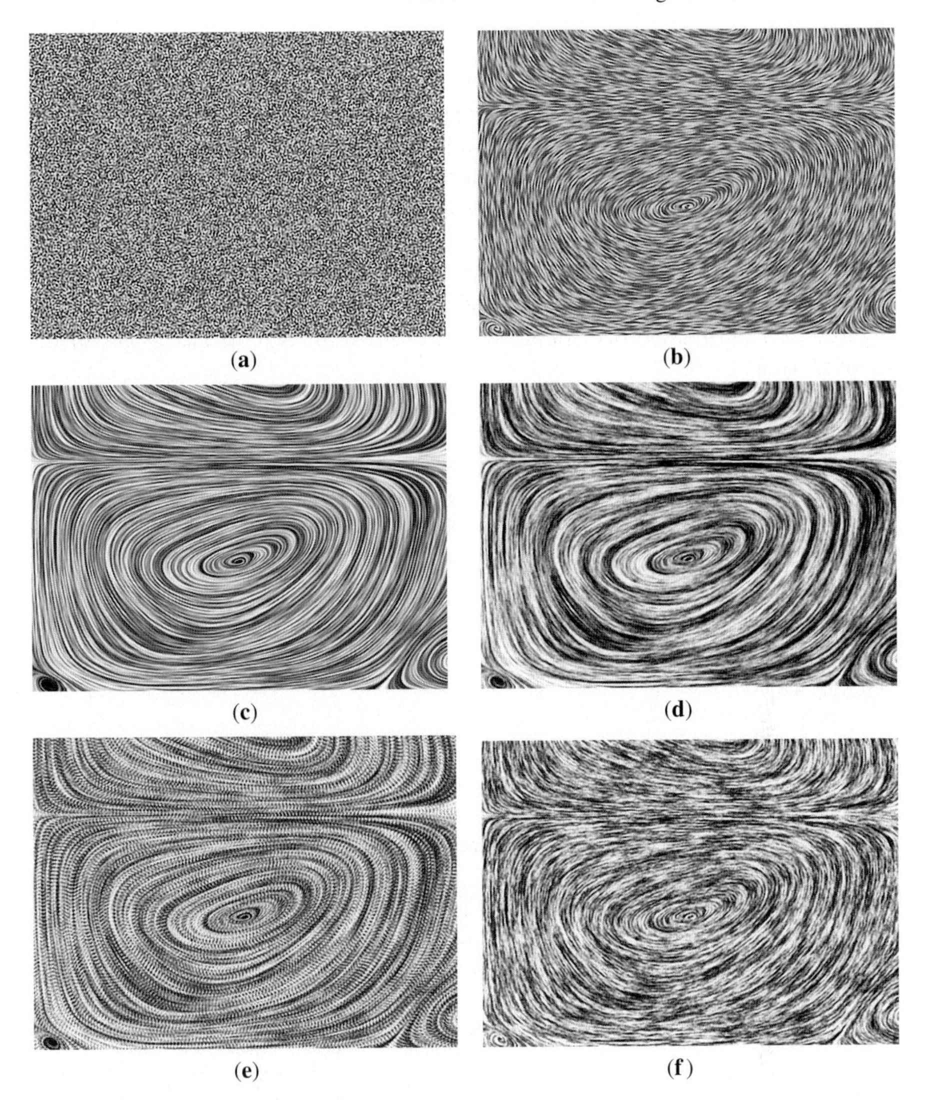

Fig. 3. Dense vector field visualization. Image (**a**) shows the input noise, (**b**) the result of the first stage of twofold convolution, and (**c**) the final visualization generated by twofold convolution. Pictures (**d**)–(**f**) are produced by purely iterative alpha blending. Image (**d**) uses a small sampling distance along with a wide exponential filter to resemble (**c**), picture (**e**) uses a larger sampling distance, and (**f**) employs a narrow exponential filter with a small sampling distance

by purely iterative alpha blending. As before, Fig. 4d uses a small sampling distance along with a modified exponential filter to resemble the result of twofold convolution. Figure 4e employs the same sampling distance as for the second stage of twofold convolution, which leads to aliasing. Short streaks are generated by a narrow exponential

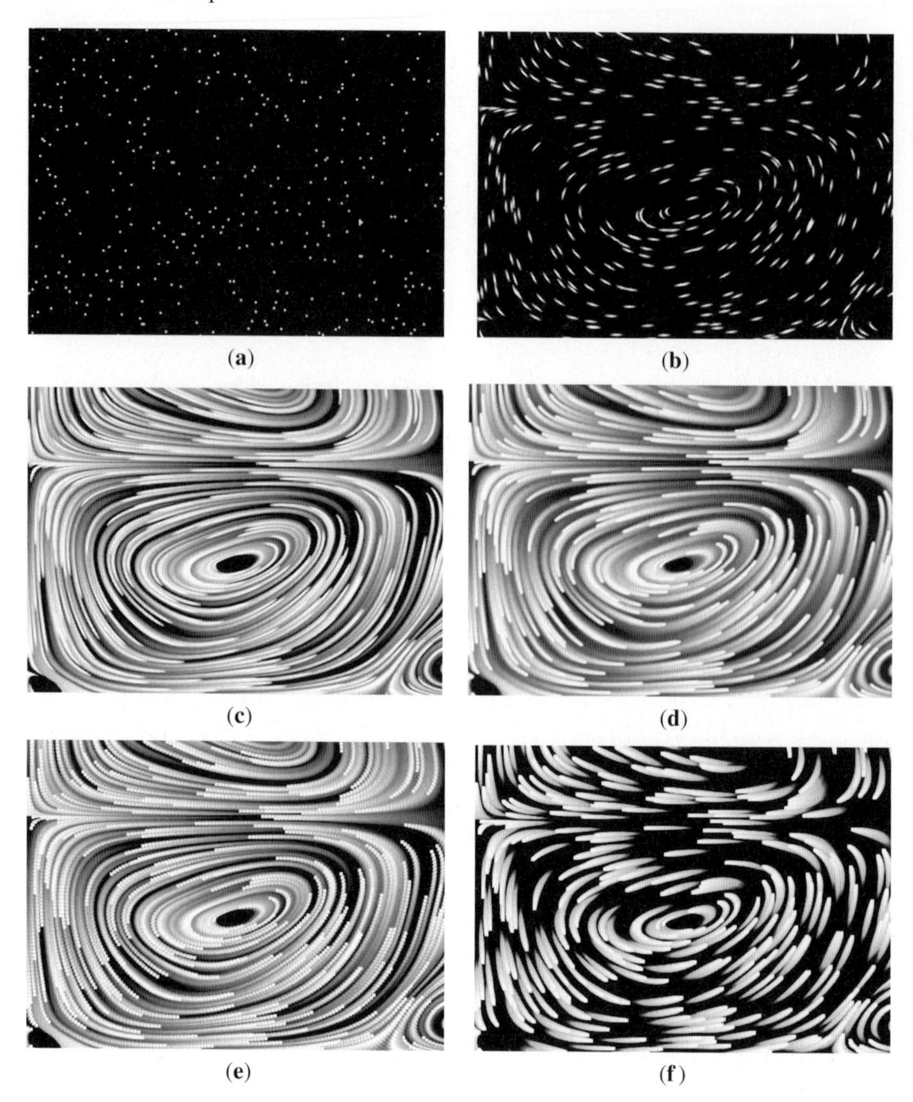

Fig. 4. Moderately sparse vector field visualization. Image (**a**) shows the input image, (**b**) the result of the first stage of twofold convolution, and (**c**) the final visualization generated by twofold convolution. Pictures (**d**)–(**f**) are produced by purely iterative alpha blending. Image (**d**) uses a small sampling distance along with a wide exponential filter to resemble (**c**), picture (**e**) uses a larger sampling distance, and (**f**) employs a narrow exponential filter with a small sampling distance

filter with small sampling distance in Fig. 4f. Again, a comparison between twofold convolution and purely iterative alpha blending should be based on Figs. 4c,d. Numerical diffusion caused by semi-Lagrangian texture advection is a most noticeable artifact in Fig. 4d, while Fig. 4c provides clearly defined streaks.

Fig. 5. Dense vector field visualization on a curved surface, using twofold convolution

Figure 5 illustrates dense vector field visualization on curved surfaces. The vector field originates from an industrial 3D simulation of air flow around an automobile. The tangential component of this vector field is extracted at the surface to obtain input data for surface flow visualization. The visualization is based on twofold convolution, where the first convolution stage uses a Bartlett window with sampling at 2×14 positions.

Figure 6 compares different methods for vector field visualization on surfaces. Here, the same data set is shown as in Fig. 5 – only the base surface is added and the viewpoint is changed. Figure 6a shows the intermediate result of the first stage of twofold convolution. Here, a Bartlett window is applied with sampling at 2×14 positions. Figure 6b presents the final result of twofold convolution. For comparison, Figs. 6c,d show purely iterative blending with the same alpha factor as in (b). Figure 6c is affected by aliasing artifacts because the same large sampling distance is used as for the second convolution stage in (b). Figure 6d is based on a smaller sampling distance, which results in short streaks.

Table 1 shows performance measurements for different GPU-based 2D vector field visualization methods and various parameter settings. The measurements were conducted on a Windows XP machine with ATI Radeon 9800 XT GPU (256 MB). The first convolution stage makes use of a Bartlett window that is evaluated at either 2×10 or 2×20 sampling positions. The row "twofold convolution" describes the overall performance for twofold convolution, including the two convolution stages and the final display with histogram equalization. The line "iterative blending (raw)" shows the performance of purely iterative alpha blending with semi-Lagrangian texture advection, including the final display. Here, the performance is independent of the filter length because only a single texture-advection step is executed.

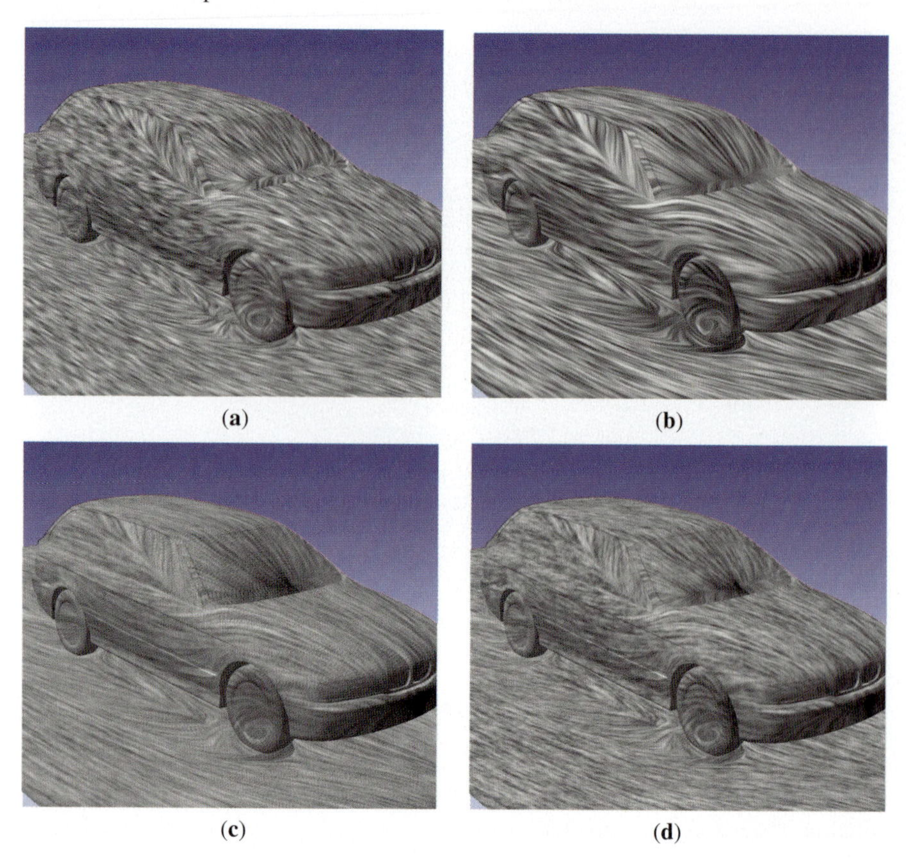

Fig. 6. Dense vector field visualization on curved surfaces. Image (**a**) shows the intermediate result of the first stage of twofold convolution, image (**b**) the final result of twofold convolution after the second convolution stage. Pictures (**c**) and (**d**) are produced by purely iterative alpha blending. Image (**c**) uses the same blending factor and the same large sampling distance as for the second convolution stage in (**b**), which leads to aliasing. Image (**d**) is based on a smaller sampling distance and the same blending factor as in (**b**), resulting in short streaks

Table 1. Performance of GPU-based 2D vector field visualization in frames per second on an ATI Radeon 9800 XT GPU

Domain size	512^2		1024^2	
Filter length	2×10	2×20	2×10	2×20
Twofold convolution	64.9	41.8	16.9	10.7
Iterative blending (raw)	688.0	688.0	192.9	192.9
Iterative blending (normalized)	137.6	68.8	38.6	19.3

As discussed in Sect. 6, purely iterative alpha blending has to be performed $(n_{\text{direct}}/4)$ times to cover the same distance as twofold advection. This fact is taken into account in the row "iterative blending (normalized)." A fair comparison between twofold convolution and iterative blending should consider this normalized performance of alpha blending. Then, the performance numbers roughly show a 1:2 ratio for twofold convolution vs. purely iterative blending, which is even more balanced than the theoretical estimate from Sect. 6. The deviation from the estimate can be primarily explained by the additional overhead for the display of intermediate results in purely iterative blending. At large, the performance measurements demonstrate that twofold convolution achieves a rendering speed comparable to the fastest known vector field visualization methods.

Similarly to the Cartesian 2D case, the performance of twofold convolution on curved surfaces is primarily determined by the costs for the first convolution stage. As an example, the visualization from Fig. 6b is rendered with 5.1 frames per second on a 950^2 viewport (ATI Radeon X800 XT Platinum Edition GPU).

8 Conclusion

Iterative twofold convolution has been introduced as an efficient high-quality two-stage filtering method for dense texture-based vector field visualization. The first stage applies a user-specified compact filter kernel, based on Lagrangian particle tracing. This stage facilitates a flexible design of filters and avoids numerical diffusion. The second stage applies iterative alpha blending to implement a large-scale exponential filter.

A discussion of frequency-space properties and adequate sampling rates has shown that this order of convolution operations is beneficial because it facilitates large integration step sizes. It has been demonstrated that the idea of twofold convolution can be used for steady and unsteady vector fields, dye and noise advection, as well as vector fields on Cartesian 2D domains and on curved surfaces. Finally, the proposed approach fits nicely into existing GPU-based visualization techniques and therefore facilitates interactive vector field visualization.

In future work, twofold convolution could be incorporated in existing GPU-based 3D vector field visualization methods [24].

A Exponential Filter and Iterative Alpha Blending

This section shows that a first-order approximation of the normalized exponential filter $\beta \exp(-\beta x) H(x)$ corresponds to iterative alpha blending, where $H(x)$ is the Heaviside function. A closely related derivation can be found in an article by Erlebacher et al. [4].

A scenario with straightened streamlines is assumed, as laid out in Sect. 3. This setting implies a vector field with normalized magnitude, pointing along the positive x axis. Applying the exponential filter to the convolution equation (3),

$$d(x) = \int_{-\infty}^{\infty} k(x-\xi)t(\xi)\,\mathrm{d}\xi = \int_{-\infty}^{\infty} k(\xi)t(x-\xi)\,\mathrm{d}\xi = \int_{0}^{\infty} \beta e^{-\beta\xi} t(x-\xi)\,\mathrm{d}\xi \quad ,$$

leads to the Riemann sum

$$d(x) \approx \sum_{j=0}^{\infty} \beta e^{-\beta j \Delta\xi} t(x - j\Delta\xi)\,\Delta\xi \quad ,$$

after sampling with equidistant rate $\Delta\xi$. Based on this approximation, the image is evaluated at the subsequent sampling position downstream the vector field,

$$\begin{aligned} d(x+\Delta\xi) &= \sum_{j=0}^{\infty} \beta e^{-\beta j \Delta\xi} t(x-(j-1)\Delta\xi)\,\Delta\xi \\ &= \sum_{j=-1}^{\infty} \beta e^{-\beta(j+1)\Delta\xi} t(x - j\Delta\xi)\,\Delta\xi \quad , \end{aligned}$$

to obtain the recursion relation

$$d(x+\Delta x) = \alpha t(x+\Delta x) + e^{-\alpha} d(x) \quad , \text{ with } \alpha = \beta\Delta x \quad .$$

First-order approximation of the exponential function leads to the alpha blending equation

$$d(x+\Delta x) = \alpha t(x+\Delta x) + (1-\alpha) d(x) \quad .$$

Either one of the two previous blending equations can be the basis for a recursive computation of exponential filtering. It is not possible to evaluate the underlying infinite sum because $d(x)$ is not available for $x \to -\infty$. In fact, the sum is truncated by setting a recursion beginning $d(x_0) = 0$ at starting point x_0. Nevertheless, for $x > x_0$, the truncated computation of $d(x)$ converges to the infinite sum.

B Fourier Transform of the Exponential Filter

In this section, the Fourier transform of the exponential filter is computed. To be more specific, the Fourier transforms of the functions

$$f : \mathbb{R} \longrightarrow \mathbb{R}\,, \quad x \longmapsto f(x) = e^{-2\pi\nu_0 |x|}$$

and

$$g : \mathbb{R} \longrightarrow \mathbb{R}\,, \quad x \longmapsto g(x) = e^{-2\pi\nu_0 x} H(x)$$

are considered. Starting with f, we have to evaluate

$$\begin{aligned} F(\nu) &= \int_{-\infty}^{\infty} e^{-2\pi\nu_0|x|} e^{-2\pi \mathrm{i}\nu x}\,\mathrm{d}x \\ &= \int_{-\infty}^{\infty} e^{-2\pi\nu_0|x|} \left(\cos(2\pi\nu x) - \mathrm{i}\sin(2\pi\nu x)\right)\,\mathrm{d}x \\ &= \int_{0}^{\infty} e^{-2\pi\nu_0 x} \left(\cos(2\pi\nu x) + \mathrm{i}\sin(2\pi\nu x)\right)\mathrm{d}x \\ &\quad + \int_{0}^{\infty} e^{-2\pi\nu_0 x} \left(\cos(2\pi\nu x) - \mathrm{i}\sin(2\pi\nu x)\right)\mathrm{d}x \\ &= 2\int_{0}^{\infty} e^{-2\pi\nu_0 x} \cos(2\pi\nu x)\,\mathrm{d}x \quad . \end{aligned}$$

The integral of the exponentially damped cosine function yields

$$\int e^{-\Omega x}\cos(\omega x)\,\mathrm{d}x = \frac{e^{-\Omega x}}{\omega^2+\Omega^2}\left[\omega\sin(\omega x) - \Omega\cos(\omega x)\right]$$

after integration by parts. Accordingly, the definite integral is

$$\int_{0}^{\infty} e^{-\Omega x}\cos(\omega x)\,\mathrm{d}x = \frac{\Omega}{\omega^2+\Omega^2} \quad .$$

Therefore, the Fourier transform of f is the Lorentz function

$$F(\nu) = \frac{1}{\pi}\frac{\nu_0}{\nu^2+\nu_0^2} \quad .$$

The Fourier transform of g is similar:

$$G(\nu) = \int_{0}^{\infty} e^{-2\pi\nu_0 x} e^{-2\pi \mathrm{i}\nu x}\,\mathrm{d}x = \int_{0}^{\infty} e^{-2\pi\nu_0 x}\left(\cos(2\pi\nu x) - \mathrm{i}\sin(2\pi\nu x)\right)\mathrm{d}x \quad .$$

The cosine term, once again, leads to a Lorentz function.Analogously, the integral of the exponentially damped sine function yields

$$\int e^{-\Omega x}\sin(\omega x)\,\mathrm{d}x = -\frac{e^{-\Omega x}}{\omega^2+\Omega^2}\left[\Omega\sin(\omega x) + \omega\cos(\omega x)\right] \quad ,$$

with the definite integral

$$\int_{0}^{\infty} e^{-\Omega x}\sin(\omega x)\,\mathrm{d}x = \frac{\omega}{\omega^2+\Omega^2} \quad .$$

Finally, the Fourier transform of g is

$$G(\nu) = \frac{1}{2\pi}\frac{\nu_0}{\nu^2+\nu_0^2} + \frac{-\mathrm{i}}{2\pi}\frac{\nu}{\nu^2+\nu_0^2} \quad .$$

Acknowledgments

The data set used for Figs. 5 and 6 was provided by the BMW Group. Thanks to Simon Stegmaier for fruitful discussions on texture advection. Special thanks to Bettina A. Salzer for proof-reading.

References

1. D. Brook and R. J. Wynne. *Signal Processing: Principles and Applications*. Edward Arnold, London, 1988.
2. B. Cabral and L. C. Leedom. Imaging vector fields using line integral convolution. In *Proc. ACM SIGGRAPH 1993*, pages 263–270, 1993.
3. F. C. Crow. Summed-area tables for texture mapping. *Computer Graphics (Proc. ACM SIGGRAPH 1984)*, 18(3):207–212, 1984.
4. G. Erlebacher, B. Jobard, and D. Weiskopf. Flow textures: High-resolution flow visualization. In C. D. Hansen and C. R. Johnson, editors, *The Visualization Handbook*, pages 279–293. Elsevier, Amsterdam, 2005.
5. T. Frankel. *The Geometry of Physics: An Introduction*. Cambridge University Press, New York, 2001.
6. G. Gorla, V. Interrante, and G. Sapiro. Texture synthesis for 3D shape representation. *IEEE Transactions on Visualization and Computer Graphics*, 9(4):512–524, 2003.
7. P. S. Heckbert. Filtering by repeated integration. *Computer Graphics (Proc. ACM SIGGRAPH 1986)*, 20(4):315–321, 1986.
8. P. S. Heckbert. Survey of texture mapping. *IEEE Computer Graphics and Applications*, 6(11):56–67, 1986.
9. H.-C. Hege and D. Stalling. Fast LIC with piecewise polynomial filter kernels. In H.-C. Hege and K. Polthier, editors, *Mathematical Visualization*, pages 295–314. Springer, Heidelberg, 1998.
10. B. Jobard, G. Erlebacher, and M. Y. Hussaini. Lagrangian-Eulerian advection of noise and dye textures for unsteady flow visualization. *IEEE Transactions on Visualization and Computer Graphics*, 8(3):211–222, 2002.
11. R. S. Laramee, H. Hauser, H. Doleisch, B. Vrolijk, F. H. Post, and D. Weiskopf. The state of the art in flow visualization: Dense and texture-based techniques. *Computer Graphics Forum*, 23(2):143–161, 2004.
12. R. S. Laramee, J. J. van Wijk, B. Jobard, and H. Hauser. ISA and IBFVS: Image space based visualization of flow on surfaces. *IEEE Transactions on Visualization and Computer Graphics*, 10(6):637–648, 2004.
13. N. Max and B. Becker. Flow visualization using moving textures. In *Proc. ICASW/LaRC Symposium on Visualizing Time-Varying Data*, pages 77–87, 1995.
14. A. Okada and D. Lane. Enhanced line integral convolution with flow feature detection. In *Proc. IS&T/SPIE Electronic Imaging 1997*, pages 206–217, 1997.
15. A. Sanna, B. Montrucchio, and P. Montuschi. A survey on visualization of vector fields by texture-based methods. *Recent Res. Devel. Pattern Rec.*, 1:13–27, 2000.
16. J. Shen, W. Shen, S. Castan, and T. Zhang. Sum-box technique for fast linear filtering. *Signal Processing*, 82(8):1109–1126, 2002.
17. D. Stalling and H.-C. Hege. Fast and resolution independent line integral convolution. In *Proc. ACM SIGGRAPH 1995*, pages 249–256, 1995.

18. J. J. van Wijk. Spot noise – texture synthesis for data visualization. *Computer Graphics (Proc. ACM SIGGRAPH 1991)*, 25(4):309–318, 1991.
19. J. J. van Wijk. Image based flow visualization. *ACM Transactions on Graphics*, 21(3): 745–754, 2002.
20. R. Wegenkittl, E. Gröller, and W. Purgathofer. Animating flow fields: Rendering of oriented line integral convolution. In *Computer Animation 1997*, pages 15–21, 1997.
21. D. Weiskopf. Dye advection without the blur: A level-set approach for texture-based visualization of unsteady flow. *Computer Graphics Forum (Proc. Eurographics 2004)*, 23(3):479–488, 2004.
22. D. Weiskopf, G. Erlebacher, and T. Ertl. A texture-based framework for spacetime-coherent visualization of time-dependent vector fields. In *Proc. IEEE Visualization 2003*, pages 107–114, 2003.
23. D. Weiskopf and T. Ertl. A hybrid physical/device-space approach for spatio-temporally coherent interactive texture advection on curved surfaces. In *Proc. Graphics Interface*, pages 263–270, 2004.
24. D. Weiskopf, T. Schafhitzel, and T. Ertl. Real-time advection and volumetric illumination for the visualization of 3D unsteady flow. In *Proc. Eurovis (EG/IEEE TCVG Symposium on Visualization)*, pages 13–20, 2005.

Constructing 3D Elliptical Gaussians for Irregular Data

Wei Hong, Neophytos Neophytou, Klaus Mueller, and Arie Kaufman

Center for Visual Computing and Department of Computer Science, Stony Brook University
{weihong, nneophyt, mueller, ari}@cs.sunysb.edu

Summary. Volumetric datasets obtained from scientific simulation and partial differential equation solvers are typically given in the form of non-rectilinear grids. The splatting technique is a popular direct volume rendering algorithm, which can provide high quality rendering results, but has been mainly described for rectilinear grids. In splatting, each voxel is represented by a 3D kernel weighted by the discrete voxel value. While the 3D reconstruction kernels for rectilinear grids can be easily constructed based on the distance among the aligned voxels, for irregular grids the kernel construction is significantly more complicated. In this paper, we propose a novel method based on a 3D Delaunay triangulation to create 3D elliptical Gaussian kernels, which then can be used by a splatting algorithm for the rendering of irregular grids. Our method does not require a resampling of the irregular grid. Instead, we use a weighted least squares method to fit a 3D elliptical Gaussian centered at each grid point, approximating its Voronoi cell. The resulting 3D elliptical Gaussians are represented using a convenient matrix representation, which allows them to be seamlessly incorporated into our elliptical splatting rendering system.

1 Introduction

Direct volume rendering is an important technology in the fields of computer graphics, as well as scientific and medical visualization. It allows the user to comprehend and visualize a volumetric dataset directly, without requiring the generation of a polygonal iso-surface. Volumetric datasets are commonly classified as *rectilinear* or *non-rectilinear*, according to their grid structure. Here, both the curvilinear and the unstructured grids belong to the class of non-rectilinear grids, while cubic grids are the simplest case of rectilinear grids. The volumetric datasets obtained from scientific simulation and partial differential equation solvers are typically given in the form of non-rectilinear grids.

The straightforward method to visualize non-rectilinear grids is to resample them into a rectilinear grid [1], where usual rendering methods readily apply. However, as a non-rectilinear grid may consist of cells of drastically different sizes, the resampling approach may either cause a loss of important information or result in a huge dataset. Thus, several techniques have been developed for the direct volume rendering of

T. Möller et al. (eds.), *Mathematical Foundations of Scientific Visualization, Computer Graphics, and Massive Data Exploration*, Mathematics and Visualization,
DOI: 10.1007/978-3-540-49926-8,

non-rectilinear grids, i.e., ray casting, cell projection, and splatting. Ray casting is the most popular direct volume rendering technique where volume rendered images are generated by casting rays from the viewer's eye, through the screen pixels, into a 3D volume, and compositing the contributions of all sample points taken along each ray into the corresponding screen pixel. Many algorithms for the ray casting of non-rectilinear grids have been developed [2–4]. Since a non-rectilinear grid may be composed of cells of drastically different sizes, sampling with a constant interval along a ray is not desirable. Therefore, sample points are usually taken at the intersections of rays and cells, which tends to be very time-consuming. In the cell projection technique [5], a cell in a volume is projected onto the screen, and its contribution to the pixels under its projection extent is calculated and composited with the contributions from the previously projected primitives. Cell projection algorithms need to obtain the proper cell visibility ordering to generate the correct compositing result. Here, the cell visibility ordering itself is not trivial and can be rather time consuming.

The splatting technique has become quite popular for directly rendering volumetric datasets of various grid structures. The original algorithm, first proposed by Westover [6] for rectilinear grids, projects each voxel onto the image plane and composites the result into an accumulation image. As each voxel is projected onto the image plane, the voxel's energy is spread over the image raster using the 2D projection of a 3D reconstruction kernel, which is centered at the voxel's projection point. For regular grids, the 3D reconstruction kernel, also called a *splat*, is spherically symmetric and centered at a voxel. Since the splat is reconstructed into a 2D image raster, it can be implemented as a 2D reconstruction kernel called a "footprint function," containing the integration of the 3D kernel along the projection direction. By ways of 2D texture mapping, rectilinear grids can be quickly rendered with a single polygon per voxel and using a single Gaussian kernel for all voxels. The direct extension of this technique to non-rectilinear grids is not straightforward, because the appropriate kernels for non-rectilinear grids are not easy to calculate. In this case, the splats are arbitrary ellipsoidal kernels, with their shape being defined by the local grid structure.

Both ray-casting and cell projection algorithms have been extended for the volume rendering of non-rectilinear grids. Recently, graphics hardware has been used to accelerate ray-casting [7] and cell projection [8, 9] algorithms for irregular grids. However, both of these modalities have some limitations. For the cell projection algorithm, the piecewise linear interpolation may result in banding at cell boundaries, degrading the quality of the resulting image. In addition, cell projection approaches are limited by the cell visibility sorting, which prevents the current graphics hardware from running at full capacity. For ray-casting algorithms, the ray-cell intersection test, the identification of the face of the cell through which a ray exists, and the interpolation from the surrounding grid points are very expensive operations. Even the hardware accelerated ray-casting algorithm [7] can not achieve interactive rendering speed.

In an attempt to overcome these problems, we propose a new approach that utilizes splatting, in conjunction with arbitrarily shaped elliptical Gaussians, for the rendering of irregularly gridded data. Our splatting approach offers the following advantages: (1) its smooth and overlapping kernel functions will reconstruct a smooth representation of the grid-sampled signal, without the artifacts of the piecewise linear representations of the cell projection approaches; (2) it promises to be more efficient than ray casting due to the ease of footprint rasterization, especially when implemented in hardware; (3) it is also more efficient than other splatting approaches for irregular grids, since the space-filling kernels are only required at the grid points, and thus the rendering complexity matches that of the grid. Finally, apart from non-rectilinear gridded data, our method also supports collections of scattered data points.

The main topic of this paper is the method for constructing arbitrarily oriented elliptical Gaussians from irregular grid topologies. Once the 3D reconstruction kernels are found, the software rendering is straightforward. We can either use the sheet buffer algorithm for composited rendering [10], or we can just splat the whole kernel for X-ray type rendering.

2 Previous Work

Only a limited amount of work has been done so far on how to use the splatting algorithm for the rendering of irregular grids. Meredith and Ma [11] proposed a spherical Gaussian splat-based rendering method for irregular data. In this method, they use an octree with roughly the same number of data points stored at each leaf node. No connectivity information is stored for the data points. For any given viewing parameters, they calculate the projected size of any octree node on the screen. Then they divide the screen area based on the number of data points within that octant to calculate the approximate kernel size. This method can only give a rough estimate of the kernel size.

Mao et al. [12, 13] presented a method that resamples irregular grids with a set of new points whose energy support extents in the 3D physical space can be approximated by spheres or ellipsoids. To approximate the scalar field represented in the original grid as accurately as possible without using too many sample points, an adaptive 3D stochastic sampling method called Poisson sphere/ellipsoid sampling is employed. Then, after the new splat distribution has been calculated, the original splatting algorithm can be used to render the irregular grid. The disadvantage of this method is that the original grid must be resampled to compute the scalar value for the new sample points. The error caused by this non-regular resampling potentially degrades the quality of the resulting images. In addition, this method also generates considerably more splats than the original number of grid points. For example, the NASA Blunt Fin dataset with resolution of $40 \times 32 \times 32$ is resampled with 79,971 sphere points and $5,041$ ellipsoid points, which more than doubles the number of points. Moreover, this method cannot be used to render scattered data.

Jang et al. [14] developed a procedure, based on the Voronoi cell describing the region around a point, to place and orient Gaussian kernels to give more uniform coverage in non-uniform cells. However, they did not specify how they construct these splats. This method also need to resample the original non-uniform cells. Jang et al. [15] performed a global optimization method to fit radial basis functions (RBFs) to irregular data. In this paper, we propose a method to construct 3D ellipsoidal kernels for irregular grids without the need for an error prone resampling of the original grid, since our 3D reconstruction kernels are still centered at the original grid points. Instead, a weighted least squares method is used to fit a single ellipsoidal kernel to the Voronoi cell at each grid point, which is a local optimization method. In our method, we do not interpolate any additional data points.

3 Creating 3D Elliptical Gaussian

The shape of a 3D elliptical Gaussian kernel centered at the origin can be modeled via the implicit equation of an ellipsoid:

$$Ax^2 + By^2 + Cz^2 + 2Dxy + 2Exz + 2Fyz - 1 = 0 \tag{1}$$

This equation has six unknowns and represents a quadric surface. The quadric surface can be represented by using matrix notation, giving rise to a 3×3 symmetric quadric matrix Q:

$$Q = \begin{vmatrix} A & D & E \\ D & B & F \\ E & F & C \end{vmatrix} \tag{2}$$

The quadric surface represented by Q can be easily translated, scaled, and rotated by multiplying it with a transformation matrix. Given a 3×3 affine transformation matrix M, the transformed quadric surface $Q^{'}$ is given by:

$$Q^{'} = (M^{-1})^t \cdot Q \cdot M^{-1} \tag{3}$$

With this representation we can create an arbitrarily oriented elliptical Gaussian by applying the scaling and rotation transformations contained in matrix $S = \{a, b, c\}$ and R on a unit sphere, respectively, as described in the following equation:

$$Q = (R^{-1})^t \cdot (S^{-1})^t \cdot I \cdot S^{-1} \cdot R^{-1} = R \cdot (S^{-1})^2 \cdot R^t \tag{4}$$

Here I is the identity matrix which represents the unit sphere, and $(S^{-1})^2 = \{1/a^2, 1/b^2, 1/c^2\}$ is a diagonal matrix, which can be thought of as a scaling matrix. It represents an axis aligned ellipsoid. The rotation matrix R is an orthogonal matrix representing the ellipsoid orientation, which can be defined by three rotation angles α, β, and γ along the three axes. Instead of directly fitting an ellipsoid using (1), we fit the scaling matrix S and rotation matrix R separately, using (4). S and R are decided by the three scaling factors and the three rotation angles, respectively. Due to this matrix representation, the resulting ellipsoidal kernel can easily be incorporated into our rendering algorithm, which represents the elliptical splats using a rotation and a scaling matrix.

The irregular grids are always described in terms of their grid structure. However, in our algorithm we are only interested in the grid points. In that respect, we treat an irregular grid as a volumetric point cloud. Our algorithm only uses these grid points as input for generating the 3D ellipsoidal kernels. In the following sections, we describe our approach to fit the 3D ellipsoidal kernel using the matrix representation.

3.1 Guide Points

As is well known, the dual of the Delaunay triangulation is the Voronoi diagram, which consists of cells around the data points such that any location in a particular cell is closer to that cell's generating point than to any other. Thus, the shape of the Voronoi cell can give us a clue about the shape of the reconstruction kernel. The main idea of our algorithm is to fit elliptical Gaussian kernels to the grid points by approximating their Voronoi cells. We show a 2D example in Fig. 1, in which the Voronoi cell of grid point V_0, shown in red, is approximated with an ellipse, shown in blue. The Voronoi cell of V_0 is obtained by connecting the circumcenters between pairs of Delaunay triangles that are adjacent and both incident to V_0.

As the first step of our algorithm, we apply the 3D Delaunay triangulation algorithm to the input grid points. Through the 3D Delaunay triangulation, we obtain for each grid point a list of tetrahedra incident to it. The circumcenters of these tetrahedra are the vertices of the Voronoi cell generated for that grid point, i.e., the cell's generating point. In the ideal case, each circumcenter is shared by four reconstruction kernels, with each of these contributing 25% to it. This would mean that the elliptical Gaussian kernel passes through these circumcenters with the 0.25-valued iso-contour. Furthermore, in this ideal case, the 0.5-valued iso-contour of the Gaussian kernel should pass through the midpoints of the edges joining the cell's generating point. However, in the general case these edge midpoints do not capture the shape of the Voronoi cell as well as the circumcenters. This is illustrated in Fig. 2, where

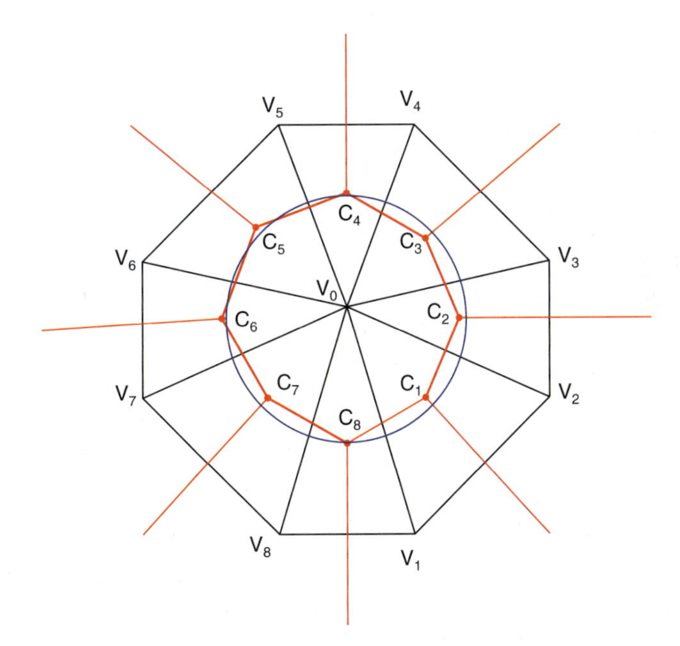

Fig. 1. A Voronoi cell is approximated by an ellipse shown in two-dimensions. V_0 is the grid point for which the Voronoi cell is constructed, $V_1, \ldots, V_8$ are neighboring grid points, and C_i is the circumcenter of triangle $V_0 V_i V_{i+1}$

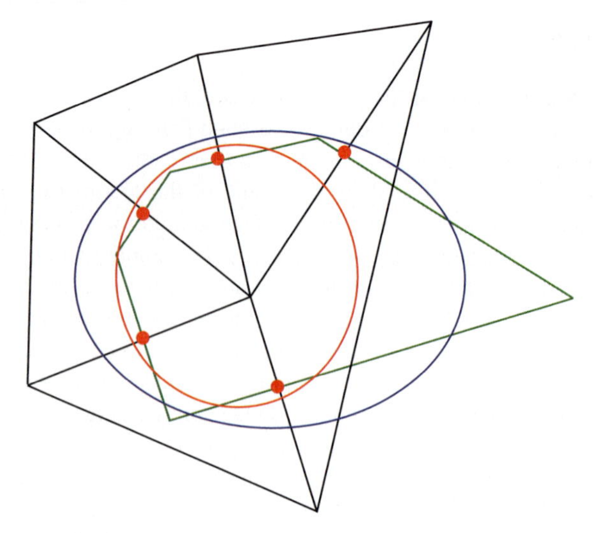

Fig. 2. In more general cases, as shown here, the edge midpoints do not capture the shape of a Voronoi cell as well as the circumcenters

the edge midpoints are shown as red dots, and the ellipse fitted from these midpoints is also shown in red. The ellipse fitted from the circumcenters is shown in blue. From this example, we can see that the blue ellipse approximates the Voronoi cell, shown in green, much better than the red ellipse. Therefore, we use the circumcenters of the incident tetrahedra as guide points for fitting the ellipsoid that approximates the 0.25-valued iso-contour of an elliptical Gaussian kernel.

If a tetrahedron is almost flat, its circumcenter is located far away from this tetrahedron. In this case, the Voronoi cell is an inferior shape for fitting the grid points kernel. Thus, if the circumcenter is too far away from the center grid point, we use the circumcenter of the triangle opposite to the center point as the contour guide point.

For each such guide point, we use its corresponding solid angle in the tetrahedron as the weight. Thus, for each grid point we have a list of weighted guide points associated with it. This list of the weighted guide points are then fed to a weighted least squares algorithm to fit the elliptical Gaussian kernels.

3.2 Initial Guess

Before we use the least squares method to fit an ellipsoid at each grid point based on the generated weighted guide points, we analyze the guide points using the Principal Component Analysis (PCA) [16] method to estimate the ellipsoid defined by the guide points. A PCA analysis of the guide points performs an eigen-decomposition of the covariance matrix of the guide points. This produces three eigenvalues and corresponding eigenvectors, which in 3D define a local orthogonal coordinate system related to the ellipsoid induced by the guide points.

Suppose the given grid point is v and the related N guide points are $u_i, i = 1, 2, \ldots, N$. We use the following equation to compute the covariance matrix M:

$$M = \sum_{i=1}^{N} (u_i - v)(u_i - v)^t \tag{5}$$

where M is a 3×3 matrix. From this 3×3 covariance matrix, we can compute the three eigenvalues and the corresponding eigenvectors. We use the three eigenvalues as the initial guess for the three scaling factors. The corresponding eigenvectors form a rotation matrix, which yields the initial guess for the three rotation angles. The PCA analysis makes the minimization process convergence faster by providing a good guess of the ellipsoid.

3.3 Energy Function

Given a set of weighted guide points $(w_i, u_i), i = 1, 2, \ldots, N$, in order to use the weighted least squares method to fit them with (4), we need to design an energy function for minimization. For this purpose, we use the sum of the weighted distances from the guide points to the quadric surface Q. This yields the energy function:

$$E(a, b, c, \alpha, \beta, \gamma) = \sum_{i=1}^{N} (w_i \times d_i^2) \tag{6}$$

where w_i is the weight of the guide point u_i, and d_i is the distance from point u_i to the ellipsoid Q, which is the length of the shortest line segment connecting u_i to any point on Q. For a given guide point $u_i = (x_i, y_i, z_i)$, Hart [17] proposed an algorithm to compute the closest point $u_i' = (x_i', y_i', z_i')$ on an axis-aligned ellipsoid defined by equation $f(x, y, z) = (x/a)^2 + (y/b)^2 + (z/c)^2 - 1 = 0$. As we know, the vector $\overrightarrow{u_i' u_i}$ is normal to the surface defined by $f(x, y, z)$ at u_i', which satisfies the following equation:

$$x_i - x = t\frac{x}{a^2},\ y_i - y = t\frac{y}{b^2},\ z_i - z = t\frac{z}{c^2} \tag{7}$$

Plugging this equation into $f(x, y, z)$ confines the point to the ellipsoid, producing:

$$\frac{a^2 x_i^2}{(t + a^2)^2} + \frac{b^2 y_i^2}{(t + b^2)^2} + \frac{c^2 z_i^2}{(t + c^2)^2} = 1 \tag{8}$$

This equation is equivalent to a sixth degree polynomial, obtained by multiplying through by the denominators. The largest root of this polynomial corresponds to the closest point on the ellipsoid. There are no closed formulas for the roots of such polynomials. We use a Newton's iteration method to find the largest root. When we obtain the largest root t_0 of this polynomial, the closest point $u_i' = (x_i', y_i', z_i')$ on the surface of the ellipsoid is obtained by plugging t_0 into (7), which yields the following equation:

$$x_i' = \frac{a^2 x_i}{t_0 + a^2},\ y_i' = \frac{b^2 y_i}{t_0 + b^2},\ z_i' = \frac{c^2 z_i}{t_0 + c^2} \tag{9}$$

Then, the distance from the point u_i to the ellipsoid is exactly the distance between u_i and u_i'.

To compute the distance from guide point u_i to an arbitrary oriented ellipsoid $Q = R \cdot (S^{-1})^2 \cdot R^t$, we transform Q and u_i to $Q' = (S^{-1})^2$ and $u'_i = R^{-1}u_i$ respectively by applying matrix R^{-1}, where Q' is an axis aligned ellipsoid centered on the origin. Then, the distance from u_i to Q is the distance from u'_i to Q' in the new coordinate system, which can be computed using the above equations. Next, we employ an iterative method to compute the minimum of E.

3.4 Minimization

The energy function of (6) is a very common unconstrained minimization problem. Powell [18] proposed a minimization method to solve this kind of problem without calculating derivatives. Powell's method ensures convergence in a finite number of steps, for a positive definite quadratic function, by making use of some properties of conjugate directions. However, this method sometimes results in search directions that become linearly dependant. The simplest way to avoid linear dependance of the search directions with Powell's basic procedure, retaining quadratic convergence, is to reset the search directions to the columns of the identity matrix after every n or $n+1$ iterations, where n is number of unknowns in the system. However, the restarting may slow down convergence, because information built up about the function is periodically thrown away. Thus, we use a modification of Powell's basic procedure proposed by Brent in [19] to solve the minimization problem. In consequence, we obtain the scaling matrix S and the rotation matrix R of the 0.25-valued iso-contour for each elliptical Gaussian kernel, which give the shape and orientation of the elliptical Gaussian kernel.

4 Evaluation

The straightforward way to evaluate the resulting 3D elliptical Gaussian kernel configuration is to resample the irregular grid data into a $N \times N \times N$ regular grid R. In the ideal case, if the grid point is inside one of the tetrahedron, the contributions from all kernels to this point sum to one. In practice, the contributions from all kernels to a grid point do not always sum to one. Therefore, the volume rendering image generated with the splatting algorithm may look blotchy. Normalizing the reconstructed value at each grid point by the contribution of reconstruction kernels can alleviate the problem.

The sum of the contributions of all kernels to each grid point can be used to evaluate the quality of the fitted 3D elliptical Gaussian kernels. Suppose the set of regular grid points inside the tetrahedra mesh is V. The standard deviation is computed as follows:

$$S = \sqrt{\frac{\sum_{v \in V}(C_v - 1.0)^2}{|V|}} \tag{10}$$

where C_v is the sum of the contributions from the reconstruction kernels to grid point v. The standard deviation S with a small value indicates a better quality of the reconstruction kernel ensemble, constructed via our fitting procedure.

5 Rendering

Our rendering system uses the sheet-buffered image aligned splatting algorithm introduced in [10, 20] and further refined in [21] for the 4D case. The system was extended in order to rasterize ellipsoids of varying size and orientation. Following this method, the elliptical kernels of the volume are sliced into image aligned sheet buffers. The slices are then shaded per-pixel and composited front-to-back onto the final image. The ellipsoids are defined by a 3×3 rotation matrix and a diagonal 3×3 scaling matrix, as produced by the fitting algorithm described above. Similar to 3D and 4D regular splatting where this method was used, it produces crisp fully shaded images. The process, however, is slightly more demanding when rendering unstructured grids using elliptical splats, because the produced sheet buffers have to be normalized before shading and compositing.

6 Implementation and Results

In this section, we present some implementation details and testing results. Our algorithm is implemented using C++ on the Windows platform, and the CGAL C++ library (`www.cgal.org`) is used to perform the 3D Delaunay triangulation in the preprocessing step. All of the experiments have been conducted on a 3.0 GHz Intel Pentium IV PC running Windows XP with 1 GB RAM. We list the datasets used in the experiments, the kernel fitting time, and the standard deviation, and max weight in Table 1. Our fitting algorithm can, on average, fit 1,300 points per minute.

We use the NASA Blunt Fin dataset with 40,960 grid points in our first experiment. To perform the numerical comparisons, we use the fitted 3D elliptical Gaussian kernels to resample it into a regular grid. One slice of the resampled regular grid is shown in Fig. 3 with two images: (a) the weight image of that slice, and (b) the density image with normalization applied. Both the weight image and the density image look smooth, but somewhat fuzzy at the boundary. The weight image is the key for quality evaluation. The more homogeneous the quality of resulting kernels is the better. We show the 3D elliptical Gaussian kernels in Fig. 4a using a surface rendering method. Figure 4b is the volume rendered image using our software splatting algorithm for elliptical splats. We observe that there are some large elliptical Gaussian kernels located at the boundary, which cause the fuzziness of the volume rendering at the boundary.

Table 1. Kernel fitting times (in minutes), standard deviation, and max weight for two different datasets

Dataset	Points	Tetrahedra	Fitting time	Standard deviation	Max weight
Blunt Fin	40,960	187,395	27.8	0.57	2.85
Combustion	47,025	215,040	40.8	0.48	3.15

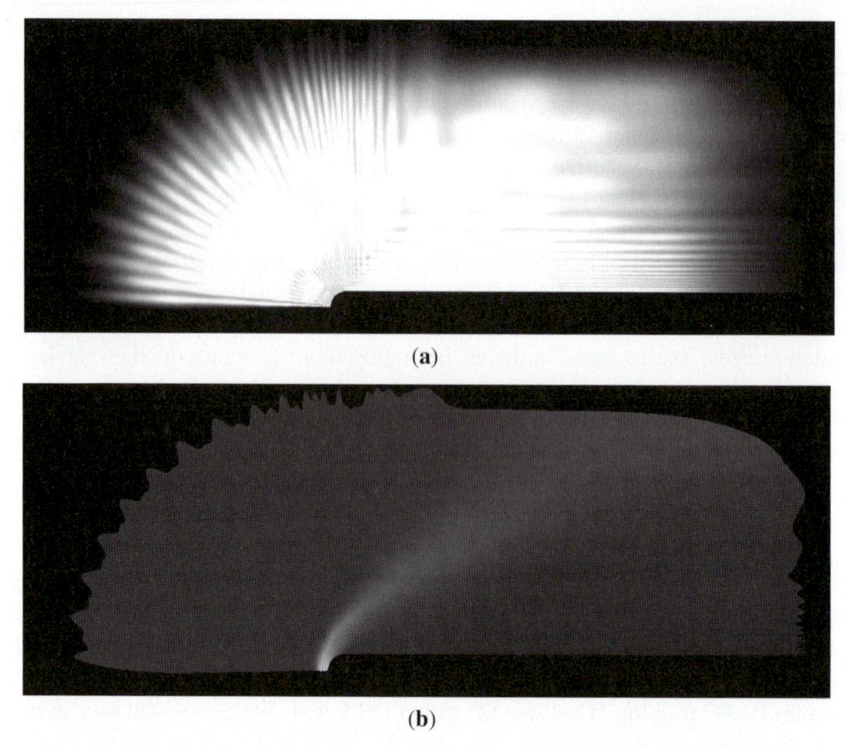

Fig. 3. Blunt Fin dataset: (a) Weight image and (b) density image for one slice of the regular grid samples evaluated using the fitted 3D elliptical Gaussian kernels of the corresponding irregular grid

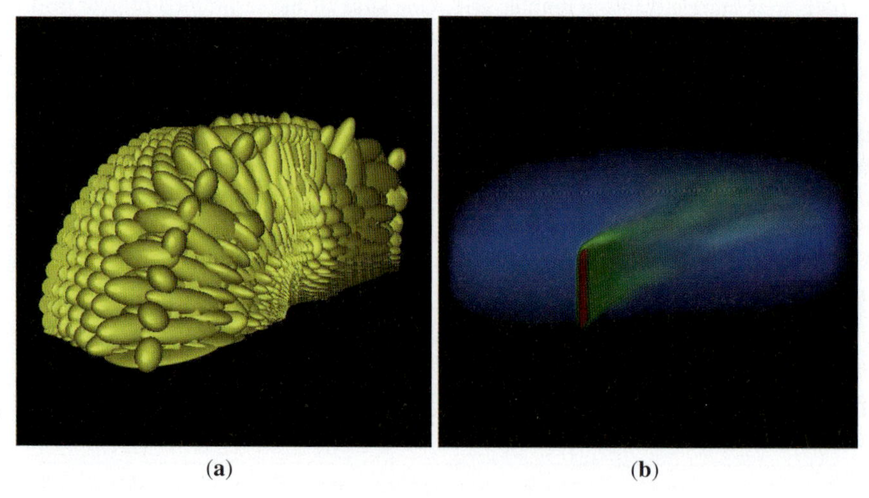

Fig. 4. Blunt Fin dataset: (a) The ensemble of fitted 3D elliptical Gaussian kernels, and (b) a volume rendered image using the elliptical splatting algorithm

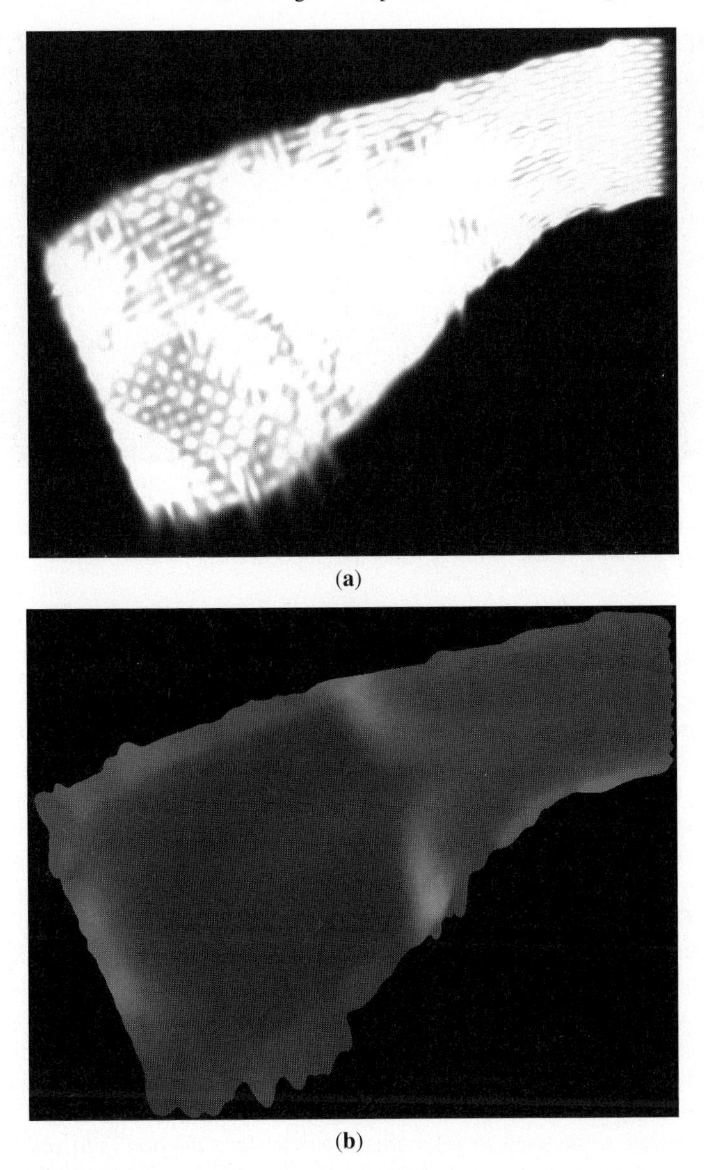

(a)

(b)

Fig. 5. Combustion Chamber dataset: (**a**) Weight image and (**b**) density image for one slice of the regular grid samples evaluated using the fitted 3D elliptical Gaussian kernels of the corresponding irregular grid

The Combustion Chamber dataset is from the Visualization Toolkit (Vtk). It consists of 47,025 grid points. One slice of the resampled regular grids is shown in Fig. 5. The resulting elliptical Gaussian kernels and volume rendered image with the splatting algorithm for the Combustion Chamber are shown in Fig. 6.

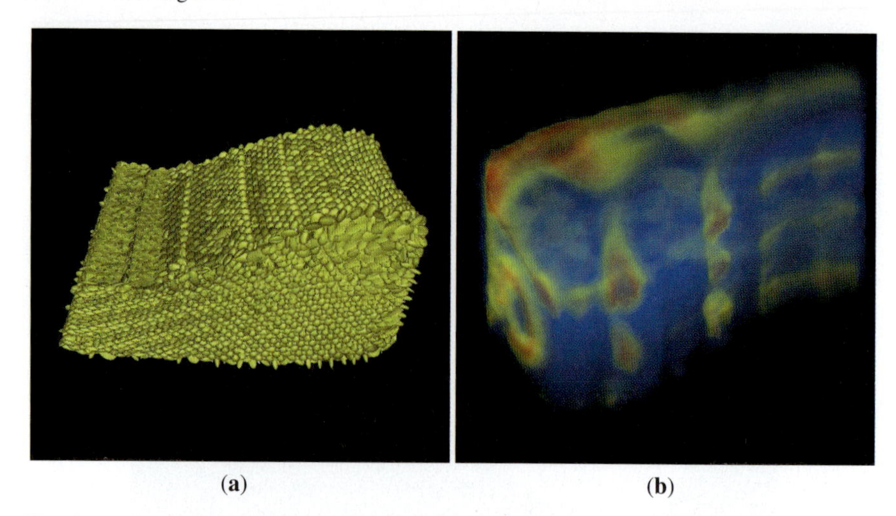

Fig. 6. Combustion Chamber dataset: (**a**) The ensemble of fitted 3D elliptical Gaussian kernels, and (**b**) a volume rendered image using the elliptical splatting algorithm

7 Conclusion and Future Work

In this paper, we have presented a method to construct an ensemble of 3D elliptical Gaussian kernels for irregular data. Our method does not resample the irregular grids to generate regular grids. Instead, we create the 3D elliptical kernels centered on the original grid points using a weighted least squares method to fit ellipsoids. We perform PCA on the guide points to provide an initial guess for the minimization process to obtain faster convergence. The resulting kernels are arbitrarily oriented elliptical Gaussians modeled via a matrix representation. The kernels are seamlessly incorporated into our splatting rendering system.

Our method has some limitations. The quality of the resulting kernels is affected by the shape of the Voronoi cells. It would require some amount of resampling in locations that are not covered well by the resulting kernels. In this case, the global optimization methods will do better. Our experiment results show that the boundaries of the irregular grids are not preserved well, appearing somewhat fuzzy. Two methods are possible to be used to solve this problem. One method is to subdivide the boundary tetrahedron. But this would require the resampling of more points and thus more splats would be generated. Another method is to add a layer of "ghost splats" outside the tetrahedra mesh to solve this problem. In future work, we would like to study where to place these ghost splats in order to preserve the boundary well. In our current implementation, the rendering is implemented using software. In subsequent, we would like to exploit the power of GPUs to accelerate the rendering. Here, the main feature of floating point blending will be highly beneficial.

References

1. Fruhauf, T.: Raycasting of nonregularly structured volume data. Eurographics, 13(3), C294–C303 (1994)
2. Garrity, M.: Raytracing irregular volume. Comput. Graph., 24(5), 35–40 (1990)
3. Ramamoorthy, S. and Wilhelms, J.: An analysis of approaches to ray-tracing curvilinear grids. Tech. Report UCSC-CRL-92-07, University of California, Santa Cruz (1992)
4. Farias, R. and Silva, T.C.: Out-of-core of large, unstructured grids. IEEE Comput. Graph. Appl., 21(4), 42–50 (2001)
5. Max, N., Hanrahan, P., and Crawfis, R.: Area and volume coherece for efficient visualization of 3D scalar functions. Comput. Graph., 24(5), 27–33 (1990)
6. Westover, L.: Footprint evaluation for volume rendering. Comput. Graph., 24(4), 367–376 (1990)
7. Weiler, M., Kraus, M., Merz, M., and Ertl, T.: Hardware-based ray casting for tetrahedral meshes. In: Proceedings of IEEE Visualization, 333–340 (2003)
8. Röttger, S., Kraus, M., and Etrl, T.: Hardware-accelerated volume and isosurface rendering based on cell-projection. In: Proceedings of IEEE Visualization, 109–116 (2000)
9. Weiler, M., Kraus, M., and Ertl, T.: Hardware-based view independant cell projection. In: Proceedings of IEEE Symposium on Volume Visualization, 13–22 (2002)
10. Mueller, K., Möller, T., and Crawfis, R.: Splatting without the blur. In: Proceedings of IEEE Visualization, 363–371 (1999)
11. Meredith, J. and Ma, K.L.: Multiresolution view-dependent splat based volume rendering of large irregular data. In: Proceedings of the IEEE symposium on parallel and large-data visualization and graphics (2001)
12. Mao, X., Hong, L., and Kaufman, A.: Splatting of curvilinear volumes. In: Proceedings of IEEE Visualization, 61–68 (1995)
13. Mao, X.: Splatting of non rectilinear volumes through stochastic resampling. IEEE Trans. Vis. Comput. Graph., 2(2), 156–170 (1996)
14. Jang, J., Shaw, C., Ribarsky, W., and Faust, N.: View-dependent multiresolution splatting of non-uniform data. In: Eurographics–IEEE Visualization Symposium, 125–132 (2002)
15. Jang, Y., Weiler, M., Hopf, M., Huang, J., Ebert, D.S., Gaither, K.P., and Ertl, T.: Interactively visualizing procedurally encoded scalar fields. In: Joint Eurographics–IEEE TCVG Symposium on Visualization (2004)
16. Jolliffe, I.T.: Principal component analysis. Springer, New York, (1986)
17. Hart, J.: Computing distance betwwen point and ellipsoid. Graphics Gems IV. Academic, Boston, MA, 113–119 (1994)
18. Powell, M.J.D.: An efficient method for finding the minimum of a function of several variables without calculating derivatives. Comp. J. 7, 303–307 (1964)
19. Brent, R.P.: Algorithms for minimization without derivatives. Dover, Mineola, NY (1973)
20. Mueller, K. and Crawfis, R.: Eliminating popping artifacts in sheet buffer-based splatting. In: Proceedings of IEEE Visualization, 239–245 (1998)
21. Neophytou, N. and Mueller, K.: Space–time points: Splatting in 4D. In: Symposium on Volume Visualization and Graphics, 97–106 (2002)

From Sphere Packing to the Theory of Optimal Lattice Sampling

Alireza Entezari, Ramsay Dyer, and Torsten Möller

Simon Fraser University, Burnaby, BC, Canada
{aentezar, rhdyer, torsten}@cs.sfu.ca

Summary. In this paper we introduce reconstruction kernels for the 3D optimal sampling lattice and demonstrate a practical realisation of a few. First, we review fundamentals of multidimensional sampling theory. We derive the optimal regular sampling lattice in 3D, namely the Body Centered Cubic (BCC) lattice, based on a spectral sphere packing argument. With the introduction of this sampling lattice, we review some of its geometric properties and its dual lattice. We introduce the ideal reconstruction kernel in the space of bandlimited functions on this lattice. Furthermore, we introduce a family of box splines for reconstruction on this sampling lattice. We conclude the paper with some images and results of sampling on the BCC lattice and contrast it with equivalent samplings on the traditionally used Cartesian lattice. Our experimental results confirm the theory that BCC sampling yields a more accurate discrete representation of a signal comparing to the commonly used Cartesian sampling.

1 Introduction

With the advent of the theory of digital signal processing various fields in science and engineering have been dealing with discrete representations of continuous phenomena. As scientific computing algorithms mature and find applications in a variety of scientific, medical and engineering fields, the question of the accuracy of the discrete representations gains an enormous importance. The theory of optimal sampling deals with this issue: given a fixed number of samples, how can one capture the most information from the underlying continuous phenomena. Such a sampling pattern would constitute the most accurate discrete representation.

While virtually all image and volume processing algorithms are based on the Cartesian sampling, it has been well known that this sampling lattice is suboptimal. Yet, only recently advances have been made by introducing reconstruction filters for the 2D optimal lattice (e.g., the Hexagonal lattice). Our paper introduces novel reconstruction filters for the Body Centered Cubic (BCC) lattice, the analogous optimal sampling lattice in 3D, that are based on the geometric structure of the underlying lattice. This should pave the way for a more mainstream adaption of the BCC lattice for the discrete representation and processing of three-dimensional phenomena.

T. Möller et al. (eds.), *Mathematical Foundations of Scientific Visualization, Computer Graphics, and Massive Data Exploration*, Mathematics and Visualization,
DOI: 10.1007/978-3-540-49926-8,

An introduction to multidimensional sampling theory can be found in Dudgeon and Mersereau [5]. A lattice can be viewed as a periodic sampling pattern. Periodic sampling of a function in the spatial domain gives rise to a periodic replication of the spectrum in the Fourier domain. The lattice that describes the centers of the replicas in the Fourier domain is called the *dual*, *reciprocal*, or *polar* lattice. Reconstruction in the spatial domain amounts to eliminating the replicas of the spectrum in the Fourier domain while preserving the primary spectrum. Therefore, the ideal reconstruction function is the inverse Fourier transform of the characteristic function of the Voronoi cell of the dual lattice.

In Sect. 2 we will give a rigorous introduction to multidimensional sampling theory and derive the relationship between the sampling pattern in the spatial and the frequency domain. This will allow us to derive the notion of the optimal sampling lattice in Sect. 3. Section 4 will discuss and derive geometric aspects of the BCC and the FCC lattices, setting the stage for deriving nearest neighbor, linear and cubic reconstruction filters in Sect. 5. A practical implementation of the linear reconstruction filter is derived in Sects. 6 and 7 discusses our experimental evaluation. Finally, Sects. 8 and 9 summarize our contributions and point to some open problems, respectively.

2 Multidimensional Sampling Theory

Let $f \in L_2(\mathbb{R}^n)$ be a multivariate function for which the Fourier transform exists and let $\hat{f} : \mathbb{R}^n \to \mathbb{C}$ be its Fourier transform:

$$\hat{f}(\boldsymbol{\omega}) = \int f(\boldsymbol{x}) e^{-2\pi i \boldsymbol{\omega} \cdot \boldsymbol{x}} d\boldsymbol{x}$$

Given the fact that $\hat{f} \in L_2(\mathbb{R}^n)$ also, the inversion formula

$$f(\boldsymbol{x}) = \int \hat{f}(\boldsymbol{\omega}) e^{2\pi i \boldsymbol{\omega} \cdot \boldsymbol{x}} d\boldsymbol{\omega}$$

recovers the original function f almost everywhere.[1] If the original function is also continuous, the reconstruction equality holds everywhere [7].

We are interested in the regular sampling of a function and its reconstruction from the discrete set of samples. In this paper we shall refer to reconstruction in the space of functions with a compact support in their Fourier representations (i.e., bandlimited functions).

The sampling operation is defined over the space of square integrable functions ($L_2(\mathbb{R})$) equipped with the usual inner product:

$$\langle f, g \rangle = \int f(\boldsymbol{x}) g(\boldsymbol{x}) d\boldsymbol{x}.$$

[1] Reconstruction takes the mean value of the left and the right limit at the points of discontinuity.

Assuming sample values are produced by a sampling device which is characterized by a function, g, called its impulse response.[2] The sampling operation, takes a function that is an element of L_2 and returns a number. This operation can be modeled by the following functional:

$$L_2 \mapsto \mathbb{R} : f \mapsto \langle f, g \rangle .$$

The ideal impulse response (i.e., sampling function) is referred to as Dirac's delta (generalized) function which is the point evaluation functional defined by the following functional equation:

$$\delta[f] = f(0) \tag{1}$$

for all continuous functions f. Formally this symbol in an integral behaves as the limit of integrals of a sequence of integrable functions K_r that have the properties:

$$\int K_r(\boldsymbol{x})d\boldsymbol{x} = 1 \text{ for all } r > 0$$

$$\lim_{r \to 0} K_r(\boldsymbol{x}) = 0 \text{ for all } \boldsymbol{x} \neq 0.$$

Examples of such kernels consist of Dirichlet, Fejér, Gaussian and Poisson kernels. It is customary to say that in the limit these kernels behave like the delta function:

$$\delta[f] = \lim_{r \to 0} \int K_r(\boldsymbol{x}) f(\boldsymbol{x}) d\boldsymbol{x} = f(0)$$

for all continuous functions f. Therefore the behavior of the functional in (1) can be considered as the behavior of the above limit. As a notational convenience, the operation of δ on a function f is defined as:

$$\int \delta(\boldsymbol{x}) f(\boldsymbol{x}) d\boldsymbol{x} \triangleq \delta[f]$$

even though, such a function $\delta(\boldsymbol{x})$ does not exist. Since Dirac's generalized function is not a function in the classical setting, the symbolic introduction of Dirac's delta function is merely for the ease of notation.

A regular sampling pattern can be viewed as a point lattice. An n-dimensional *point lattice* is characterized by a set of n basis vectors $\{\boldsymbol{T}_j\}_{1 \leq j \leq n}$. A point is on the lattice if and only if it is described by a linear combination with integer coefficients of the basis vectors. The matrix, $\boldsymbol{T} = [\boldsymbol{T}_1 \boldsymbol{T}_2 \ldots \boldsymbol{T}_n]$, whose columns are the basis vectors is called the *sampling matrix* and any lattice point $\boldsymbol{t}$ is given by $\boldsymbol{t} = \boldsymbol{T}\boldsymbol{p}$ for some $\boldsymbol{p} \in \mathbb{Z}^n$. Figure 1 illustrates a two-dimensional lattice. The impulse response of such a sampling lattice is:

$$Ш_{\boldsymbol{T}}(\boldsymbol{x}) = \sum_{\boldsymbol{k} \in \mathbb{Z}^n} \delta(\boldsymbol{x} - \boldsymbol{T}\boldsymbol{k}) \tag{2}$$

[2] In the medical imaging community the impulse response is sometimes referred to as the excitation function.

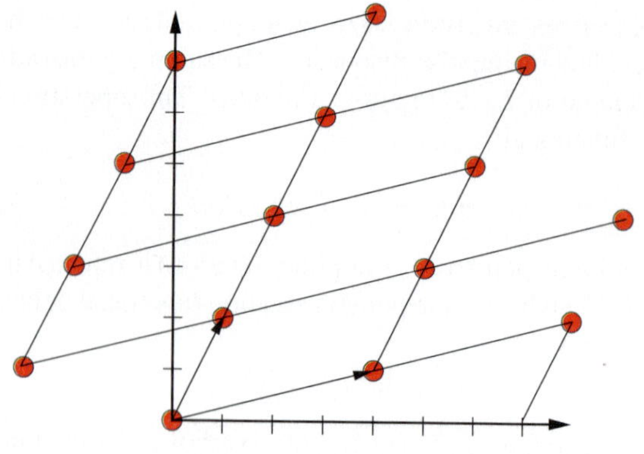

Fig. 1. A two-dimensional lattice with $\boldsymbol{T} = [\boldsymbol{T}_1 \boldsymbol{T}_2]$, where $\boldsymbol{T}_1 = [4, 1]^\top$ and $\boldsymbol{T}_2 = [1, 2]^\top$

This equation is again a symbolic equation that eases the notation. The corresponding definition, then, is:

$$\int Ш_{\boldsymbol{T}}(\boldsymbol{x}) f(\boldsymbol{x}) d\boldsymbol{x} \triangleq \lim_{r \to 0} \int \sum_{\boldsymbol{k} \in \mathbb{Z}^n} K_r(\boldsymbol{x} - \boldsymbol{T}\boldsymbol{k}) f(\boldsymbol{x}) d\boldsymbol{x} \tag{3}$$

for all continuous functions f with bounded support.
The corresponding functional that defines the $Ш_{\boldsymbol{T}}$ is:

$$\int Ш_{\boldsymbol{T}}(\boldsymbol{x}) f(\boldsymbol{x}) d\boldsymbol{x} = \sum_{\boldsymbol{k} \in \mathbb{Z}^n} f(\boldsymbol{T}\boldsymbol{k}) \tag{4}$$

Therefore the function that is obtained by the sampling device is:

$$f_s(\boldsymbol{x}) = Ш_{\boldsymbol{T}}(\boldsymbol{x}) f(\boldsymbol{x}) \tag{5}$$

We observe that $Ш_{\boldsymbol{T}}$ is a periodic function with period $\boldsymbol{T}$: $Ш_{\boldsymbol{T}}(\boldsymbol{x} + \boldsymbol{T}\boldsymbol{m}) = Ш_{\boldsymbol{T}}(\boldsymbol{x})$ for $\boldsymbol{m} \in \mathbb{Z}^n$.

In order to study the effect of the sampling operator on the Fourier representation of the underlying function, we need to derive the Fourier transform (FT) of $Ш_{\boldsymbol{T}}$. Without claiming any of the convergence properties of the Fourier transform for the shah function, we transform the shah as:

$$\hat{Ш}_{\boldsymbol{T}}(\boldsymbol{\omega}) = \int Ш_{\boldsymbol{T}}(\boldsymbol{x}) e^{-2\pi i \boldsymbol{\omega} \cdot \boldsymbol{x}} d\boldsymbol{x}.$$

Since the exponential functions are continuous, (4) yields:

$$\hat{Ш}_{\boldsymbol{T}}(\boldsymbol{\omega}) = \sum_{\boldsymbol{k} \in \mathbb{Z}^n} e^{-2\pi i \boldsymbol{\omega} \cdot (\boldsymbol{T}\boldsymbol{k})}.$$

Note that $\hat{Ш}_{\boldsymbol{T}}$ is periodic with the periodicity matrix $\tilde{\boldsymbol{T}} = \boldsymbol{T}^{-\top}$, since for any $\boldsymbol{m} \in \mathbb{Z}^n$:

$$\begin{aligned}
\hat{Ш}_{\boldsymbol{T}}(\boldsymbol{\omega} + \boldsymbol{T}^{-\top}\boldsymbol{m}) &= \sum_{\boldsymbol{k}\in\mathbb{Z}^n} e^{-2\pi i(\boldsymbol{\omega}+\boldsymbol{T}^{-\top}\boldsymbol{m})\cdot(\boldsymbol{T}\boldsymbol{k})} \\
&= \sum_{\boldsymbol{k}\in\mathbb{Z}^n} e^{-2\pi i\left[\boldsymbol{\omega}\cdot(\boldsymbol{T}\boldsymbol{k})+(\boldsymbol{T}^{-\top}\boldsymbol{m})\cdot(\boldsymbol{T}\boldsymbol{k})\right]} \\
&= \sum_{\boldsymbol{k}\in\mathbb{Z}^n} e^{-2\pi i\left[\boldsymbol{\omega}\cdot(\boldsymbol{T}\boldsymbol{k})+(\boldsymbol{m}^{\top}\boldsymbol{T}^{-1}\boldsymbol{T}\boldsymbol{k})\right]} \\
&= \sum_{\boldsymbol{k}\in\mathbb{Z}^n} e^{-2\pi i\left[\boldsymbol{\omega}\cdot(\boldsymbol{T}\boldsymbol{k})+(\boldsymbol{m}^{\top}\boldsymbol{k})\right]} \\
&= \sum_{\boldsymbol{k}\in\mathbb{Z}^n} e^{-2\pi i\boldsymbol{\omega}\cdot(\boldsymbol{T}\boldsymbol{k})} e^{-2\pi i\boldsymbol{m}\cdot\boldsymbol{k}} \\
&= \sum_{\boldsymbol{k}\in\mathbb{Z}^n} e^{-2\pi i\boldsymbol{\omega}\cdot(\boldsymbol{T}\boldsymbol{k})} \text{ since } e^{-2\pi i\boldsymbol{m}\cdot\boldsymbol{k}} = 1 \text{ for all } \boldsymbol{m}, \boldsymbol{k} \in \mathbb{Z}^n \\
&= \hat{Ш}_{\boldsymbol{T}}(\boldsymbol{\omega}).
\end{aligned}$$

Theorem 1. *The Fourier transform of a Shah function* $Ш_{\boldsymbol{T}}(\boldsymbol{x})$ *over a lattice* $\boldsymbol{T}$ *as defined in (2) can be described as another Shah function* $Ш_{\tilde{\boldsymbol{T}}}(\boldsymbol{\omega})$ *over the dual lattice* $\tilde{\boldsymbol{T}} = \boldsymbol{T}^{-\top}$ *with* $\hat{Ш}_{\boldsymbol{T}}(\boldsymbol{\omega}) = Ш_{\tilde{\boldsymbol{T}}}(\boldsymbol{\omega})$

Proof. To prove this theorem we note that the $\hat{Ш}_{\boldsymbol{T}}$ is $\tilde{\boldsymbol{T}}$ periodic. Let $\Omega = \tilde{\boldsymbol{T}}[-\frac{1}{2}, \frac{1}{2})^n$ be one period of the domain $\mathbb{R}^n$. Therefore, all we need to show is that:

$$\int_{\Omega} \hat{Ш}_{\boldsymbol{T}}(\boldsymbol{\omega})g(\boldsymbol{\omega})d\boldsymbol{\omega} = g(0) = \int \delta(\boldsymbol{\omega})g(\boldsymbol{\omega})d\boldsymbol{\omega} \tag{6}$$

for all continuous $g : \mathbb{R}^n \mapsto \mathbb{C}$ with bounded support.

For the choice of the kernel K_r in (3), we resort to the Poisson kernel. The family of functions $P_r : \mathbb{R}^n \mapsto \mathbb{C},\ 0 < r < 1$ defined by:

$$P_r(\boldsymbol{\omega}) = \sum_{\boldsymbol{k}\in\mathbb{Z}^n} r^{\|\boldsymbol{k}\|} e^{2\pi i\boldsymbol{\omega}\cdot(\boldsymbol{T}\boldsymbol{k})}$$

where $\|\boldsymbol{k}\| = \sum_{i=1}^n |k_i|$ for $\boldsymbol{k} = [k_1 \dots k_n]^{\top}$.

Expanding the right-hand side of the above equation we have:

$$\begin{aligned}
P_r(\tilde{\boldsymbol{T}}\boldsymbol{\omega}) &= \sum_{\boldsymbol{k}\in\mathbb{Z}^n} r^{\|\boldsymbol{k}\|} e^{2\pi i\boldsymbol{\omega}\cdot\boldsymbol{k}} \\
&= \sum_{k_1,k_2,\dots,k_n\in\mathbb{Z}} r^{|k_1|}r^{|k_2|}\dots r^{|k_n|}e^{2\pi i\omega_1 k_1}e^{2\pi i\omega_2 k_2}\dots e^{2\pi i\omega_n k_n} \\
&= (\sum_{k_1\in\mathbb{Z}} r^{|k_1|}e^{2\pi i\omega_1 k_1})(\sum_{k_2\in\mathbb{Z}} r^{|k_2|}e^{2\pi i\omega_2 k_2})\dots(\sum_{k_n\in\mathbb{Z}} r^{|k_n|}e^{2\pi i\omega_n k_n}) \\
&= P_r(\omega_1)P_r(\omega_2)\dots P_r(\omega_n)
\end{aligned}$$

In other words, the multidimensional Poisson kernel is a separable kernel and therefore we can use the following one-dimensional results from [1]:

$$P_r(\omega) = \sum_{k\in\mathbb{Z}} r^{|k|} e^{2\pi i \omega k}$$
$$P_r(\omega) \geq 0$$
$$\int_{[-\frac{1}{2},\frac{1}{2})} P_r(\omega) d\omega = 1$$

Furthermore, [1]:

$$\lim_{r\to 1} \int P_r(\omega) g(\omega) d\omega = g(0) = \int \delta(\omega) g(\omega) d\omega$$

Since $P_r(\omega)$ is positive and bounded, $P_r(\boldsymbol{\omega}) \geq 0$ and by the Fubini theorem we have:

$$\int_{[-\frac{1}{2},\frac{1}{2})^n} P_r(\tilde{\boldsymbol{T}}\boldsymbol{\omega}) d\boldsymbol{\omega} = \int_{-\frac{1}{2}}^{\frac{1}{2}} P_r(\omega_1) d\omega_1 \int_{-\frac{1}{2}}^{\frac{1}{2}} P_r(\omega_2) d\omega_2 \ldots \int_{-\frac{1}{2}}^{\frac{1}{2}} P_r(\omega_n) d\omega_n = 1.$$

Moreover, for any continuous function $g : \mathbb{R}^n \mapsto \mathbb{C}$ we have:

$$\begin{aligned}
&\lim_{r\to 1} \int_{\Omega} P_r(\boldsymbol{\omega}) g(\boldsymbol{\omega}) d\boldsymbol{\omega} = \\
&\quad = \lim_{r\to 1} \int_{\omega_1} \cdots \int_{\omega_n} P_r(\omega_1, \ldots, \omega_n) g(\omega_1, \ldots, \omega_n) d\omega_n \ldots d\omega_1 \\
&\quad = \lim_{r\to 1} \int_{\omega_1} \cdots \int_{\omega_{n-1}} P_r(\omega_1, \ldots, \omega_{n-1}) g(\omega_1, \ldots, \omega_{n-1}, 0) d\omega_{n-1} \ldots d\omega_1 \\
&\quad \vdots \\
&\quad = \lim_{r\to 1} \int_{\omega_1} P_r(\omega_1) g(\omega_1, 0, \ldots, 0) d\omega_1 = g(0)
\end{aligned}$$

Hence, we conclude by the dominated convergence theorem that:

$$\lim_{r\to 1} \int_{\Omega} P_r(\boldsymbol{\omega}) g(\boldsymbol{\omega}) d\boldsymbol{\omega} = \int_{\Omega} \hat{Ш}_{\boldsymbol{T}}(\boldsymbol{\omega}) g(\boldsymbol{\omega}) d\boldsymbol{\omega} = g(0)$$

Since $\hat{Ш}_{\boldsymbol{T}}$ is $\tilde{\boldsymbol{T}}$ periodic, we have:

$$\hat{Ш}_{\boldsymbol{T}}(\boldsymbol{\omega}) = Ш_{\tilde{\boldsymbol{T}}}(\boldsymbol{\omega}).$$

This equality is again a symbolic equality and its meaning is only defined under an integral:

$$\int \hat{Ш}_{\boldsymbol{T}}(\boldsymbol{\omega}) f(\boldsymbol{\omega}) d\boldsymbol{\omega} = \int Ш_{\tilde{\boldsymbol{T}}}(\boldsymbol{\omega}) f(\boldsymbol{\omega}) d\boldsymbol{\omega} \tag{7}$$

for all continuous functions f with bounded support.

In conclusion, the Fourier transform of $Ш_{\boldsymbol{T}}$ is yet another shah function on the reciprocal lattice $Ш_{\tilde{\boldsymbol{T}}}$. □

In order to find the Fourier transform of the sampled function f_s as in (5), one can use the Convolution-Multiplication theorem to show:

$$\hat{f}_s(\omega) = (\hat{\text{Ш}}_T * \hat{f})(\omega) = (\text{Ш}_{\tilde{T}} * \hat{f})(\omega)$$

The important observation from this result is that the two lattices represented by $\boldsymbol{T}$ and $\tilde{\boldsymbol{T}}$ are duals of each other through the Fourier transform.

3 The Optimal Lattice Sampling

The main result of the previous section was that sampling a function f on a lattice $\boldsymbol{T}$, brings about the replication of the Fourier Transform of f on the dual lattice $\tilde{\boldsymbol{T}} = \boldsymbol{T}^{-\top}$. Due to this reciprocal relationship, the sparsest sampling matrix $\boldsymbol{T}$ will have to produce the densest packing of the replicas of the spectrum on the dual lattice $\tilde{\boldsymbol{T}}$. Therefore, in order to distribute the samples in the spatial domain in the most economical (sparse) fashion, the dual lattice $\tilde{\boldsymbol{T}}$ needs to be as densely packed as possible.

In the typical three-dimensional case, usually there is no knowledge of a direction of preferred resolution for sampling the underlying function f and the function is assumed to be qualitatively isotropic. This means that f has a spherically uniform spectrum. With this assumption, the dense packing of the spectra in the Fourier domain can be addressed by the sphere packing problem. Consequently the best sampling lattice in 3D is dual to the lattice that attains the highest sphere packing density.

The sphere packing problem [3] can be traced back to the early seventeenth century. Finding the densest packing of spheres is known as the Kepler problem. The fact that the face centered cubic (FCC) packing attains the highest density of lattice packings was first proven by Gauß in 1831 [3]. Further, the Kepler conjecture – that the FCC packing is an optimal packing of spheres in 3D even when the lattice condition is not imposed – was not proven until 1998 by a lengthy computer-aided proof [6].

In two dimensions however, the hexagonal packing structure can be easily shown to attain the optimal density of packing. Since the two-dimensional hexagonal lattice is self-dual, the optimal sampling in 2D is a hexagonal lattice. Consequently, by sampling a 2D function f on a hexagonal lattice, its Fourier domain representation is replicated on the dual hexagonal lattice; similarly by sampling a function on the commonly used Cartesian lattice, its Fourier domain representation is replicated on the dual Cartesian lattice.

Figure 2 illustrates the optimality of hexagonal sampling vs. Cartesian sampling. An equivalent spatial domain sampling density is used for both the Cartesian and the hexagonal sampling lattice. The Fourier domain replication is shown for the Cartesian lattice in Fig. 2a and for the hexagonal lattice in Fig. 2b. It is apparent that the area of the main spectrum (in red) that is captured in the hexagonal case is larger than that of the Cartesian case. This means that with the equivalent sampling density in the spatial domain, the hexagonal sampling captures more of the frequency content

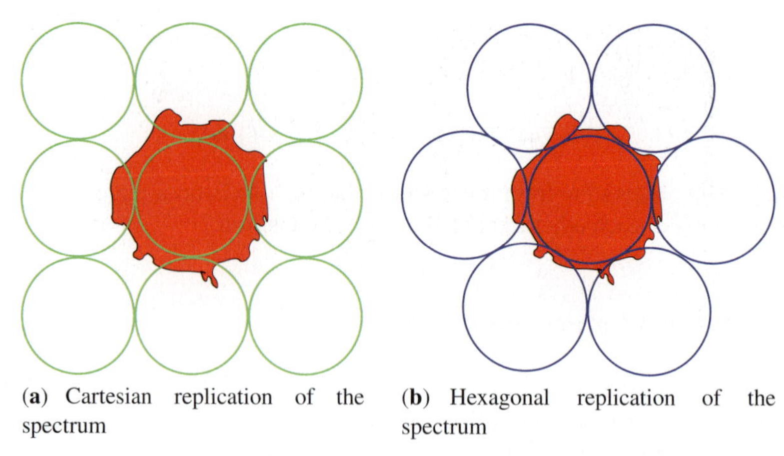

(**a**) Cartesian replication of the spectrum

(**b**) Hexagonal replication of the spectrum

Fig. 2. Hexagonal sampling captures higher frequencies with equal sampling density

of the spectrum of f and in the process of band limiting the underlying signal for sampling, we can allow a larger baseband to be captured. This means that more information can be captured with the same number of samples. The increased efficiency of the optimal sampling lattice in 2D is about 14% and in 3D it is about 30%.

While the optimal regular sampling theory is attractive for its theoretical advantages, it hasn't been widely employed in practice due to the lack of signal processing theory and tools to handle such a sampling lattice.

4 The BCC Lattice

A *lattice* can be viewed as an infinite array of points in which each point has surroundings identical to those of all the other points [2]. In other words, every lattice point has the same Voronoi cell and we can refer to the Voronoi cell of the lattice without ambiguity. The lattice points form a group under vector addition in the Euclidean space.

The BCC lattice is a sublattice of the Cartesian lattice. The BCC lattice points are located on the corners of the cube with an additional sample in the center of the cube as illustrated in Fig. 3. An alternative way of describing the BCC lattice is to start with a Cartesian lattice (i.e., $\mathbb{Z}^3$) and retain only those points whose coordinates have identical parity.

The simplest interpolation kernel on any lattice is the characteristic function of the Voronoi cell of the lattice. This is usually called nearest neighbor interpolation. More sophisticated reconstruction kernels involve information from the neighboring points of a given lattice point. With this in mind, we focus in the next section on the geometry and the polyhedra associated with the BCC lattice.

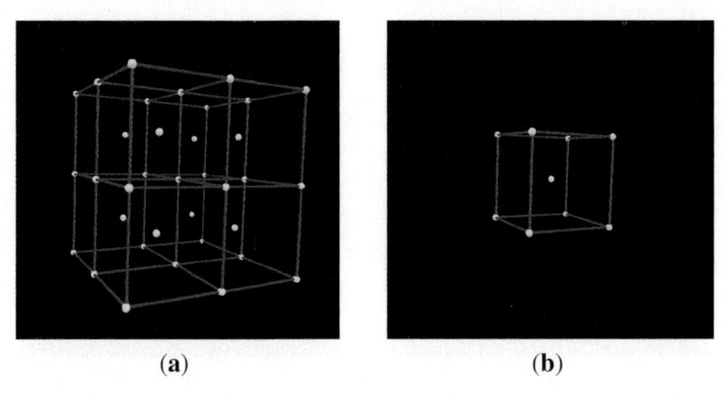

Fig. 3. The BCC lattice. A neighborhood of 35 points is displayed on the *left*, while a simple neighborhood of nine points is displayed on the *right*

4.1 Polyhedra Associated with the BCC Lattice

Certain polyhedra arise naturally in the process of constructing interpolation filters for a lattice. The Voronoi cell of the lattice is one such example. The Voronoi cell of the Cartesian lattice is a cube and the Voronoi cell of the BCC lattice is a truncated octahedron as illustrated in Fig. 4a.

We are also interested in the cell formed by the immediate neighbors of a lattice point. The first neighbors of a lattice point are defined by the Delaunay tetrahedralization of the lattice; a point $\boldsymbol{q}$ is a *first neighbor* of $\boldsymbol{p}$ if their respective Voronoi cells share a (nondegenerate) face. The *first neighbors cell* is the polyhedron whose vertices are the first neighbors. Again, this cell is the same for all points on the lattice.

For example, by this definition there are six first neighbors of a point in a Cartesian lattice; the first neighbors cell for the Cartesian lattice is the octahedron. For the BCC lattice there are fourteen first neighbors for each lattice point. The first neighbor cell is a rhombic dodecahedron as illustrated in Fig. 4b.

The geometry of the dual lattice is of interest when we consider the spectrum of the function captured by the sampling operation. The Cartesian lattice is self dual. However, the dual of the BCC lattice is the FCC lattice. The FCC lattice is a sublattice of $\mathbb{Z}^3$ and is often referred to as the $\boldsymbol{D}_3$ lattice [3]. In fact $\boldsymbol{D}_3$ belongs to a general family of lattices, $\boldsymbol{D}_n$, sometimes called checkerboard lattices. The checkerboard property implies that the sum of the coordinates of the lattice sites is always even. We will use this property to demonstrate the zero crossings of the frequency response of the reconstruction filters at the FCC lattice sites.

The Voronoi cell of the FCC lattice is the rhombic dodecahedron as illustrated in Fig. 4c. Its characteristic function is the frequency response of the ideal reconstruction filter for the BCC lattice. Figure 4d shows the first neighbors cell of the FCC lattice; the cuboctahedron.

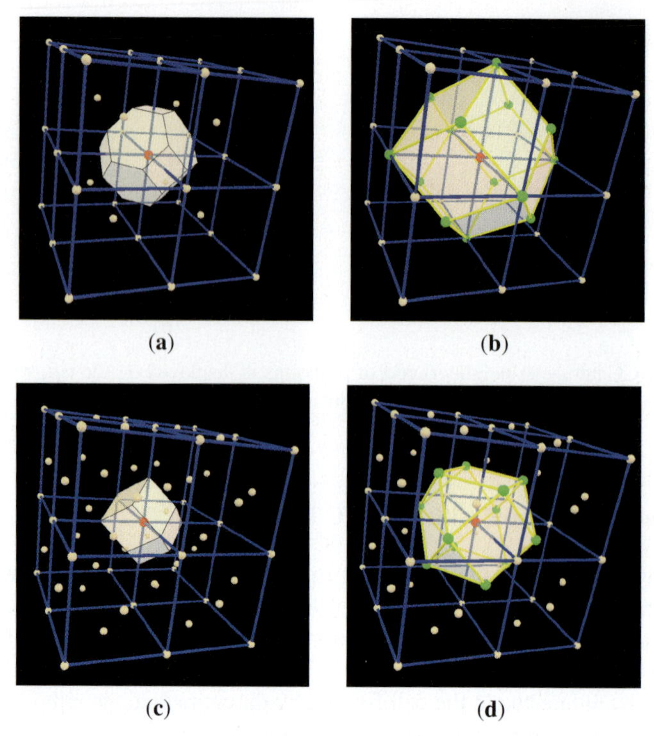

Fig. 4. The Voronoi cell of the BCC lattice is the truncated octahedron (**a**), and its first neighbor cell is the rhombic dodecahedron (**b**). For the FCC lattice, the rhombic dodecahedron is the Voronoi cell (**c**), and the cuboctahedron is the first neighbor cell (**d**)

5 Reconstruction Filters

The kernel for the nearest neighbor interpolation in 1D is the Box function. It is the characteristic function of the Voronoi cell of the samples on the real line. The nearest neighbor interpolation on the BCC lattice is similarly defined in terms of the Voronoi cell of the lattice which is a truncated octahedron (Fig. 4a). In this scheme, a point in space is assigned the value of the sample in whose Voronoi cell it is located. Since the Voronoi cell tiles the space, its characteristic function induces an interpolation scheme for that lattice. Based on the fact that the periodic tiling of the Voronoi cell yields the constant function in the spatial domain, Van De Ville et al. [12] proves by means of the Poisson summation formula that the frequency response of such a kernel does in fact vanish at the aliasing frequencies.

5.1 Ideal Interpolation

As noted earlier, sampling a function on a periodic lattice replicates the spectrum of the function in the Fourier domain on the dual lattice. When the space of bandlimited functions is the space of choice for reconstruction, the ideal interpolation function is the one that removes the replicates of the spectrum in the Fourier domain. This proves that the Fourier transform of the ideal interpolation function is the characteristic function of the Voronoi cell of the dual lattice; hence, convolving by the ideal interpolation function, leaves out the main spectrum and eliminates all of the replicas.

The ideal reconstruction function for the Cartesian lattice has a Fourier transform that is the characteristic function of a cube and the one for the BCC lattice has Fourier transform which is the characteristic function of a rhombic dodecahedron. Therefore, in order to find the ideal interpolation function for the BCC lattice we need to find a function whose Fourier transform is constant on the rhombic dodecahedron in Fig. 4c and is zero everywhere else.

As it is not easy to derive this function directly, to construct an explicit function we decompose the rhombic dodecahedron into simpler objects that are easy to construct in the dual domain. Figure 5 illustrates the decomposition of the rhombic dodecahedron into four three-dimensional parallelepipeds. These parallelepipeds share the origin and each are formed by three vectors from the origin. For a rhombic dodecahedron oriented as in Fig. 4c we define the set of vectors:

$$\xi_1 = \begin{bmatrix} -1 \\ -1 \\ 1 \end{bmatrix}, \xi_2 = \begin{bmatrix} -1 \\ 1 \\ -1 \end{bmatrix}, \xi_3 = \begin{bmatrix} 1 \\ -1 \\ -1 \end{bmatrix}, \xi_4 = \begin{bmatrix} 1 \\ 1 \\ 1 \end{bmatrix}.$$

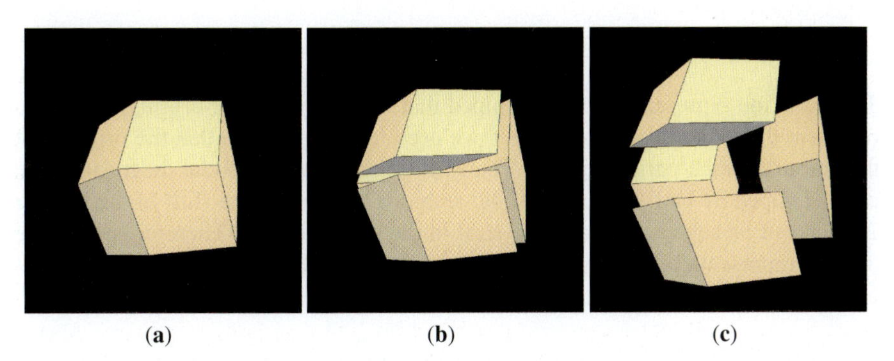

(a) (b) (c)

Fig. 5. The rhombic dodecahedron, the Voronoi cell of the FCC lattice, can be decomposed into four parallelepipeds

Any three of these four vectors form one of the parallelepipeds in the decomposition of the rhombic dodecahedron illustrated in Fig. 5. One can observe that:

$$\boldsymbol{\xi}_1 + \boldsymbol{\xi}_2 + \boldsymbol{\xi}_3 + \boldsymbol{\xi}_4 = \begin{bmatrix} 0 \\ 0 \\ 0 \end{bmatrix} \tag{8}$$

that is attributed to the symmetries of the rhombic dodecahedron.

Since a parallelepiped can be constructed by a linear transform from a cube, we can start by constructing a cube in the Fourier domain:

$$\mathcal{F}\{\text{sinc}(x)\text{sinc}(y)\text{sinc}(z)\} = \mathcal{B}(\omega_x)\mathcal{B}(\omega_y)\mathcal{B}(\omega_z)$$

where $\text{sinc}(x) = \frac{\sin(\pi x)}{\pi x}$. Rewriting the above equation in terms of a 3D extension of sinc:

$$\mathcal{F}\{\text{sinc}_{3D}(\boldsymbol{x})\} = \mathcal{B}_{3D}(\boldsymbol{\omega}).$$

Let $\boldsymbol{\xi}_i, \boldsymbol{\xi}_j, \boldsymbol{\xi}_k$ denote the vectors forming a parallelepiped. Then the matrix $\boldsymbol{T} = [\boldsymbol{\xi}_i|\boldsymbol{\xi}_j|\boldsymbol{\xi}_k]$ transforms the unit cube to the parallelepiped. If $\chi_{\boldsymbol{T}}$ denotes the characteristic function of the parallelepiped formed by the columns of $\boldsymbol{T}$, then:

$$\chi_{\boldsymbol{T}}(\boldsymbol{\omega}) = \mathcal{B}_{3D}(\boldsymbol{T}^{-1}\boldsymbol{\omega}).$$

In order to get $\chi_{\boldsymbol{T}}$ in the Fourier domain, we use the multidimensional scaling lemma in the Fourier transform:

$$\mathcal{F}\{\text{sinc}_{3D}(\boldsymbol{T}^\top \boldsymbol{x})\} = \det \boldsymbol{T}\, \chi_{\boldsymbol{T}}(\boldsymbol{\omega}).$$

That means the spatial domain form of a constant parallelepiped formed by $\boldsymbol{T}$ in the Fourier domain is:

$$\mathcal{F}\{\text{sinc}(\boldsymbol{\xi}_i \cdot \boldsymbol{x})\text{sinc}(\boldsymbol{\xi}_j \cdot \boldsymbol{x})\text{sinc}(\boldsymbol{\xi}_k \cdot \boldsymbol{x})\} = \det \boldsymbol{T}\, \chi_{\boldsymbol{T}}(\boldsymbol{\omega}). \tag{9}$$

This equation represents a parallelepiped that is centered at the origin; in order to represent the parallelepipeds in Fig. 5 we need to shift them so that the origin is at the corner of each parallelepiped. The shift is along the antipodal diagonal of the parallelepiped by half the length of the antipodal diagonal. The shift in the Fourier domain can be achieved by a phase shift in the space domain. Therefore, the space domain representation of a parallelepiped formed by $\boldsymbol{T}$ with its corner at the origin is

$$\mathcal{F}\{e^{-2\pi i \frac{1}{2}(\boldsymbol{\xi}_i + \boldsymbol{\xi}_j + \boldsymbol{\xi}_k)\cdot \boldsymbol{x}} \text{sinc}(\boldsymbol{\xi}_i \cdot \boldsymbol{x})\text{sinc}(\boldsymbol{\xi}_j \cdot \boldsymbol{x})\text{sinc}(\boldsymbol{\xi}_k \cdot \boldsymbol{x})\} = \det \boldsymbol{T}\, \chi^o_{\boldsymbol{T}}(\boldsymbol{\omega}).$$

where $\chi^o_{\boldsymbol{T}}$ is the characteristic function of the parallelepiped with its corner at the origin.

Now we can write the space domain representation of the rhombic dodecahedron in Fig. 4c.

$$\begin{aligned}
\mathrm{sincBCC}(\boldsymbol{x}) = & \\
& e^{\pi i \xi_4 \cdot \boldsymbol{x}} \mathrm{sinc}(\xi_1 \cdot \boldsymbol{x})\mathrm{sinc}(\xi_2 \cdot \boldsymbol{x})\mathrm{sinc}(\xi_3 \cdot \boldsymbol{x}) + \\
& e^{\pi i \xi_3 \cdot \boldsymbol{x}} \mathrm{sinc}(\xi_1 \cdot \boldsymbol{x})\mathrm{sinc}(\xi_2 \cdot \boldsymbol{x})\mathrm{sinc}(\xi_4 \cdot \boldsymbol{x}) + \\
& e^{\pi i \xi_2 \cdot \boldsymbol{x}} \mathrm{sinc}(\xi_1 \cdot \boldsymbol{x})\mathrm{sinc}(\xi_3 \cdot \boldsymbol{x})\mathrm{sinc}(\xi_4 \cdot \boldsymbol{x}) + \qquad (10)\\
& e^{\pi i \xi_1 \cdot \boldsymbol{x}} \mathrm{sinc}(\xi_2 \cdot \boldsymbol{x})\mathrm{sinc}(\xi_3 \cdot \boldsymbol{x})\mathrm{sinc}(\xi_4 \cdot \boldsymbol{x}) \\
= & \sum_{j=1}^{4} e^{\pi i \xi_j \cdot \boldsymbol{x}} \prod_{k \neq j} \mathrm{sinc}(\xi_k \cdot \boldsymbol{x}).
\end{aligned}$$

Claim. $\mathrm{sincBCC}(\boldsymbol{x})$ is a real valued function

Proof. In order to show that $\mathrm{sincBCC}(\boldsymbol{x})$ is a real valued function we subtract it from its conjugate:

$$\begin{aligned}
& \overline{\mathrm{sincBCC}(\boldsymbol{x})} - \mathrm{sincBCC}(\boldsymbol{x}) = \\
& \sum_{j=1}^{4} (e^{-\pi i \xi_j \cdot \boldsymbol{x}} - e^{\pi i \xi_j \cdot \boldsymbol{x}}) \prod_{k \neq j} \mathrm{sinc}(\xi_k \cdot \boldsymbol{x}) = \\
& \sum_{j=1}^{4} (2i \sin(\pi \xi_j \cdot \boldsymbol{x})) \prod_{k \neq j} \mathrm{sinc}(\xi_k \cdot \boldsymbol{x}) = \\
& \sum_{j=1}^{4} (2\pi i (\xi_j \cdot \boldsymbol{x}) \mathrm{sinc}(\xi_j \cdot \boldsymbol{x})) \prod_{k \neq j} \mathrm{sinc}(\xi_k \cdot \boldsymbol{x}) = \\
& \sum_{j=1}^{4} 2\pi i (\xi_j \cdot \boldsymbol{x}) \prod \mathrm{sinc}(\xi_k \cdot \boldsymbol{x}) = \\
& 2\pi i ((\xi_1 + \xi_2 + \xi_3 + \xi_4) \cdot \boldsymbol{x}) \prod \mathrm{sinc}(\xi_k \cdot \boldsymbol{x}) = 0
\end{aligned}$$

due to symmetries of the rhombic dodecahedron illustrated in (8).

As a corollary to this claim, using the fact that $\mathrm{sincBCC}(\boldsymbol{x}) = \Re\{\mathrm{sincBCC}(\boldsymbol{x})\} = \frac{1}{2}(\mathrm{sincBCC}(\boldsymbol{x}) + \overline{\mathrm{sincBCC}(\boldsymbol{x})})$ we simplify the $\mathrm{sincBCC}(\boldsymbol{x})$ to:

$$\mathrm{sincBCC}(\boldsymbol{x}) = \sum_{\mathrm{j}=1}^{4} \cos(\pi \xi_{\mathrm{j}} \cdot \boldsymbol{x}) \prod_{\mathrm{k} \neq \mathrm{j}} \mathrm{sinc}(\xi_{\mathrm{k}} \cdot \boldsymbol{x}). \qquad (11)$$

5.2 Linear Box Spline

de Boor et al. [4] analytically define the box splines, in n-dimensional space, by successive directional convolutions. They also describe an alternative geometric description of the box splines in terms of the projection of higher dimensional boxes (nD cubes). A simple example of a one-dimensional linear box spline is the triangle function which can be obtained by projecting a 2D box along its diagonal axis down to 1D. The resulting function (after proper scaling) is one at the origin and has a linear fall off toward the first neighbors as illustrated in Fig. 6a.

The properties and behaviors of box splines are studied in [4]. For example, the order of the box splines can be determined in terms of the difference in dimension between the higher dimensional box and the lower dimensional projection. For instance, the triangle function is a projection of a 2D cube into 1D, hence it is a first order box spline.

Our construction of box splines for the BCC lattice is guided by the fact that the rhombic dodecahedron (the first neighbors cell of the BCC lattice) is the three-dimensional shadow of a four-dimensional hypercube (tesseract) along its antipodal axis. This fact will be revealed in the following discussion. This construction is reminiscent of constructing a hexagon by projecting a three-dimensional cube along its antipodal axis; see Fig. 6b for the 2D case.

Integrating a constant tesseract of unit side length along its antipodal axis yields a function that has a rhombic dodecahedron support (see Fig. 3b), has the value two[3]

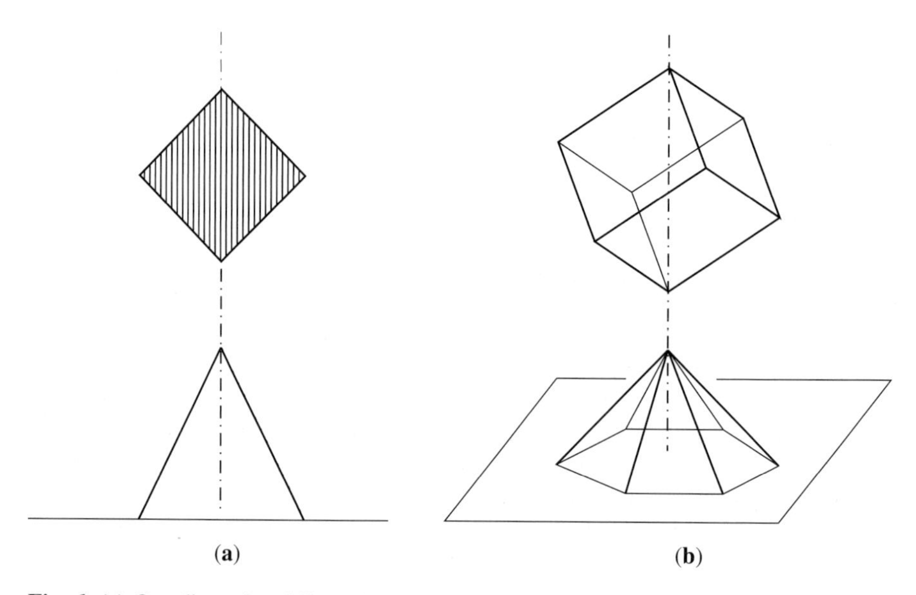

Fig. 6. (**a**) One-dimensional linear box spline (triangle function). (**b**) The two-dimensional hexagonal linear box spline

[3] Note that the BCC sampling lattice has a sampling density of two samples per unit volume.

at the center and has a linear fall off toward the 14 first neighbor vertices. Since it arises from the projection of a higher dimensional box, this filter is the first order (linear) box spline interpolation filter on the BCC lattice.

Let $\mathcal{B}$ denote the Box distribution. The characteristic function of the unit tesseract is given by a product of these functions:

$$\mathcal{T}(x, y, z, w) = \mathcal{B}(x)\mathcal{B}(y)\mathcal{B}(z)\mathcal{B}(w). \tag{12}$$

Let $\boldsymbol{v} = \langle 1, 1, 1, 1\rangle$ denote a vector along the antipodal axis. In order to project along this axis, it is convenient to rotate it so that it aligns with the w axis. Let

$$\boldsymbol{R} = \frac{1}{2}[\boldsymbol{\rho}_1\boldsymbol{\rho}_2\boldsymbol{\rho}_3\boldsymbol{\rho}_4] = \frac{1}{2}\begin{bmatrix} -1 & -1 & 1 & 1 \\ -1 & 1 & -1 & 1 \\ 1 & -1 & -1 & 1 \\ 1 & 1 & 1 & 1 \end{bmatrix} \tag{13}$$

This rotation matrix transforms $\boldsymbol{v}$ to $\langle 0, 0, 0, 2\rangle$.[4] Let $\boldsymbol{x} = \langle x, y, z, w\rangle$; now the linear kernel is given by

$$L_{RD}(x, y, z) = \int \mathcal{T}(\boldsymbol{R}^{\top}\boldsymbol{x})\, dw.$$

Substituting in (12) we get

$$L_{RD}(x, y, z) = \int \prod_{i=1}^{4} \mathcal{B}(\frac{1}{2}\boldsymbol{\rho}_i \cdot \boldsymbol{x})\, dw. \tag{14}$$

We illustrate an analytical evaluation of this integral in Sect. 6.

5.3 Cubic Box Spline

By convolving the linear box spline filter kernel with itself we double its vanishing moments in the frequency domain. Hence the result of such an operation will have a cubic approximation order [10]. As noted by de Boor et al. [4], the convolution of two box splines is again a box spline.

An equivalent method of deriving this function would be to convolve the tesseract with itself and project the resulting distribution along a diagonal axis (this commutation of convolution and projection is easy to understand in terms of the corresponding operators in the Fourier domain – see Sect. 5.4). Convolving a tesseract with itself results in another tesseract which is the tensor product of four one-dimensional triangle functions.

---

[4] By examining (13), one can see that each vertex of the rotated tesseract, when projected along the w axis, will coincide with the origin or one of the vertices of the rhombic dodecahedron: $\left\langle \pm\frac{1}{2}, \pm\frac{1}{2}, \pm\frac{1}{2}\right\rangle$, $\langle \pm 1, 0, 0\rangle$, $\langle 0, \pm 1, 0\rangle$ or $\langle 0, 0, \pm 1\rangle$.

Let Λ denote the triangle function. Then convolving the characteristic function of the tesseract yields

$$\mathcal{T}_c(x, y, z, w) = \Lambda(x)\Lambda(y)\Lambda(z)\Lambda(w). \tag{15}$$

Following the same 4D rotation as in the previous section, we obtain a space domain representation of the cubic box spline filter kernel:

$$C_{RD}(x, y, z) = \int \prod_{i=1}^{4} \Lambda(\frac{1}{2}\boldsymbol{\rho}_i \cdot \boldsymbol{x})\, dw. \tag{16}$$

Again, we will illustrate in Sect. 6 how to evaluate this integral analytically.

5.4 Frequency Response

From the construction of the rhombic dodecahedron discussed earlier, we can analytically derive the frequency response of the linear function described by (14).

From (12), it is evident that the frequency domain representation of the characteristic function of the tesseract is given by the product of four sinc functions:

$$\tilde{\mathcal{T}}(\omega_x, \omega_y, \omega_z, \omega_w) = \text{sinc}(\omega_x)\text{sinc}(\omega_y)\text{sinc}(\omega_z)\text{sinc}(\omega_w).$$

While in the previous section the origin was assumed to be at the corner of the tesseract, for the simplicity of derivation, we now consider a tesseract whose center is at the origin. The actual integral, computed in (14) or (16) will not change.

By the Fourier slice-projection theorem, projecting the tesseract in the spatial domain is equivalent to slicing $\tilde{\mathcal{T}}$ perpendicular to the direction of projection. This slice runs through the origin. Again we make use of the rotation (13) to align the projection axis with the w axis. Thus in the frequency domain we take the slice $\omega_w = 0$.

It is convenient to introduce the 3×4 matrix

$$\Xi = \frac{1}{2}[\xi_1 \xi_2 \xi_3 \xi_4] = \frac{1}{2}\begin{bmatrix} -1 & -1 & 1 & 1 \\ -1 & 1 & -1 & 1 \\ 1 & -1 & -1 & 1 \end{bmatrix} \tag{17}$$

given by the first three rows of the rotation matrix $\boldsymbol{R}$ of (13). The frequency response of the linear kernel can now be written as

$$\tilde{L}_{RD}(\omega_x, \omega_y, \omega_z) = \prod_{i=1}^{4} \text{sinc}(\frac{1}{2}\xi_i \cdot \boldsymbol{\omega}), \tag{18}$$

where $\boldsymbol{\omega} = \langle \omega_x, \omega_y, \omega_z \rangle$.

The box spline associated with this filter is represented by the Ξ matrix. The properties of this box spline can be derived based on this matrix according to the theory developed in [4]. For instance, one can verify C^0 smoothness of this filter using Ξ.

We can verify the zero crossings of the frequency response at the aliasing frequencies on the FCC lattice points. Due to the checkerboard property for every $\boldsymbol{\omega}$ on the FCC lattice, $\boldsymbol{\xi}_4 \cdot \boldsymbol{\omega} = (\omega_x + \omega_y + \omega_z) = 2k$ for $k \in \mathbb{Z}$; therefore, $\mathrm{sinc}(\frac{1}{2}\boldsymbol{\xi}_4 \cdot \boldsymbol{\omega}) = 0$ on all of the aliasing frequencies. Since $\boldsymbol{\xi}_4 \cdot \boldsymbol{\omega} = -\boldsymbol{\xi}_1 \cdot \boldsymbol{\omega} - \boldsymbol{\xi}_2 \cdot \boldsymbol{\omega} - \boldsymbol{\xi}_3 \cdot \boldsymbol{\omega}$, at least one of the $\boldsymbol{\xi}_i \cdot \boldsymbol{\omega}$ for $i = 1, 2, 3$ needs to be also an even integer and for such i we have $\mathrm{sinc}(\frac{1}{2}\boldsymbol{\xi}_i \cdot \boldsymbol{\omega}) = 0$; therefore, there is a zero of order at least two at each aliasing frequency, yielding a C^0 filter.

The cubic box spline filter can be similarly derived by projecting a tesseract composed of triangle functions. Again, the frequency response can be obtained via the Fourier slice-projection theorem.

Since convolution corresponds to multiplication in the dual domain, the frequency response of (15) is

$$\tilde{T}_c(\omega_x, \omega_y, \omega_z, \omega_w) = \mathrm{sinc}^2(\omega_x)\mathrm{sinc}^2(\omega_y)\mathrm{sinc}^2(\omega_z)\mathrm{sinc}^2(\omega_w).$$

By rotating and taking a slice as before we obtain:

$$\tilde{C}_{RD}(\omega_x, \omega_y, \omega_z) = \prod_{i=1}^{4} \mathrm{sinc}^2(\frac{1}{2}\boldsymbol{\xi}_i \cdot \boldsymbol{\omega}). \tag{19}$$

We can see that the vanishing moments of the cubic kernel are doubled from the linear kernel. We could also have obtained (19) by simply multiplying (18) with itself, which corresponds to convolving the linear 3D kernel with itself in the spatial domain.

The box spline matrix for the cubic kernel is $\Xi' = [\Xi|\Xi]$. One can verify the C^2 continuity of this box spline using Ξ' and the theory in [4].

6 Implementation

In this section we describe a method to evaluate the linear and the cubic kernel analytically.

Let $\mathcal{H}$ denote the Heaviside distribution. Using the fact that $\mathcal{B}(x) = \mathcal{H}(x) - \mathcal{H}(x-1)$ we can expand the integrand of the linear kernel ((14)) in terms of Heaviside distributions. After simplifying the product of four Box distributions in terms of $\mathcal{H}$, we get sixteen terms in the integrand. Each term in the integrand is a product of four Heaviside distributions. Since x, y, z are constants in the integral and the integration is with respect to w, we group the x, y, z argument of each $\mathcal{H}$ and call it t_i, using the fact that $\mathcal{H}(\frac{1}{2}x) = \mathcal{H}(x)$, we can write each term in the integrand as:

$$I = \int_a^b \mathcal{H}(w + t_0)\mathcal{H}(w + t_1)\mathcal{H}(w + t_2)\mathcal{H}(w + t_3)\,dw.$$

The integrand is nonzero only when all of the Heaviside distributions are nonzero and since the integrand will be constant one we have:

$$I = \max(0, b - \max(a, \max(-t_i))).$$

Similarly, for the cubic kernel in (16) we substitute $\Lambda(x) = \mathcal{R}(x) - 2\mathcal{R}(x - 1) + \mathcal{R}(x - 2)$, where $\mathcal{R}$ denotes the ramp function. We obtain eighty one terms, each of which is a product of four ramp functions. Using $\mathcal{R}(\frac{1}{2}x) = \frac{1}{2}\mathcal{R}(x)$, we can write each term in the integrand as a scalar fraction of:

$$I = \int_a^b \mathcal{R}(w + t_0)\mathcal{R}(w + t_1)\mathcal{R}(w + t_2)\mathcal{R}(w + t_3)\, dw.$$

This simplifies to a polynomial times four Heaviside distributions that we can evaluate analytically:

$$\begin{aligned} I &= \int_a^b \prod_{i=1}^{4} (w + t_i)\mathcal{H}(w + t_i)\, dw \\ &= \int_c^b \prod_{i=1}^{4} (w + t_i)\, dw. \end{aligned}$$

where $c = \min(b, \max(a, \max(-t_i)))$ and one can compute the integral of this polynomial analytically.

6.1 Simplification of the Linear Kernel

An alternative method of deriving the linear kernel can be obtained through a geometric argument.

All of the polyhedra discussed in Sect. 4 are convex and therefore may be described as the intersection of a set of half spaces. Further, each face is matched by a parallel antipodal face; this is due to the group structure of the lattice. If a point $\boldsymbol{a}$ is in the lattice and vector $\boldsymbol{b}$ takes it to a neighbor then $\boldsymbol{a} + \boldsymbol{b}$ is in the lattice; then the group property enforces $\boldsymbol{a} - \boldsymbol{b}$ be a point in the lattice as well, hence the antipodal symmetry. As a consequence the polyhedra lend themselves to a convenient description in terms of the level sets of piecewise linear functions.

Consider the rhombic dodecahedron, for example. Each of its twelve rhombic faces can be seen to lie centered on the edges of a cube such that the vector from the center of the cube to the center of its edge is orthogonal to the rhombic face placed on that edge.

So the interior of the rhombic dodecahedron that encloses the unit cube in this way can be described as the intersection of the twelve half spaces

$$\pm x \pm y \leq \sqrt{2}, \quad \pm x \pm z \leq \sqrt{2}, \quad \pm y \pm z \leq \sqrt{2}. \tag{20}$$

Now consider the pyramid with apex at the center of the polyhedron and whose base is a face f with unit outward normal $\hat{\boldsymbol{n}}_f$. Notice that for any point $\boldsymbol{p}$ within this pyramid, the scalar product $p \cdot \hat{\boldsymbol{n}}_f$ is larger than $p \cdot \hat{\boldsymbol{n}}_{f'}$, where $\hat{\boldsymbol{n}}_{f'}$ is the outward normal for any other rhombic face f'. Thus if we define a function

$$\begin{aligned} \phi &: \mathbb{R}^3 \longrightarrow \mathbb{R} \\ \phi &: \boldsymbol{p} \longmapsto \max_{\hat{\boldsymbol{n}}_f} p \cdot \hat{\boldsymbol{n}}_f, \end{aligned} \tag{21}$$

its level sets are rhombic dodecahedra. We can use the axial symmetries of the half spaces (20) to write the function (21) for the rhombic dodecahedron in the compact form

$$\phi(x, y, z) = \max(|x| + |y|, |x| + |z|, |y| + |z|).$$

For a fixed s, all the points in the space with $\phi(x, y, z) < s$ are the interior of the rhombic dodecahedron, $\phi(x, y, z) = s$ are on the rhombic dodecahedron and $\phi(x, y, z) > s$ are on the outside of the rhombic dodecahedron. Therefore for all $s \geq 0$ the function $\phi(x, y, z)$ describes concentric rhombic dodecahedra that are growing outside from the origin linearly with respect to s.

Using this fact, one can derive the function that is two at the center of the rhombic dodecahedron and decreases linearly to zero at the vertices, similar to the linear kernel described in (14), to be:

$$L_{RD}(x, y, z) = 2 \max(0, 1 - \max(|x| + |y|, |x| + |z|, |y| + |z|)). \tag{22}$$

7 Results and Discussion

The optimality properties of the BCC sampling imply that the spectrum of a Cartesian sampled volume matches the spectrum of a BCC sampled volume with 29.3% fewer samples [11]. On the other hand, given equivalent sampling density per volume, the BCC sampled volume outperforms the Cartesian sampling in terms of information captured during the sampling operation. Therefore, in our test cases, we are comparing renditions of a Cartesian sampled dataset against renditions of an equivalently dense BCC sampled volume as well as against a BCC volume with 30% fewer samples.

In order to examine the reconstruction schemes discussed in this paper, we have implemented a ray-caster to render images from the Cartesian and the BCC sampled volumetric datasets.[5] The normal estimation, needed for shading, was based on central differencing of the reconstructed continuous function both in the Cartesian and BCC case. Central Differencing is easy to implement and there is no reason to believe that it performs any better or worse than taking the analytical derivative of the reconstruction kernel [9].

---

[5] In order to ensure fair comparison of Cartesian vs. BCC sampling we should compare our new reconstruction filters with filters based on the octahedron of first neighbors cell (see

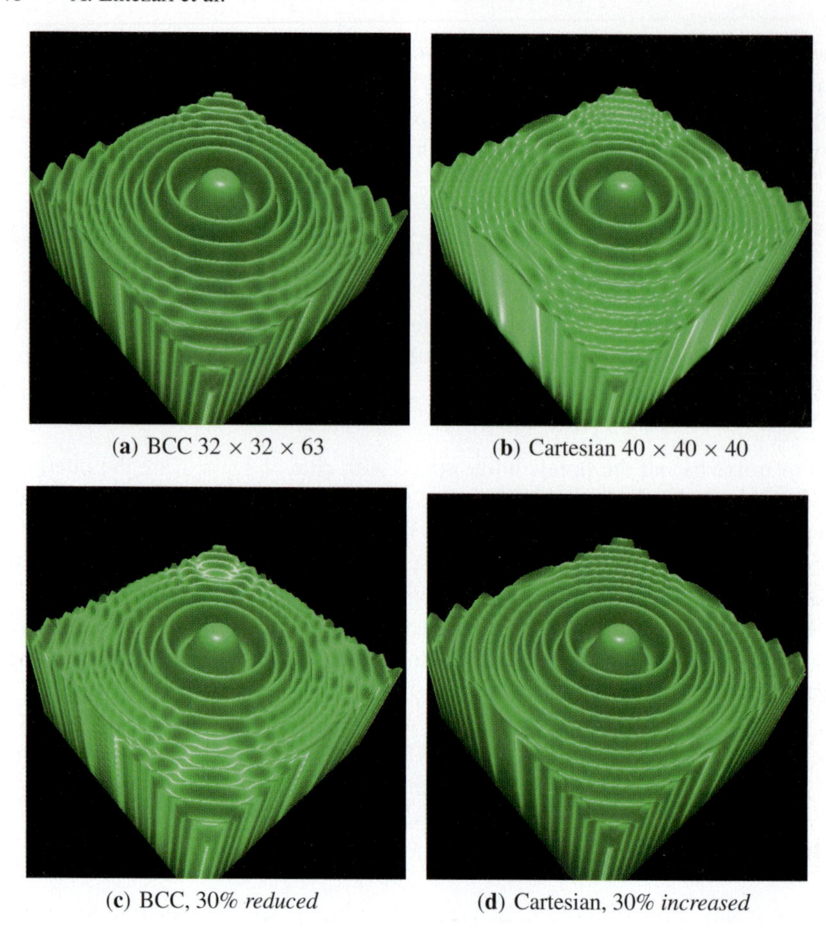

(**a**) BCC 32 × 32 × 63 (**b**) Cartesian 40 × 40 × 40

(**c**) BCC, 30% *reduced* (**d**) Cartesian, 30% *increased*

Fig. 7. Comparison of BCC and Cartesian sampling of the Marschner–Lobb data set, cubic reconstruction

We have chosen the synthetic dataset first proposed in [8] as a benchmark for our comparisons. The function was sampled at the resolution of $40 \times 40 \times 40$ on the Cartesian lattice and at an equivalent sampling on the BCC lattice of $32 \times 32 \times 63$. For the sake of comparison with these volumes a 30% reduced volume of $28 \times 28 \times 55$ samples on the BCC lattice along with a volume of 30% increased sampling resolution of $44 \times 44 \times 44$ for the Cartesian sampling was also rendered. The images in Fig. 7 are rendered using the cubic box spline on the BCC sampled datasets and the tri-cubic B-spline on the Cartesian sampled datasets. The images in Fig. 8 document

---

Sect. 4.1). However, tri-linear filtering is the common standard in volume rendering and since tri-linear filters are superior to the octahedron based filters, we will compare our new filters to the tensor-product spline family instead.

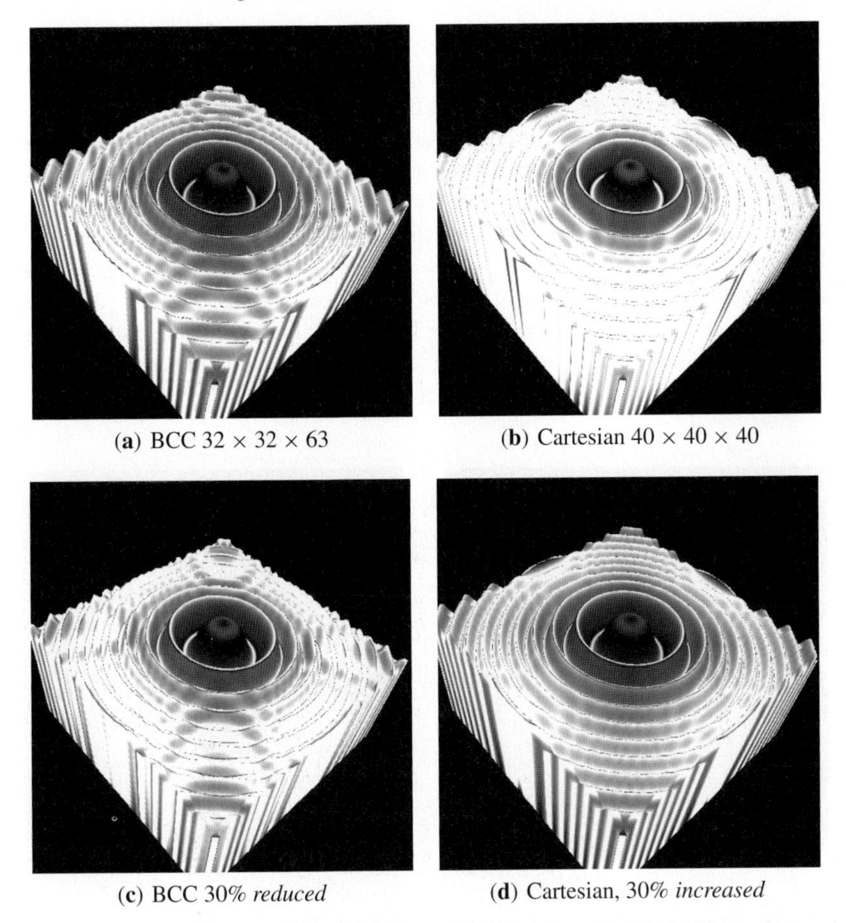

(**a**) BCC 32 × 32 × 63 (**b**) Cartesian 40 × 40 × 40

(**c**) BCC 30% *reduced* (**d**) Cartesian, 30% *increased*

Fig. 8. Angular error of the computed normal vs. the exact normal of the cubic reconstruction in Fig. 7. Angular error of 30° mapped to white

the corresponding error images that are obtained by the angular error incurred in estimating the normal (by central differencing) on the reconstructed function. The gray value of 255 (white) denotes the angular error of 30° between the computed normal and the exact normal.

The optimality of the BCC sampling is apparent by comparing the images Figs. 7a and 7b as these are obtained from an equivalent sampling density over the volume. While the lobes are mainly preserved in the BCC case, they are smoothed out in the case of Cartesian sampling. This is also confirmed by their corresponding error images in Fig. 8. The image in Fig. 7c is obtained with a 30% reduction in the sampling density over the volume of the BCC sampled data while the image in Fig. 7d is obtained with a 30% increase in the sampling density over the volume of the Cartesian sampled data. One could match the quality in Fig. 7c with Fig. 7b and the Fig. 7d with the Fig. 7a, this pattern can also be observed in the error images of Fig. 8. This matches our predictions from the theory of optimal sampling.

We also examined the quality of the linear kernel on this test function. The renditions of the test function using the linear kernel on the BCC lattice and tri-linear interpolation on the Cartesian lattice are illustrated in Fig. 9. Since 98% of the energy of the test function is concentrated below the 41st wavenumber in the frequency domain [8], this sampling resolution is at a critical sampling rate and hence a lot of aliasing appears during linear reconstruction. We doubled the sampling rate on each dimension and repeated the experiment in Fig. 10. Figure 11 demonstrates the errors in the normal estimation. Due to the higher sampling density, the errors in normal

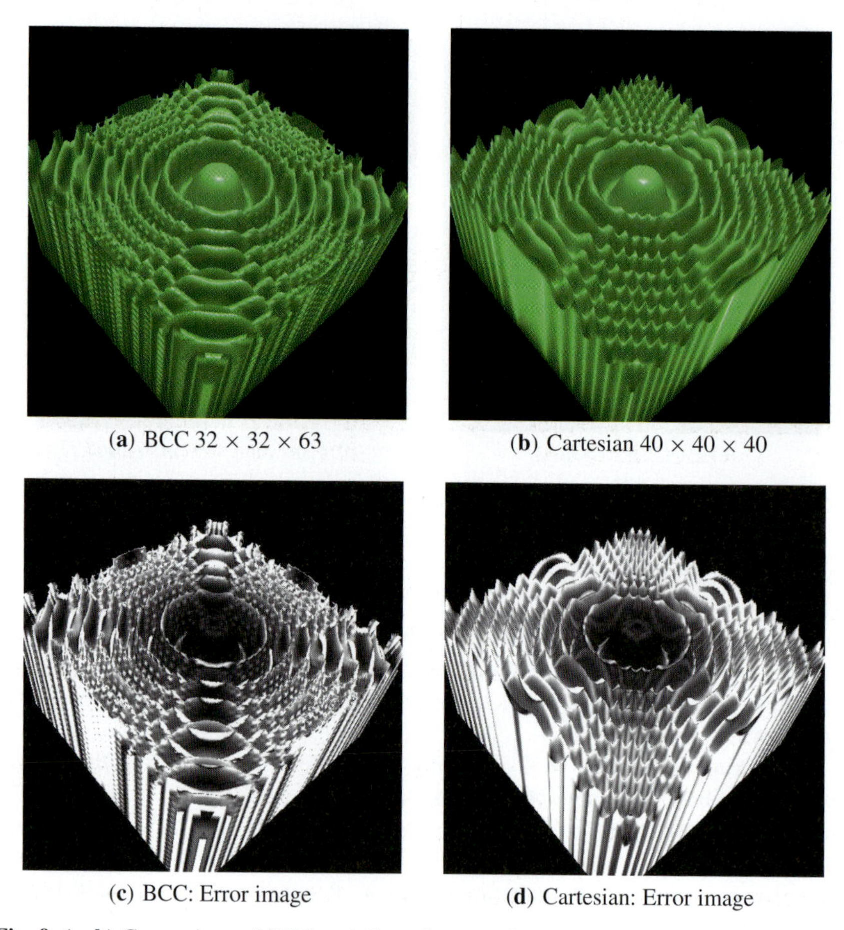

(**a**) BCC $32 \times 32 \times 63$ (**b**) Cartesian $40 \times 40 \times 40$

(**c**) BCC: Error image (**d**) Cartesian: Error image

Fig. 9. (**a**, **b**) Comparison of BCC and Cartesian sampling of the Marschner–Lobb data set, linear reconstruction. (**c**, **d**) The corresponding error images map an angular error of 30° to white

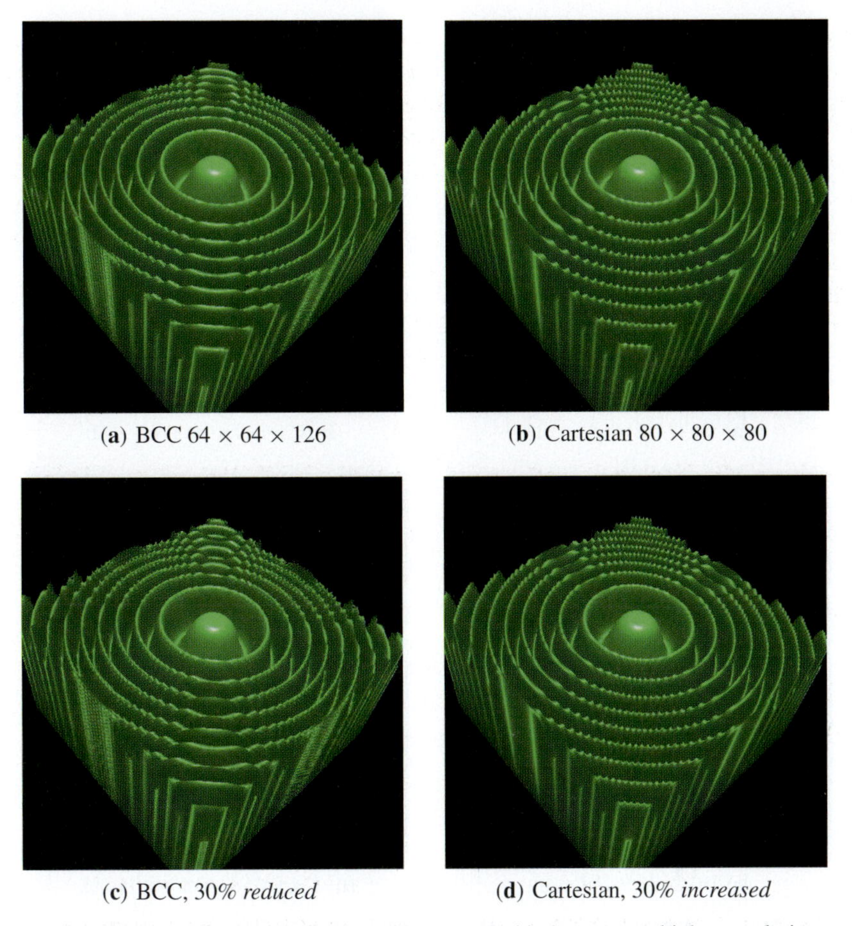

(**a**) BCC $64 \times 64 \times 126$ (**b**) Cartesian $80 \times 80 \times 80$

(**c**) BCC, 30% *reduced* (**d**) Cartesian, 30% *increased*

Fig. 10. Linear reconstruction of the Marschner–Lobb data set at a higher resolution

estimation are considerably decreased; hence we have mapped the gray value 255 (white) to $5°$ of error.

Renditions of the Marschner–Lobb function with this higher sampling resolution using cubic reconstruction and the corresponding error images are illustrated in Fig. 12 and in Fig. 13.

Throughout the images in Fig. 7 through Fig. 13, one can observe the superior fidelity of the BCC sampling compared to the Cartesian sampling.

Real volumetric datasets are scanned and reconstructed on the Cartesian lattice; there are filtering steps involved in scanning and reconstruction that tune the data according to the Cartesian sampling so the spectrum of the captured data is

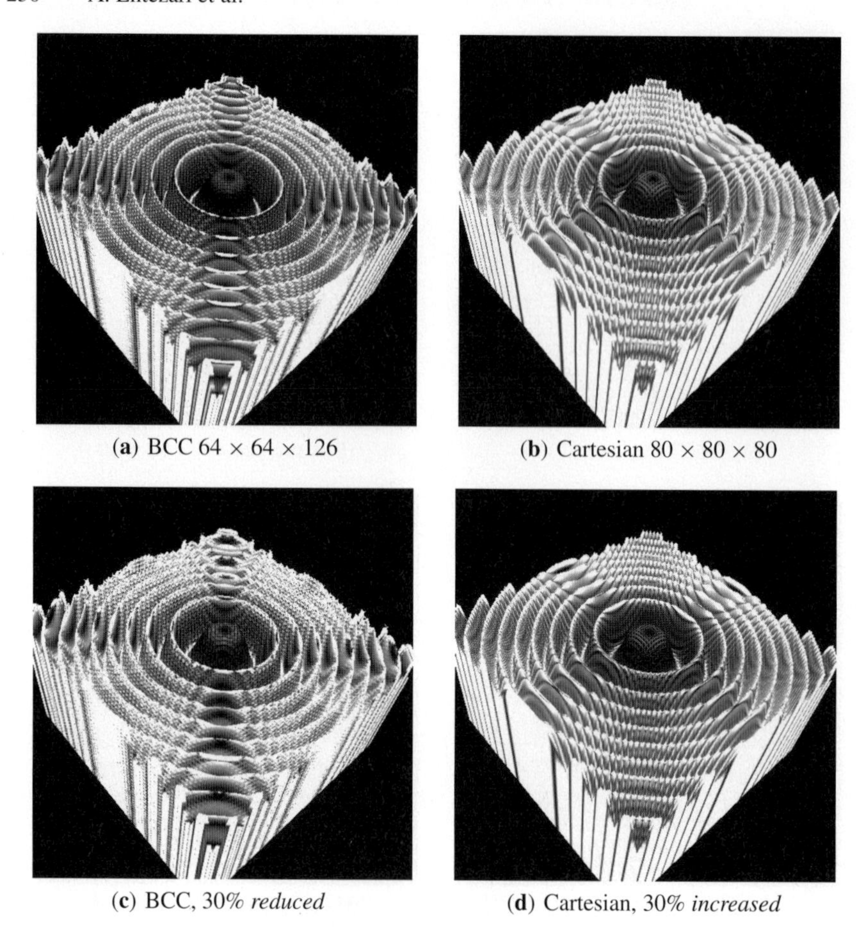

(**a**) BCC 64 × 64 × 126 (**b**) Cartesian 80 × 80 × 80

(**c**) BCC, 30% *reduced* (**d**) Cartesian, 30% *increased*

Fig. 11. Angular error images for linear reconstruction at a higher resolution as shown in Fig. 10. Angular error of 5° mapped to white (255)

antialiased with respect to the geometry of the Cartesian lattice. Therefore, the ultimate test of the BCC reconstruction can not be performed until there are optimal BCC sampling scanners available.

However, for examining the quality of our reconstruction filters on real world datasets we used a Cartesian filter to resample the Cartesian datasets on the BCC lattice. While prone to the errors of the reconstruction before resampling, we have produced BCC sampled volumes of the tooth and the UNC brain datasets with 30% reduction in the number of samples. The original tooth volume has a resolution of 160 × 160 × 160 and the BCC volume after the 30% reduction has a resolution of 113 × 113 × 226; similarly for the UNC dataset, the original Cartesian resolution of 256 × 256 × 145 was reduced by 30% to the BCC resolution of 181 × 181 × 205. The result of their rendering using the linear and the cubic box spline in the BCC

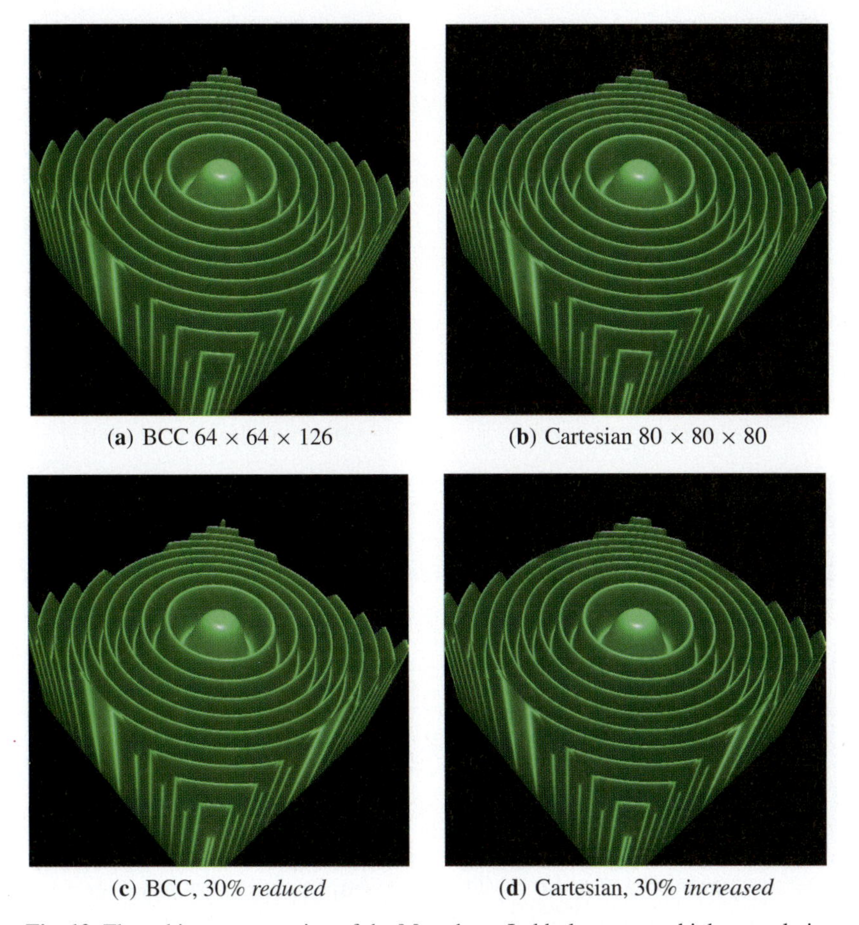

(**a**) BCC $64 \times 64 \times 126$

(**b**) Cartesian $80 \times 80 \times 80$

(**c**) BCC, 30% *reduced*

(**d**) Cartesian, 30% *increased*

Fig. 12. The cubic reconstruction of the Marschner–Lobb data set at a higher resolution

case and the tri-linear and tri-cubic B-spline reconstruction in the Cartesian case is illustrated in Figs. 14 and 15. These images were rendered at a 512^2 resolution on an SGI Altix with sixty-four 1.5 GHz Intel Itanium processors running Linux.

8 Conclusion

In this paper we have derived an analytic description of linear and cubic box splines for the body centered cubic (BCC) lattice. Using geometric arguments, we have further derived a simplified analytical form of the linear box spline in (22), which is simple and fast to evaluate (simpler than the trilinear interpolation function for Cartesian lattices).

Further we have also derived the analytical description of the Fourier transform of these novel filters and by demonstrating the number of vanishing moments we

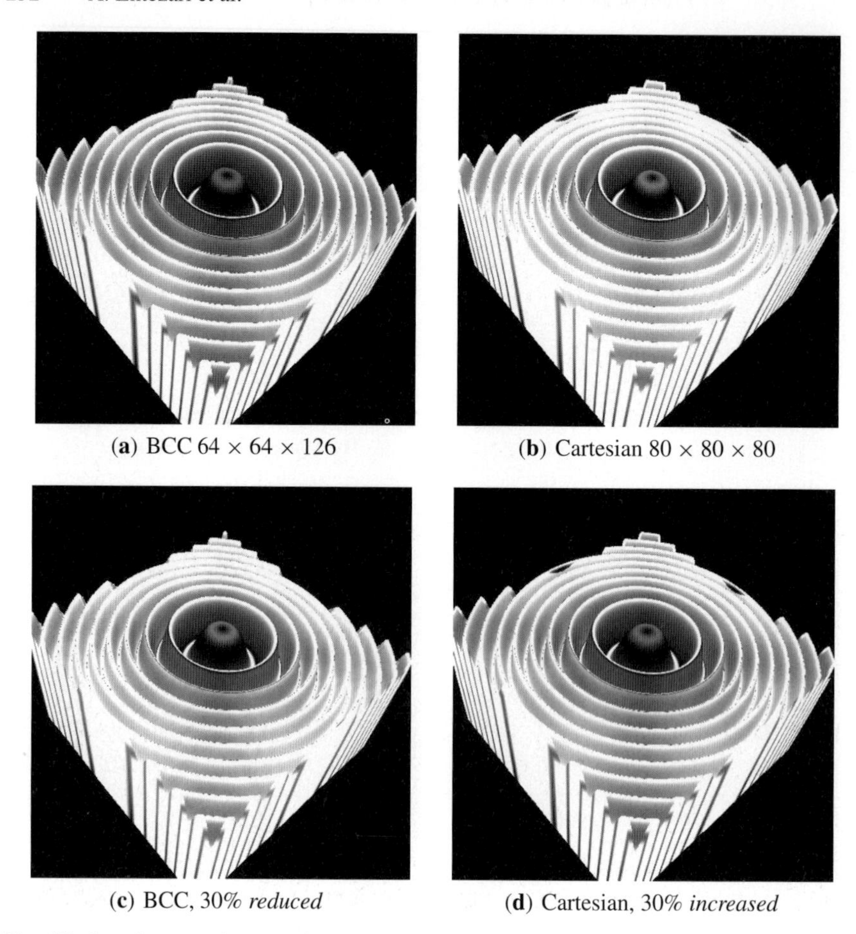

(**a**) BCC $64 \times 64 \times 126$ (**b**) Cartesian $80 \times 80 \times 80$

(**c**) BCC, 30% *reduced* (**d**) Cartesian, 30% *increased*

Fig. 13. Angular error images for cubic reconstruction at a higher resolution as shown in Fig. 12. Angular error of 5° mapped to white (255)

have established the numerical order of these filters. We believe that these filters will provide the key for a more widespread use of BCC sampled lattices.

Our images support the theoretical results of the equivalence of Cartesian lattices with BCC lattices of 30% fewer samples.

9 Future Research

As we have obtained the linear interpolation filter from projection of the tesseract, we can obtain odd order splines by successive convolutions of the linear kernel (or alternatively – projecting a tesseract which is the tensor product of higher order one-dimensional splines). However, the even order splines and their analytical forms do not seem to be easily derived. We are currently investigating this case.

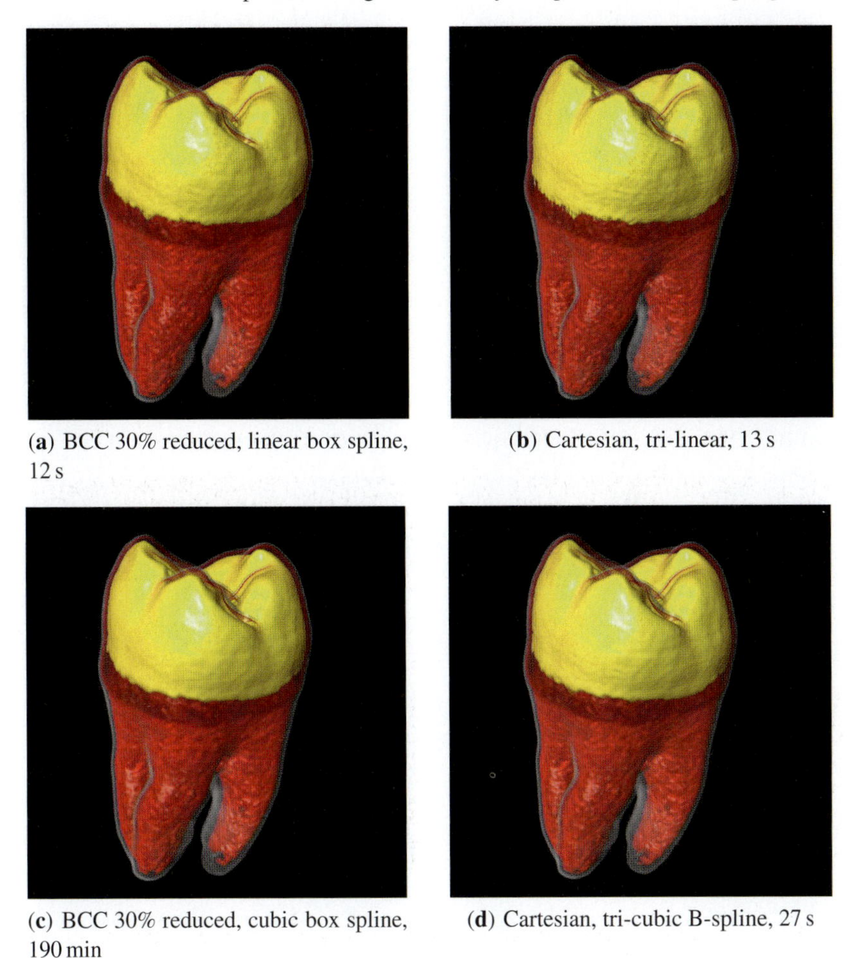

(**a**) BCC 30% reduced, linear box spline, 12 s

(**b**) Cartesian, tri-linear, 13 s

(**c**) BCC 30% reduced, cubic box spline, 190 min

(**d**) Cartesian, tri-cubic B-spline, 27 s

Fig. 14. The tooth dataset

The ease of deriving the frequency response of these interpolation filters lends itself to a thorough error analysis on this family.

Further, the computation of the cubic box spline in (16) currently entails the evaluation of 81 terms. This makes the evaluation of the cubic kernel computationally expensive. We are currently investigating simplifications similar to that of the linear kernel discussed in Sect. 6.1.

Except for the first order box spline, the spline family are approximating filters, hence research on exact interpolatory filters, similar to those of Catmull–Rom for the BCC lattice is being explored.

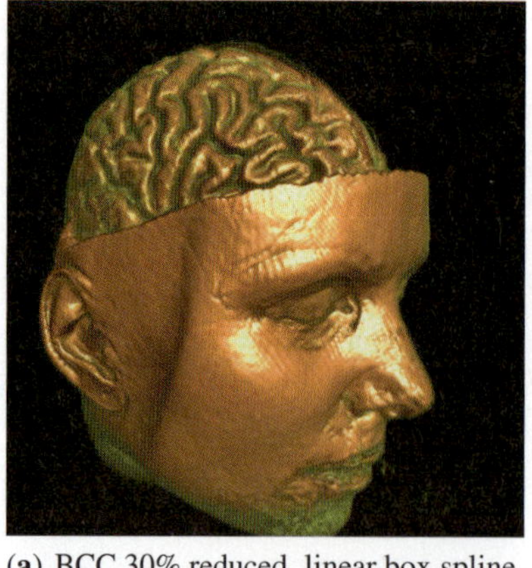

(a) BCC 30% reduced, linear box spline, 11 s

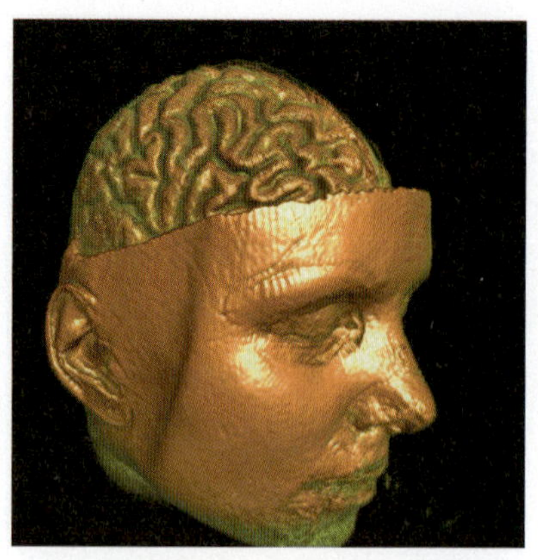

(b) Cartesian, tri-linear, 12 s

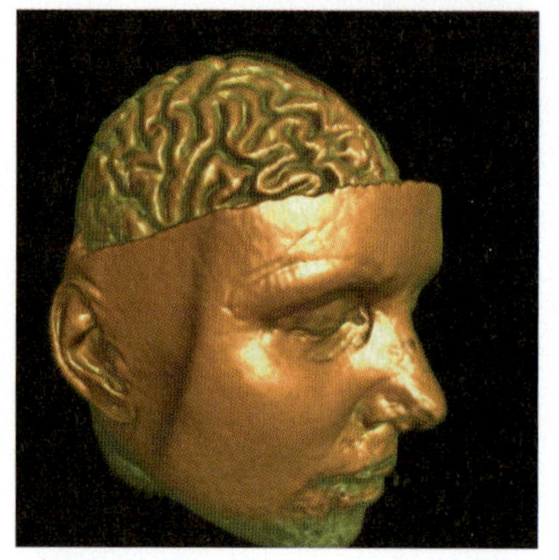

(c) BCC 30% reduced, cubic box spline, 170 min

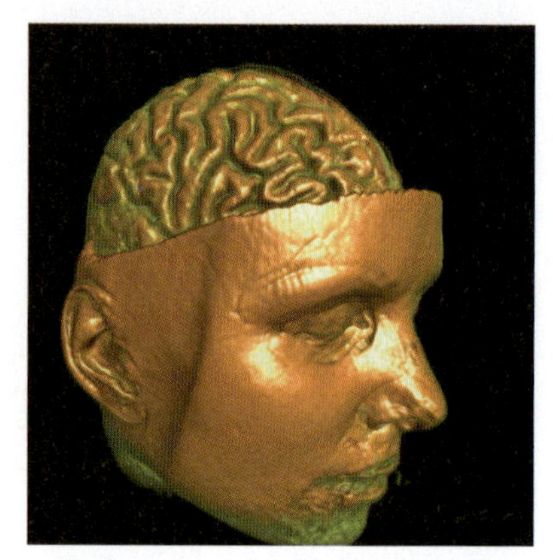

(d) Cartesian, tri-cubic B-spline, 24 s

Fig. 15. The UNC dataset

References

1. P. Brémaud. *Mathematical Principles of Signal Processing*. Springer, Berlin, 2002.
2. G. Burns. *Solid State Physics*. Academic, New York, 1985.
3. J.H Conway and N.J.A. Sloane. *Sphere Packings, Lattices and Groups*, 3rd edition. Springer, Berlin, 1999.
4. C. de Boor, K Höllig, and S Riemenschneider. *Box Splines*. Springer, Berlin, 1993.
5. D. E. Dudgeon and R. M. Mersereau. *Multidimensional Digital Signal Processing*, 1st edition. Prentice-Hall, Englewood-Cliffs, NJ, 1984.
6. T.C. Hales. Cannonballs and honeycombs. *Notices of the AMS*, 47(4):440–449, April 2000.
7. C. Lanczos. *Discourse on Fourier Series*. New York, Hafner, 1966.

8. S.R. Marschner and R.J. Lobb. An evaluation of reconstruction filters for volume rendering. In R. Daniel Bergeron and Arie E. Kaufman, editors, *Proceedings of the IEEE Conference on Visualization 1994*, pages 100–107, Los Alamitos, CA, USA, October 1994. IEEE Computer Society Press.
9. T. Möller, R. Machiraju, K. Mueller, and R. Yagel. A comparison of normal estimation schemes. In *Proceedings of the IEEE Conference on Visualization 1997*, pages 19–26, October 1997.
10. G. Strang and G.J. Fix. A Fourier analysis of the finite element variational method. In *Construct. Aspects of Funct. Anal.*, pages 796–830, 1971.
11. T. Theußl, T. Möller, and E. Gröller. Optimal regular volume sampling. In *Proceedings of the IEEE Conference on Visualization 2001*, pages 91–98, Oct. 2001.
12. D. Van De Ville, T. Blu, M. Unser, W. Philips, I. Lemahieu, and R. Van de Walle. Hex-splines: A novel spline family for hexagonal lattices. *IEEE Transactions on Image Processing*, 13(6):758–772, June 2004.

Reducing Interpolation Artifacts by Globally Fairing Contours

Martin Bertram and Hans Hagen

TU Kaiserslautern, FB Informatik, P.O. Box 3049, 67653 Kaiserslautern, Germany
{bertram, hagen}@informatik.uni-kl.de

Summary. We propose an iterative fairing method for scalar fields reducing the artifacts of bi- and trilinear interpolation. Our method reconstructs a two-dimensional scalar field from interpolation constraints optimizing the smoothness of its contours (isolines) based on variational principles. It generalizes to the trivariate case and is used to increase the quality of data sets employing a cubic B-spline representation. In contrast to filtering methods, our approach preserves the level of detail, enhancing features supported by the data while reducing interpolation artifacts.

1 Introduction

Visualization of data defined on regular grids is mostly based on (multi-)linear interpolation. Examples are view-dependent rendering of piecewise linear elevation models and volume rendering of trilinearly interpolated computer tomography data. In regions of high geometric complexity, multilinear interpolation introduces artifacts that are particularly visible in extracted contours, like elevation lines and isosurfaces. Besides of lacking smoothness, small topological features of these contours are often damaged by the interpolation method. In this work, we demonstrate that interpolation artifacts can be reduced significantly without eliminating such features. We present a contour fairing-method providing a bicubic B-spline representation of two-dimensional scalar fields. Our algorithm generalizes to the trivariate case and is used to increase the quality when resampling data sets at higher resolution.

Fairness of a scalar field can be defined reciprocal to the magnitude of its principal curvatures. Smooth interpolation based on cubic B-Splines, for example, provides fair curves, surfaces, and volumes. However, the contours of such representations are not as smooth as they could be, see Fig. 1. Many small contour components of high curvature do not represent features supported by data and could be merged to larger and smoother components. An approach is needed for fairing each individual contour of an interpolating scalar field. While fairing techniques for parametric curves and surfaces are well known, we consider the problem of fairing implicitly defined

T. Möller et al. (eds.), *Mathematical Foundations of Scientific Visualization, Computer Graphics, and Massive Data Exploration*, Mathematics and Visualization,
DOI: 10.1007/978-3-540-49926-8,

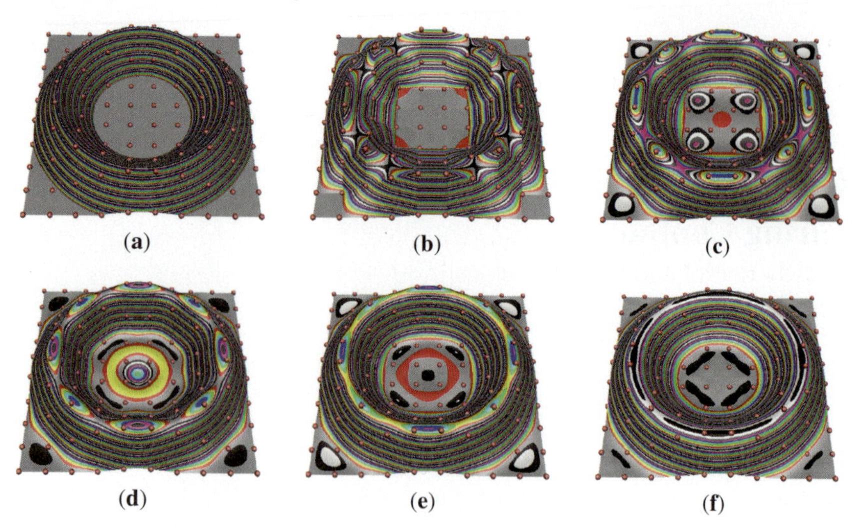

Fig. 1. Contours of an analytic function (**a**) sampled on a 10×10-grid. (**b**) Bilinear interpolation; (**c**) bicubic interpolation; (**d**) sinc interpolation; (**e**) our interpolating method (five iterations); (**f**) our approximating method (five iterations, $w = 0.5$)

geometry. In the present work, we propose a global optimization process minimizing the variation of the scalar field's gradient along all individual contours. The method can be combined with interpolation as shown in Fig. 1e or approximation, see Fig. 1f.

These are the contents of our work: In Sect. 2, we summarize related work. Section 3 presents our contour fairing approach for scalar fields, extending an earlier approach [2]. As basis functions, we use bicubic B-splines with dyadic refinement by knot insertion. In Sect. 4, we provide numerical examples and conclude our work in Sect. 5.

2 Related Work

Following the initial idea of *marching cubes* [10], a variety of different contouring schemes have been proposed for trilinear volume data. Efficient methods extract multiple contours in one pass for volume rendering purposes [7]. Isosurfaces can also be extracted from hierarchical octree representations [17], facilitating level-of-detail.

An important breakthrough is the extraction of topologically correct isosurfaces with respect to the trilinear interplant [9, 12]. Topological analysis of scalar fields provides *critical points* where the topology of contours changes when a passing a certain isovalue [14]. Unfortunately, the topology of a trilinear interplant is often different from the topology of an original scalar field prior to discretization. The question arises how to find the best reconstruction of the original shape consistent with the discrete data.

Image processing techniques like anisotropic diffusion [6, 15] are capable of recognizing local features, but they modify the data. Such approaches are mostly useful when the data is contaminated with noise. Fairing techniques of this kind are also applicable to the fairing of geometric shapes [4, 5].

Variational modeling [8, 16] provides useful fairing methods for parametric surfaces. Our challenge is to apply these methods to implicit geometries that do not have a parametric domain. Fairing a single extracted isosurface with nonlinear constraints imposed by the volume grid is feasible [11]. Besides variational principles, wavelets can be used for fairing of parameterized contours, as well [1].

In the present work, we contribute two fairing methods for the contours of a sampled scalar field based on interpolation and least-squares fitting. The methods are based on an earlier approach [2] using a local C^1-continuous basis blended into a global representation. The methods presented here are based on cubic B-splines with dyadic refinement. We provide examples for the bivariate case. An extension to three dimensions is possible by fairing different sets of slices in a volume. This has already proven to work well for a more efficient but less accurate discrete fairing method [3].

3 Fairing Contours of a Scalar Field

Parametric curves and surfaces can be smoothed based on variational modeling [8, 16]. In this section, we adapt these techniques to the fairing of implicitly defined geometries, like isolines. We obtain fair contours by minimizing their curvature within a prescribed space of basis functions representing the underlying scalar field.

3.1 Linear Optimization

Given some ordinates f_i with associated parameters (s_i, t_i), we intend to construct an interpolating scalar field $f(s, t)$, satisfying

$$f(s_i, t_i) = f_i. \tag{1}$$

A straightforward construction, such as interpolation with bicubic splines, uses the same number of (proper) basis functions ψ_i as the number of interpolation constraints. In this case, a linear system of equations,

$$\mathbf{Ac} = \mathbf{f}, \quad a_{ij} = \psi_j(s_i, t_i) \tag{2}$$

provides the coefficients representing

$$f(s, t) = \sum_i c_i \; \psi_i(s, t).$$

If the number of basis functions is greater than the number of constraints, then there exist some degrees of freedom for optimizing the scalar field's shape. A well-known optimization method is thin-plate energy minimization,

$$\|f\|_{tp}^2 = < f, f >_{tp} \longrightarrow min,$$
$$< f, g >_{tp} := \int\int f_{ss}g_{ss} + 2f_{st}g_{st} + f_{tt}g_{tt}\, ds\, dt, \tag{3}$$

where f_{ss}, f_{st}, and f_{tt} denote second order partial derivatives of f. We note that this functional is only an approximation to the energy of a thin elastic surface. Bicubic splines satisfy this optimization a priori. If other basis functions are used, the minimum within the spanned space can be found by solving a system of equations.

If it is sufficient to approximate the ordinates f_i (assuming these are not exact due to some uncertainty), then the thin-plate minimization can be combined with least-squares fitting,

$$r_f := (1-w)\sum_i |f(s_i,t_i) - f_i|^2 + w < f, f >_{tp} \longrightarrow min. \tag{4}$$

The constant $w \in (0, 1)$ allows to put more emphasis on either accuracy or smoothness. The coefficients satisfying (4) are determined by a linear system of equations,

$$((1-w)\mathbf{A}^T\mathbf{A} + w\mathbf{E})\,\mathbf{c} = (1-w)\mathbf{A}^T\mathbf{f}, \tag{5}$$

where $e_{ij} = <\psi_i, \psi_j>_{tp}$ and $\mathbf{A}$ is the (nonsquare) matrix of (2). The system of equations for such an optimization problem is typically found by the necessary constraints $\frac{\partial r_f}{\partial c_k} = 0$, where r_f is the residual to be minimized.

In cases where the ordinates f_i are exact values that need to be interpolated, these constraints can be worked into the system $\mathbf{Ec} = \mathbf{0}$ using Lagrangian multipliers. A more efficient approach uses a transformed set of basis functions, such that the interpolation constraints are a priori satisfied by the construction. The rank of the system corresponds then to the number of degrees of freedom.

In our approach, we use two sets of basis functions, Φ and Ψ, where Φ contains one basis function for every interpolation constraint and Ψ provides the remaining degrees of freedom used for fairing. We construct these bases such that

$$\begin{aligned} f(s,t) &= \sum_{i:\phi_i\in\Phi} f_i\phi_i(s,t) + \sum_{k:\psi_k\in\Psi} c_k\psi_k(s,t), \\ \phi_i(s_j,t_j) &= \delta_{ij}, \quad \text{and} \\ \psi_k(s_j,t_j) &= 0. \end{aligned} \tag{6}$$

Using this representation, the coefficients c_k minimizing (3) need to be determined. Again, the constraints $\frac{\partial}{\partial c_k} < f, f >_{tp} = 0$ necessary for optimization provide a linear system of equations,

$$\begin{aligned} \mathbf{Ec} &= -\mathbf{Gf}, \quad \text{where} \\ e_{ij} &= < \psi_i, \psi_j >, \quad \text{and} \quad g_{ij} = < \psi_i, \phi_j > . \end{aligned} \tag{7}$$

The matrix $\mathbf{E}$ is positive definite, since for any vector $\mathbf{x} \neq 0$,

$$\mathbf{xEx} = \sum_i \sum_j x_i x_j < \psi_i, \psi_j > = \|\sum_i x_i\psi_i\|^2 > 0, \tag{8}$$

provided that the functions ψ_i are linearly independent. The solution of this system provides the remaining coefficients $\mathbf{c}$ representing f in (6), minimizing (3) among all choices of $\mathbf{c}$.

The above optimization approach can be implemented for different choices of inner products, as well. In the following section, we construct an inner product whose norm minimizes the curvature of the scalar field's isolines. This construction is adopted from [2] with a few modifications.

3.2 Fairing Contours

Consider an interpolating scalar field $f(x, y)$, satisfying $f(x_i, y_i) = f_i$. Selecting a certain isovalue α, we intend to smooth the corresponding isoline composed of all points (x, y) satisfying $f(x, y) = \alpha$. We note that the following deliberations also generalize to volumes, where contours are surfaces.

Suppose that we get hold of a parametric representation of the contour associated with isovalue α, say $g_\alpha(s)$, such that

$$f(g_\alpha(s)) = \alpha. \tag{9}$$

Using this parametric form, fairing the contour can be achieved by minimizing its second derivative,

$$\int \|g''_\alpha(s)\|^2 \, ds \longrightarrow \min. \tag{10}$$

In the case of an arc-length parametrization, this is equivalent to minimizing the variation $n_s := \frac{\partial n}{\partial s}$ of the contour's normal vector n along $g_\alpha(s)$, see Fig. 2:

$$\begin{aligned} &\int \|n_s(s)\|^2 \, ds \longrightarrow \min, \\ &n(s) = \frac{\nabla f(g_\alpha(s))}{\|\nabla f(g_\alpha(s))\|} \end{aligned} \tag{11}$$

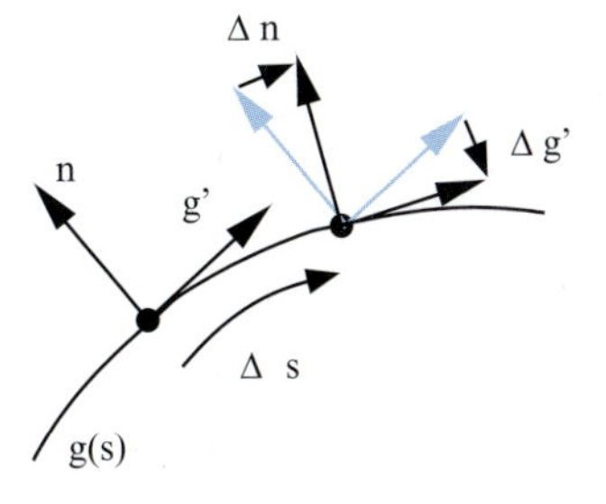

Fig. 2. Variation of g' and n are equal in absolute value

The variation of the vector n has two components, $n'_\perp = (n \cdot \nabla)n$ orthogonal and $n'_\parallel = (n \times \nabla)n$ tangential to the contour:

$$
\begin{aligned}
(n'_\perp)_i &= \sum_{k=1}^{3} n_k \frac{\partial n_i}{\partial n_k}, \\
(n'_\parallel)_{ij} &= \sum_{k=1}^{3}\sum_{l=1}^{3} \varepsilon_{jkl}\, n_k \frac{\partial n_i}{\partial n_l}, \\
\varepsilon_{jkl} &= \begin{cases} 1 & \text{if } ijk \in \{123, 231, 312\} \\ -1 & \text{if } ijk \in \{321, 213, 132\} \\ 0 & \text{else.} \end{cases}
\end{aligned}
\tag{12}
$$

In the two-dimensional case, the tangential component $n'_\parallel$ is a vector, since its components are only nonzero for $j = 3$. In three dimensions it is a matrix. In both cases, its magnitude can be defined by a norm taking the square root of the sum of all squared scalar components. Now, we substitute $n'_\parallel$ for n_s in (11) and obtain

$$
\int \|(n \times \nabla)n\|^2 \, ds \longrightarrow \min. \tag{13}
$$

To combine all contours of a scalar field into a single optimization process, we need to integrate over the isovalues α as well. To emphasize certain regions in the optimization process, a nonnegative weighting function $w(s, \alpha)$ is multiplied with the normal variation,

$$
\int\int w(s, \alpha)\, \|(n \times \nabla)n\|^2 \, ds \, d\alpha \longrightarrow \min. \tag{14}
$$

Assuming that each contour is given in arc-length parametrization, this is equivalent to

$$
\begin{aligned}
\int\int \tilde{w}(x, y)\, \|(n \times \nabla)n\|^2 \, dx \, dy &\longrightarrow \min, \\
\tilde{w}(g_\alpha(s)) &:= w(s, \alpha),
\end{aligned}
\tag{15}
$$

where x and y are parameters of the scalar field and the weighting function is now redefined for this domain.

The residual norm corresponding to this optimization problem is induced by a scalar product, which can be used in a (nonlinear) optimization algorithm. In numerical experiments, we observed that the normalization of the gradient can be skipped, improving convergence rates. Hence, we substitute ∇f for the contour normal n and obtain with

$$
\begin{aligned}
\|(n \times \nabla)\nabla f\|^2 &= \|n_1 \,(f_{xy},\ f_{yy})^T - n_2 \,(f_{xx},\ f_{xy})^T\|^2 \\
&= (n_1 f_{xy} - n_2 f_{xx})^2 + (n_1 f_{yy} - n_2 f_{xy})^2
\end{aligned}
\tag{16}
$$

the scalar product

$$< f, g >_n = \int\int w \left((n_1 f_{xy} - n_2 f_{xx})(n_1 g_{xy} - n_2 g_{xx}) + (n_1 f_{yy} - n_2 f_{xy})(n_1 g_{yy} - n_2 g_{xy})\right) dx\, dy. \tag{17}$$

If the normal field $n(x, y) = (n_1(x, y), n_2(x, y))^T$ is known from a previous estimate for the scalar field f, then above scalar product used in (7) provides a linear optimization method. Since n depends on f, an iterative algorithm with linear optimization steps is required.

3.3 Nonlinear Fairing Algorithm

Our iterative fairing algorithm works as follows: The first estimate $f^{(0)}$ is obtained by thin-plate energy minimization. To compute $f^{(i+1)}$ from $f^{(i)}$, we first sample the gradient field from the previous estimate, $n^{(i)} := \nabla f^{(i)}$ (without normalization). Then, we compute $f^{(i+1)}$ from the optimization process using

$$< f, g > := < f, g >_{n^{(i)}} + \varepsilon \int\int f_x g_x + f_y g_y \, dx\, dy, \tag{18}$$

where ε is a small number, say 0.001. We need to add a small portion of slope minimization to increase numerical stability, since our optimization method does not converge to a unique solution, otherwise. This is due to the fact that the contour geometry does not carry any information about the corresponding isovalues, and multiple solutions with same contour geometry exist, see Fig. 3. All scalar products between basis functions need to be estimated by numerical integration.

Our algorithm requires smooth basis functions with nonzero second-order derivatives. In our implementation, we used cubic B-splines with two segments per grid cell to obtain enough degrees of freedom. For a row of n grid points located at $x = 0, 1, \cdots, n-1$, the corresponding knot vector is

$$\begin{aligned} t_0 &= \cdots = t_3 = 0, \\ t_i &= (i-3)/2 \quad (i = 4, \cdots, 2n), \\ t_{2n+1} &= \cdots = t_{2n+4} = n - 1. \end{aligned} \tag{19}$$

The corresponding B-splines N_i^4, $(i = 0, \cdots, 2n)$ are depicted in Fig. 4.

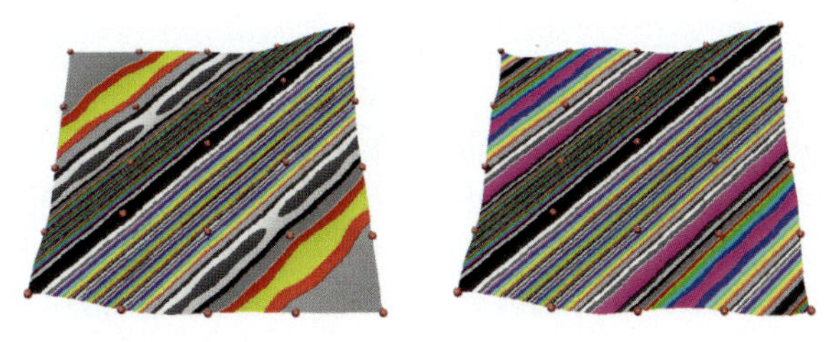

Fig. 3. Multiple solutions exist, when smoothing isolines without considering corresponding isovalues

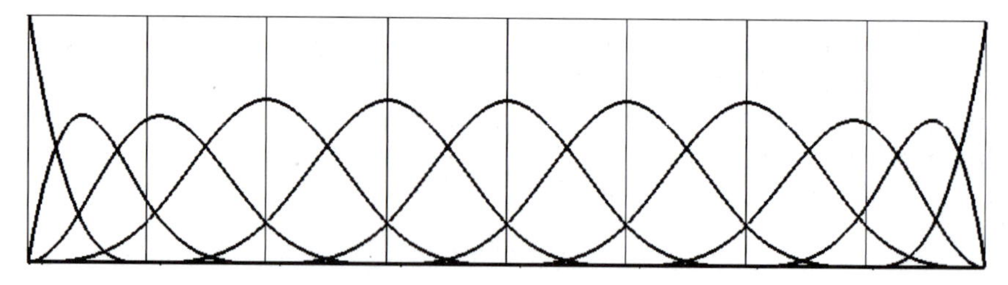

Fig. 4. B-splines N_i^4 for a row of five grid points (eight segments after refinement)

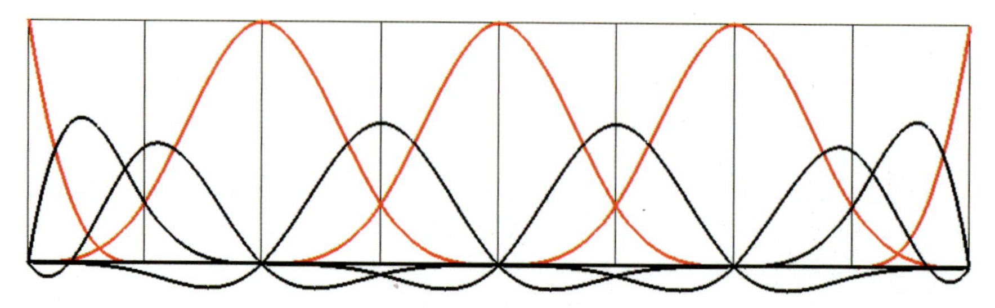

Fig. 5. Transformed basis functions Φ (*red*) and Ψ (*black*)

A simple basis transform provides the two sets Φ and Ψ used for interpolation and optimization, respectively:

$$\begin{aligned}
\phi_0 &= N_0^4, \\
\phi_i &= 1.5\, N_{2i+1}^4, \quad (i = 1, \cdots, n-2), \\
\phi_{n-1} &= N_{2n}^4, \\
\psi_0 &= N_1^4, \\
\psi_i &= N_{2i}^4 - 0.25\,(N_{2i-1}^4 + N_{2i+1}^4), \quad (i = 1, \cdots, n-1), \\
\psi_n &= N_{2n-1}.
\end{aligned} \tag{20}$$

These functions, shown in Fig. 5, have compact support and satisfy

$$\begin{aligned}
\phi_i(j) &= \delta_{ij} \quad (i, j = 0, \cdots, n-1), \\
\psi_i(j) &= 0 \quad (i = 0, \cdots, n;\ j = 0, \cdots, n-1).
\end{aligned} \tag{21}$$

In the two-dimensional case, we use the tensor-products $\Phi \times \Phi$ for interpolation and the functions in $\Phi \times \Psi$, $\Psi \times \Phi$, and $\Psi \times \Psi$ for optimization. A scalar field defined by $m \times n$ grid points will be represented by $(2m+1) \times (2n+1)$ coefficients. For the trivariate case, the method would be rather slow. Here we suggest to apply the bivariate approach to three canonical sets of slices, as described earlier for a similar method [3].

If interpolation is not required, our fairing method can be combined with least-squares fitting, based on (5). In this case, all basis functions are associated with degrees of freedom and no specific basis transform is required. In applications where the given data is not contaminated with noise or some uncertainty, however, the interpolative approach is recommended.

4 Numerical Examples

Figures 1, 6, and 7 provide a comparison of the contours obtained by standard interpolation methods and our fairing approach. Besides bilinear and bicubic interpolation, sinc-functions [13] are often used for interpolation since they

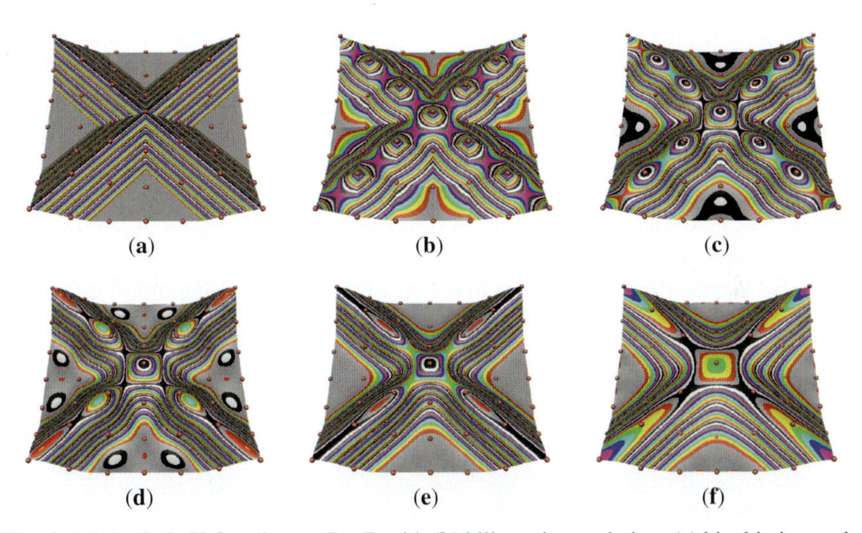

Fig. 6. (**a**) Analytic X-function on 7×7-grid; (**b**) bilinear interpolation; (**c**) bicubic interpolation; (**d**) sinc interpolation; (**e**) our interpolating method (five iterations); (**f**) our approximating method (five iterations, $w = 0.5$)

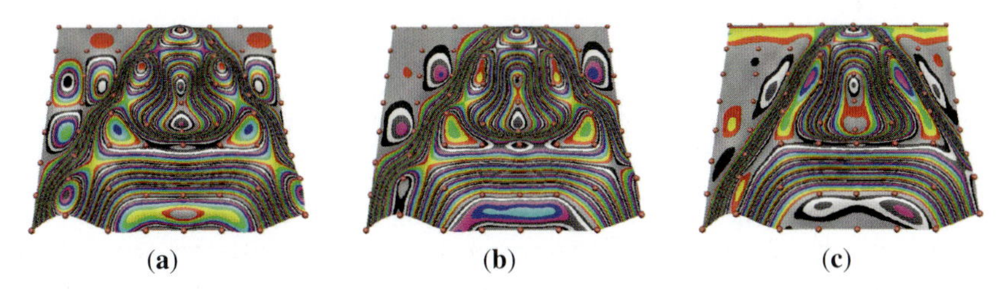

Fig. 7. (**a**) Sinc interpolation of a discrete point set, (**b**) bicubic interpolation, (**c**) optimized contours (five iterations)

correspond to box functions in Fourier domain and provide an optimal reconstruction from uniformly sampled data. The univariate sinc interplant can be defined as a convolution,

$$\begin{aligned} f_{sinc}(x) &= \sum_i f_i \; sinc((x-i)\pi), \\ sinc(x) &= \begin{cases} \frac{sin(x)}{x} & \text{if } x \neq 0, \\ 1 & \text{else.} \end{cases} \end{aligned} \tag{22}$$

Due to the infinite support of these basis functions, the sinc interplant may be computed using the Fourier transform. To avoid problems at the boundaries of our data sets, we assumed zero values at all grid points outside the domain.

The examples show that the sinc interplant mostly represents contour lines better than the bicubic interplant, despite of its energy-minimization property. In particular, diagonal features like those in Fig. 6 are better represented by sinc functions. The drawback of these is that often over- and undershoots occur, since exact localization in both frequency and time-domain is not possible (uncertainty principle). When using our contour fairing approach, these over- and undershoots are reduced to a minimum, showing that most contour lines generated are supported by the data. If we relax the interpolation constraints in order to get a least-squares approximation, the contours become even smoother.

Figure 8 shows the approximation results of the first three iterations. Computation times for data sets of different size are summarized in Table 1. These examples

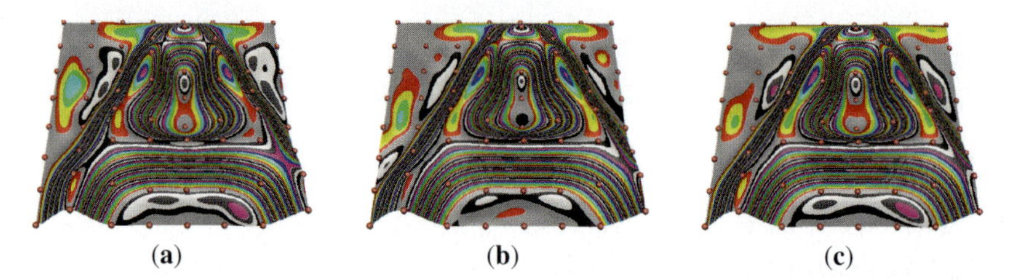

(a) (b) (c)

Fig. 8. Convergence of our iteration scheme: (**a**) first, (**b**) second, and (**c**) third iteration

Table 1. Computation times in milliseconds for thin-plate energy minimization and for each iteration of the fairing process. The total number of basis functions is the number of vertices plus the number of degrees of freedom (dofs)

No. vertices	No. dofs	Thin-plate min.	Iteration time
25	96	69	92
49	146	181	250
64	225	259	352
100	341	537	723

were computed on a Linux PC equipped with a 2.4 GHz processor, using only 16 samples per grid cell for numerical integration of the inner products. Most of the computation time is used up by this integration. Due to compact support of the basis functions, however, the time used for numerical integration is linear in the number of grid points. The systems of equations are sparse and require only small solution times. The method could be accelerated by precomputing the basis functions on the integration grid and by implementing the integration on a GPU.

We found that five iterations were sufficient to obtain good results with both methods. The computation times for the approximating method were slightly larger, since the systems are larger compared to the interpolating method (Fig. 9). The times for thin-plate minimization were given for comparison. Of course, thin-plate minimization can be obtained more efficiently by cubic spline interpolation at the coarse level, followed by knot insertion. This is due to the fact that bicubic surfaces minimize thin-plate energy a priori.

The quality of our fairing methods can be measured in the L^2-sense (taking the square root of the average squared error), in case that a continuous function representing a data set is available. To judge the quality of contour lines, we define A_r to

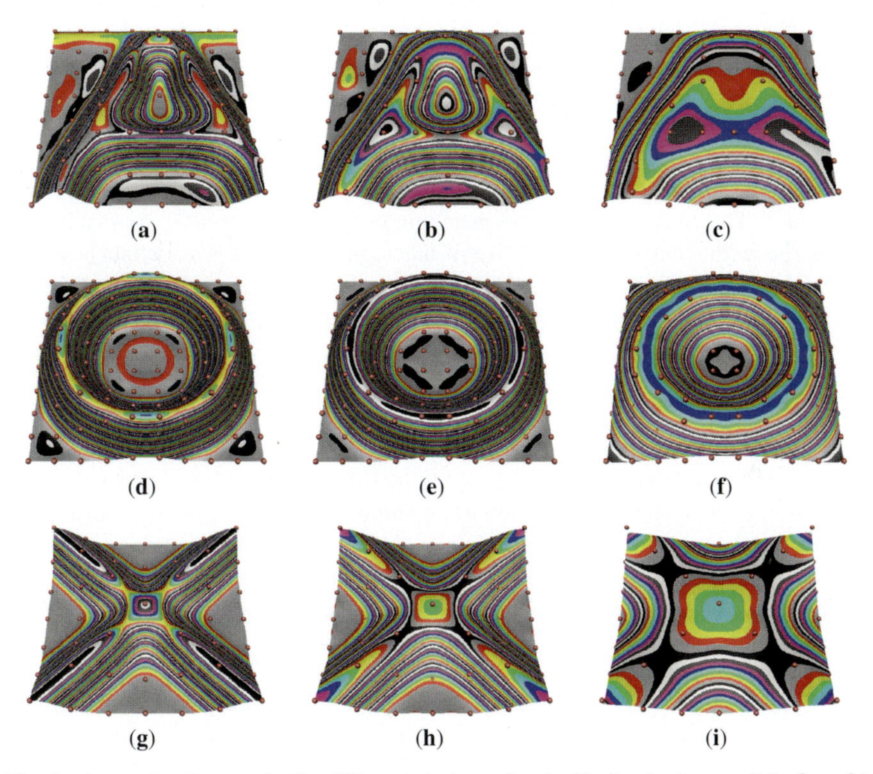

Fig. 9. Approximative results for different choices of w in (4). (**a**, **d**, **g**) $w = 0.1$; (**b**, **e**, **h**) $w = 0.5$; (**c**, **f**, **i**) $w = 0.9$. For $w \to 0$ the results equal those of the interpolating method

Table 2. Approximation results of O- and X-data sets based on three different norms. L^2 and d^2 are measured in percent of one grid unit, where a_∞ denotes the percentage of the domain area where no matching contour is found

Method	$\mathrm{O}: L^2$	$\mathrm{O}: d^2$	$\mathrm{O}: a_\infty$	$\mathrm{X}: L^2$	$\mathrm{X}: d^2$	$\mathrm{X}: a_\infty$
Bilinear	6.51	17.26	5.81	5.38	18.09	7.80
Bicubic	4.35	13.66	20.23	3.90	12.61	12.95
Sinc	2.98	10.15	22.97	4.20	12.78	26.14
Interpol. fairing	2.33	9.31	22.49	2.96	9.15	19.06
Approx. $w = 0.1$	2.29	9.59	21.71	2.70	9.52	14.08

be the subset of the domain D where the distance between corresponding isolines in the exact data f_{orig} and its reconstruction f_{reco} is no larger than a radius r:

$$A_r = \{p \in D \mid \exists q \in D : \; f_{orig}(q) = f_{reco}(p) \;\wedge\; \|p - q\| \leq r\} \qquad (23)$$

For all points in A_r, the distance between the reconstructed contour through this point and the matching original contour is bounded by r. We denote d^2 as the L^2-norm of these distances taken from all points in A_r. Let A_∞ be the complement of A_r and a_∞ denote its relative area within the domain. The radius r is chosen to be half a grid unit. The values obtained by these norms are summarized for the O- and X-data sets in Table 2.

The reconstruction results show that our interpolating approach and also the approximation based on small fairing weights, e.g., $w = 0.1$, provide small L^2 and d^2 values. The area A_∞ where no matching contours are found is mostly located in regions with small slopes. The size of this region does not provide much insight, but it can be used in combination with the d^2-values. The good performance of the bilinear interplant in terms of locating close contours may be due to the piecewise linear shape of the example data sets.

5 Conclusions and Future Work

We presented two iterative variational-modeling algorithms reducing interpolation artifacts of discretized scalar fields. Our methods find an interpolating (or approximating) scalar field for gridded data minimizing the curvature of all contours. The methods can be used for resampling or to improve the quality in regions of high geometric complexity. Of course, the size of reconstructed details is limited by the Nyquist frequency, i.e., by half the rate of discretization.

Numerical examples show that the visual quality of gridded data can be improved significantly in the two-dimensional case. The method generalizes to three dimensions, but it may be more efficient to use the bivariate approach for fairing the slices of a three-dimensional data set. When implemented on a GPU, the approach may be efficient enough to be used interactively.

References

1. M. Bertram, D.E. Laney, M.A. Duchaineau, C.D. Hansen, B. Hamann, and K.I. Joy, *Wavelet representation of contour sets*, In: Proceedings of IEEE Visualization 2001, pp. 303–310, 566.
2. M. Bertram, *Fairing scalar fields by variational modeling of contours*, In: Proceedings of IEEE Visualization 2003, pp. 387–392.
3. M. Bertram, *Volume refinement fairing isosurfaces*, In: Proceedings of IEEE Visualization 2004, pp. 449–505.
4. U. Clarenz, U. Diewald, and M. Rumpf, *Nonlinear anisotropic diffusion in surface processing*, In: Proceedings of IEEE Visualization 2000, pp. 397–405, 580.
5. M. Desbrun, M. Meyer, P. Schroeder, and A. Barr, *Implicit fairing of irregular meshes using diffusion and curvature flow*, In: Proceedings of ACM Siggraph, 1999, pp. 317–324.
6. U. Diewald, T. Preußer, and M. Rumpf, *Anisotropic diffusion in vector field visualization on euclidean domains and surfaces*, IEEE Transactions on Visualization and Computer Graphics, vol. 6, no. 2, 2000, pp. 139–149.
7. T. Gerstner, *Fast multiresolution extraction of multiple transparent isosurfaces*, In: Proceedings of VisSym'01, Joint Eurographics and IEEE TCVG Symposium on Visualization, 2001, pp. 35–44, 336.
8. H. Hagen, G. Brunnett, and P. Santarelli, *Variational principles in curve and surface design*, Surveys on Mathematics for Industry, vol. 3, no. 1, 1993, pp. 1–27.
9. A. Lopes and K. Brodlie, *Improving the robustness and accuracy of the marching cubes algorithm for isosurfacing*, IEEE Transactions on Visualization and Computer Graphics, vol. 9, no. 1, 2003, pp. 16–29.
10. W.E. Lorensen and H.E. Cline, *Marching cubes: a high resolution 3D surface construction algorithm*, In: Proceedings of ACM Siggraph, 1987, pp.163–169.
11. G.M. Nielson, G. Graf, R. Holmes, A. Huang, and M. Phielipp, *Shrouds: optimal separating surfaces for enumerated volumes*, In: Proceedings of VisSym'03, Joint Eurographics and IEEE TCVG Symposium on Visualization, 2003, pp. 75–84, 287.
12. G.M. Nielson, *On marching cubes*, IEEE Transactions on Visualization and Computer Graphics, vol. 9, no. 3, 2003, pp. 283–297.
13. F. Stenger, *Numerical methods based on sinc and analytic functions*, Springer, New York, 1993.
14. G.H. Weber, G. Scheuermann, and B. Hamann, *Detecting critical regions in scalar fields*, In: Proceedings of VisSym'03, Joint Eurographicsand IEEE TCVG Symposium on Visualization, 2003, pp. 85–94, 288.
15. J. Weickert, *Anisotropic diffusion in image processing*, ECMI Series, Teubner, Stuttgart, 1998.
16. W. Welch and A. Witkin, *Variational surface modeling*, In: Proceedings of ACM Siggraph, 1992, pp.157–166.
17. R. Westermann, L. Kobbelt, and T. Ertl, *Real-time exploration of regular volume data by adaptive reconstruction of isosurfaces*, The Visual Computer, vol. 15, 1999, pp. 100–111.

Time- and Space-Efficient Error Calculation for Multiresolution Direct Volume Rendering

Attila Gyulassy[1], Lars Linsen[1,2], and Bernd Hamann[1]

[1] Institute for Data Analysis and Visualization (IDAV), University of California, Davis, Davis, CA 95616, USA
{aggyulassy, hamann}@cs.ucdavis.edu

[2] Department of Mathematics and Computer Science, Ernst-Moritz-Arndt-Universität Greifswald, Greifswald, Germany
linsen@uni-greifswald.de

Summary. Multiresolution data representations are crucial for viewing large volumetric datasets interactively. When data is too large to fit into texture memory, or into main memory, a "cut" must be made through the multiresolution data hierarchy to attain a subset of the data that satisfies the memory requirements. Ideally, a subset is chosen such that the error made when visualizing the subset (compared to a visualization of the full data set) is smaller than that of any other subset of the same size. For real-time applications it is computationally too expensive to calculate the exact error during runtime. Further, computing error in a preprocessing step is usually not practical due to a large number of possible different configurations each requiring its own error computation. For example, when coupling a multiresolution representation with a direct volume rendering technique, screen-space error depends on the transfer function and viewing direction, making impossible its precomputation. We present an algorithm that stores an intermediate form of the error, which allows us to approximate screen-space error efficiently. The input for our algorithm is any spatially subdivided multiresolution representation of grid-aligned scalar or multivariate volume data. We focus on octree- and wavelet-based multiresolution techniques. For each level in the multiresolution hierarchy, the algorithm estimates screen-space error "on the fly," with respect to the current transfer function and viewing direction. The error is approximated by means of a two-dimensional histogram of error pairs. We have extended previous methods by presenting an approach that balances computational and memory costs with approximation quality of the error estimate.

1 Introduction

Visualization of volumetric datasets is a common task used in many fields, including medicine, physics, and other sciences. Complexity is introduced in this task by the fact that datasets sometimes are too large to fit into texture memory, or main memory. Therefore, data reduction schemes are required to enable interactive exploration and visualization. We refer to the subset that is selected as a "cut" of the data structure. Three steps summarize the main tasks involved in the visualization of large data

T. Möller et al. (eds.), *Mathematical Foundations of Scientific Visualization, Computer Graphics, and Massive Data Exploration*, Mathematics and Visualization,
DOI: 10.1007/978-3-540-49926-8,

using multiresolution approximation methods: (1) Compress the original data to a more manageable size; (2) select the optimal cut such that it minimizes screen-space error while maintaining interactive exploration; and (3) render the cut in an efficient way. For interactive visualization, frame-rates of at least ten frames per second are desirable. Therefore, the algorithm that selects the cut must be efficient. Also, the algorithm must not have a large memory overhead, since the primary goal is to use available memory to attain as high-quality a representation of the original data as possible. Furthermore, interactive modification of the transfer function is desirable.

Several studies have shown that it is possible to compress large datasets and thereby reduce the I/O and memory footprint. Nguyen and Saupe [8] showed that it is possible to attain quality compression of volumetric datasets using blockwise wavelet representations. Guthe and Straßer [2] demonstrated that using such compression methods and graphics hardware, it is possible to render large datasets at interactive frame rates. Unlike these methods, however, our algorithm can be applied to any data representation, including compressed representations, as long as the representation is hierarchical/nested, i.e., it maintains the property that high-resolution levels are spatially contained in low-resolution levels. Such representations include tree- and wavelet-based structures.

Selection of the cut is important when using multiresolution representations, since different cuts yield different screen-space errors when applying visualization methods. If the error associated with a node in the representation is known, then it is possible to make a decision about the importance of refining the resolution of that node. We present an algorithm that estimates screen-space error efficiently and supports interactive modification of the transfer function. The key observation motivating the algorithm presented in this paper is that there exists an intermediate form of the error that can be exploited to make possible lazy evaluation of the actual screen-space error. The intermediate form of the error can be computed independently of the chosen transfer function and viewing direction. Specifically, instead of storing the actual error (in color space) associated with each node in the data hierarchy, we store a precalculated histogram of values and deviations, such that the error can be reconstructed for any transfer function and viewing direction without processing the entire dataset again.

We show that with our algorithm one can attain substantially higher framerates for a guaranteed error bound by minimizing the size of the cut required for that error. Alternately, given a fixed amount of memory, we show that our algorithm selects a near-optimal cut for minimizing screen-space error.

2 Previous Work

The algorithm presented here combines techniques described by Guthe and Straßer [2] and LaMar et al. [6]. LaMar et al. presented an algorithm that makes use of the fact that there are fewer unique error pairs in a large data set than occurrences of such pairs. An error pair (a, b) is a pair of values out of the range of the considered data field. (An error pair (a, b) occurs when value a is used instead of the correct value b.) Their algorithm considers byte datasets that have the property that there exist only 256 distinct values and $256^2 = 2^{16}$ possible error pairs, whereas in a data set over a uniform rectilinear grid, e.g., of size 512^3, there are already $512^3 = 2^{27}$ entries of data values. Therefore, storing a 2D table for each nonleaf node in the representation, where the table contains for each error pair (a, b) entries

$$\text{“}Q(a, b) = \text{number of times error pair } (a, b) \text{ occurs,”}$$

makes calculation of the actual error at runtime faster. Several optimizations were introduced to reduce the size of this table, including halving the size of the table by reflecting with respect to the table's diagonal, and run-length encoding. Even though this method provides a fast method for recalculating error, it introduces large storage overhead, as each nonleaf node in the hierarchy has to contain a large data structure representing this table. In addition, there is a high cost of calculating the error at each node, since the entire table must be traversed. Also, this method is restricted to byte datasets. Previous studies [4, 5] developed the error metrics necessary for error calculation.

Guthe and Straßer [2] took a different approach to calculating the error for each node in the hierarchy. In their algorithm, each node in the data hierarchy stores only a small histogram of the maximum deviation for each value in that node. Furthermore, the method bins values into eight groups. Therefore, instead of dealing with a 256^2 table, a single eight-entry array is used. In reconstructing the error, however, all possible combinations of the values must be considered to find a conservative estimate of the error. In practice, the n^2 complexity of this operation can be avoided by storing another small table of the “maximum” and “minimum colors” and opacities for each bin. While this approach is computationally inexpensive and memory-efficient, it creates an overly conservative estimation of error. Indeed, the maximum color is compared with the minimum color of every other element, and then scaled by the number of items binned, which greatly overshoots the actual error. Therefore, this method overestimates the error associated with a node in such a way that it could produce an inferior cut of the multiresolution representation.

We present a new, hybrid approach, which uses a histogram similar to the one used by LaMar et al. [6], but also applies a binning procedure, such as described by Guthe and Straßer [2], see Fig. 1. In this way, we obtain a closer approximation of the actual error, while keeping the data structure and computation overheads small. Furthermore, our binning approach enables us to work on any multivariate dataset, not only on byte data.

Other methods [1, 3, 11] can be used to calculate the error associated with levels in a hierarchy. However, these methods rely on a fixed transfer function. Therefore, whenever the transfer function is modified, the entire dataset must be traversed. This characteristic prohibits interactive modification of the transfer function.

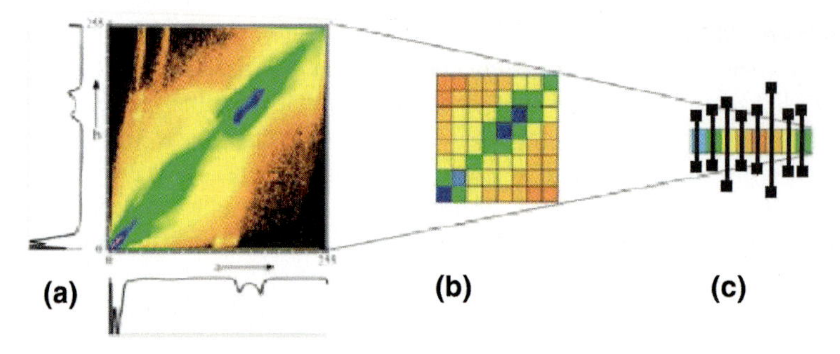

Fig. 1. For each node in the data hierarchy, (**a**) shows the table generated by LaMar et al. [6], (**c**) shows the binning method of Guthe and Straßer [2], and (**b**) shows how our algorithm combines the two methods

3 Error Estimation

Our algorithm presented here utilizes a conservative error estimation to select the cut through the multiresolution representation of a dataset. The cut consists of the nodes that are kept in memory for the purpose of rendering. The error can be measured by the root-mean-square (RMS) difference between the images generated by rendering the actual and approximating data, called the screen-space error. The error in a single pixel is defined as

$$error_{color}(x) = \left| \int_0^d Opacity_h(x) * Color(Value_h(x))\, dx - \int_0^d Opacity_l(x) * Color(Value_l(x))\, dx \right| .$$

The function $Opacity_h$ refers to the function that defines the progressive visibility along the ray, parameterized by x for the high-resolution data representation; $Value_h$ refers to the value returned by interpolation of the high-resolution data. Similarly, $Opacity_l$ and $Value_l$ refer to the corresponding functions for the low-resolution data. The integral is evaluated over the interval $(0, d)$, where d is the far cut plane.

In our conservative estimate, we approximate this value by the integral of the errors along the ray projected from that pixel convolved with the progressive opacity function, i.e.,

$$error_{approx}(x) = \int_0^d \left| Opacity_h(x) * Color(Value_h(x)) - Opacity_l(x) * Color(Value_l(x)) \right| dx .$$

Considering the triangle inequality, $error_{color}(x) \leq error_{approx}(x)$.

We further simplify this equation by using the opacity in each node instead of the progressive opacity, so that the error contributed by each node is view-independent. Therefore, a conservative estimate of the error contributed by each node is sufficient to compute a conservative estimate of the final screen-space error of the image. The error of a particular node in the cut is also useful for determining whether or not to refine the cut at that node.

3.1 View-Independent Error at a Node

For simplicity, we first only consider a piecewise-constant interpolation of the data. Also, we consider only one color channel at a time so that the transfer function is scalar-valued. We combine the error contributions of each channel after they have been computed independently. As is done by LaMar et al. [6], we can use two different error norms for calculating the absolute error at a node in the data representation. The L_∞-error defines the error as the maximum of the errors under that node, i.e.,

$$error_{L_\infty} = \max_{p \in B}\{|Color(Value_h(p)) - Color(Value_l(p))|\} \,, \tag{1}$$

where p refers to points of the original data that reside inside node B of the hierarchy. This error calculates the largest deviation in color that can occur inside a node. The RMS error averages the errors under that node, i.e.,

$$error_{RMS} = \sqrt{\frac{1}{n}\sum_{p \in B}(Color(Value_h(p)) - Color(Values_l(p)))^2} \,, \tag{2}$$

where n is the total number of points $p \in B$. For simplicity, we limit our discussion to $error_{RMS}$. To calculate this error, however, we need to use the entire data structure. We can rewrite this error as

$$error_{RMS} = \sqrt{\frac{1}{n}\sum_{i}(Color(a_i) - Color(b_i))^2 * Q(a_i, b_i)} \,, \tag{3}$$

where $\{(a_i, b_i)\} = \{(Value_h(p_i), Value_l(p_i))|p_i \in B\}$ and $Q(a_i, b_i)$ is the number of times that the error pair (a_i, b_i) appears inside node B. In the case of byte data, it is possible to represent Q explicitly as a table that is significantly smaller than n for blocked data, and we can efficiently compute the error at a node using this histogram. However, for real-valued data, no guarantees can be made concerning the number of unique error pairs (a_i, b_j). Therefore, the size of Q is only bounded by n. Straightforward storage of the table Q would be inefficient. Instead, we fix a histogram size for Q, binning error pairs. Q represents a table of bins, each bin (a_i, b_j) counting the number of occurrences of error pairs in its range. We reconstruct a conservative error efficiently from this histogram.

At each node in the hierarchy, we store such a table Q of error pairs. We define the table Q at node B as $Q(a_i, b_j) = m$, where m is the number of points $p \in B$ such that $Value_l(p) \in [a_i, a_{i+1})$ and $Value_h(p) \in [b_j, b_{j+1})$. Each bin in the histogram stores the number of occurrences of error pairs whose values fall within a range of values. Figure 2 shows how this table is created for the lowest nonleaf node in the multiresolution hierarchy.

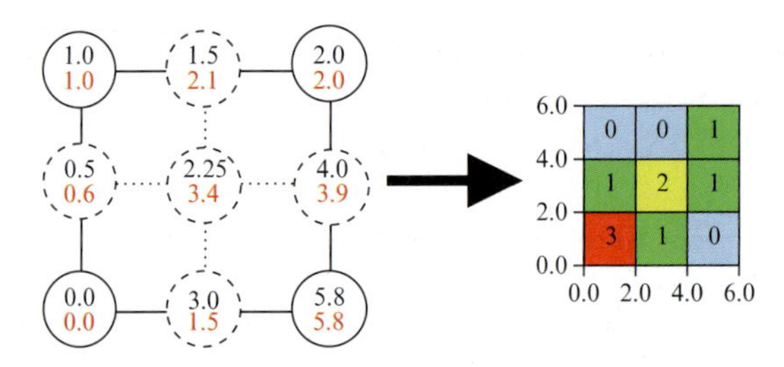

Fig. 2. Two levels in a quad-tree hierarchy. Data points (*left*) display low-resolution $Value_l(p)$ on *top*, and high-resolution $Value_h(p)$ on the *bottom*, and the two values together form an error pair. A 3 × 3 histogram (*right*) associated with the low-resolution node stores the number of occurrences of error pairs. We assume piecewise-constant interpolation, with the data ranging in value in the interval [0.0, 6.0]

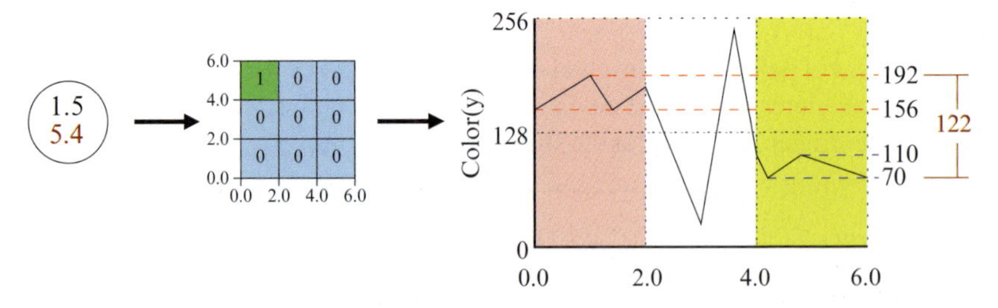

Fig. 3. Error pairs (*left*) are used to populate the histogram (*middle*). To reconstruct the error associated with bin [0.0, 2.0) × [4.0, 6.0) we determine the maximally possible difference in those ranges in the transfer function (*right*), *MaxError*() = 122. We assume piecewise-constant interpolation of the data

To compute the error at a given node B, we reference the table Q of error pairs associated with that node. We use the form of the error given by (3) to generate our error formula as

$$error_{RMS} = \sqrt{\frac{1}{n} \sum_{i,j} (MaxError([a_i, a_{i+1}), [b_j, b_{j+1})))^2 * Q(a_i, b_j)}\,, \qquad (4)$$

where $MaxError([a_i, a_{i+1}), [b_j, b_{j+1}))$ is the largest possible error in color that can occur in the intervals $[a_i, a_{i+1})$ and $[b_j, b_{j+1})$. We apply the transfer function to compute the maximum and minimum colors for each interval. We define *MaxError* as

$$\begin{aligned} MaxError([a, b), [c, d)) = {} & \max(Color([a, b)) \cup Color([c, d))) \\ & - \min(Color([a, b)) \cup Color([c, d)))\,. \end{aligned} \qquad (5)$$

The variables a, b, c, and d represent values in the domain of the transfer function. Here, $Color([a, b))$ returns a range of color values in the interval $[a, b)$. Figure 3 illustrates *MaxError*. The transfer function is piecewise-linear. Therefore, all extrema occur either at the endpoints of the intervals, or at data points of the transfer function. We extract extreme values by sweeping through the transfer function inside the intervals $[a_i, a_{i+1})$ and $[b_j, b_{j+1})$. The maximal error is computed as the difference between the maximum color value and the minimum color value. We scale the maximum error in a node by the maximum distance inside the node to obtain a conservative estimate of the error. The maximum distance in a node in a standard rectilinear grid is the diagonal length $\sqrt{3}$.

The error formulation in (6) is defined for piecewise-constant interpolation applied to a dataset. To maintain a conservative error estimate when using trilinear interpolation, the histogram Q of a node B must be modified. $Q(a_i, b_j)$ represents the number of nodes B_i contained in B, where the minimum function value is in the range a_i, and the maximum function value is in the range b_j. To reconstruct the error, *MaxError* now returns the maximum color difference in the range $[min(a_i, b_j), max(a_i, b_j)]$. When the dimension of Q is $N \times 1$, this method reduces to the one presented in [2]. For our results, we have used this formulation of error.

The calculation of the maximum error of a bin is an expensive operation. However, a major improvement can be made when each table Q has the same range and size at every node. In this case, a single 2D table can be calculated representing the maximum difference between colors for each bin in the histogram Q. Therefore, the (i, j)-th element holds the maximum color difference for a bin (a_i, b_j) in table Q. This drastically reduces the amount of calculation necessary for each node, since this only needs to be calculated once whenever the transfer function is modified. The result is stored in a table for subsequent look-ups.

Increasing histogram size improves the accuracy of the error estimation, as it reduces the range of each bin, and therefore the maximum error associated with that bin. Table 1 shows the space overhead associated with different histogram sizes. The maximum acceptable histogram size at each node depends on the size of the blocks, and also on data size. A large histogram can lead to severe storage overhead. Indeed, for efficient estimation of the error, it is important to keep the error estimation structure in memory. Therefore, the size of the histogram must be balanced with performance considerations.

Table 1. Increasing histogram size leads to larger memory overhead associated with multiple block sizes. The numbers are the percentages of the total memory used to store histograms

Histogram size	2^2 (%)	8^2 (%)	64^2 (%)	256^2 (%)
32^3	0.012	0.195	12.500	200.000
64^3	0.002	0.024	1.560	25.000
128^3	<0.001	0.003	0.195	3.120
256^3	<0.001	<0.001	0.024	0.391

3.2 View-Dependent Error

Another consideration in calculating error is visibility of a node. Two factors contribute to visibility: projected solid angle and opacity.

Projected solid angle refers to the amount of screen space that a node and its subhierarchy occupy when projected. When a node occupies less than a pixel of screen space, its screen-space error is very small. Conversely, when the screen space of a node is large, even smaller errors are noticed. Therefore, a new view-dependent error function can be defined as

$$error_{node} = error_{RMS} \cdot \phi \ ,$$

where ϕ is the projected solid angle of the node. Since the actual value of the projected solid angle is expensive to calculate, we approximate it by using the distance d from the camera to the node, i.e.,

$$error_{node} = error_{RMS} \cdot \alpha \cdot \frac{r^2}{d^2} \ ,$$

where r is the maximum radius of the node and α is a constant.

Opacity is difficult to calculate efficiently, especially since it relies on the transfer function and on the viewing direction. The error methods discussed so far are used to select the nodes that in the working set, needed for rendering. To calculate opacity, a front-to-back calculation must take place to eliminate nodes that are occluded. One possible way to perform this task, without calculating the entire working set, is to process nodes front-to-back and subdivide them in that order. Therefore, a node is only considered for subdivision according to the previously defined error metric once all nodes in front of it satisfy a particular error condition.

Following this approach, only the nodes in the final working set are considered. Unfortunately, the size of this set is no longer bounded, as there is no limit on the number of subdivision levels of nodes in the front. As a result, the memory bound for the working set may be exceeded with a suboptimal selection of nodes. Due to the complexity involved in calculating opacity while selecting a cut of the multiresolution representation, occlusion culling is usually employed only during rendering, once the cut has been selected. Guthe and Straßer [2] used such a method to avoid rendering occluded blocks.

3.3 Cut Selection

Assuming that the error for each node is known, the overall procedure for selecting nodes and rendering performs these steps:

Preprocessing:

1. Initialize multiresolution data structure.
2. Calculate histograms in bottom-up manner

Runtime:

1. Initialize the cut with the top level node
2. While space left in memory:
   - Find node with largest error
   - Subdivide this node and add children to the cut
3. Render the cut

The greedy algorithm selects the node with the largest error, and subdivides it. As a termination condition we either use the amount of free memory left or a particular error threshold not to be exceeded. In either case, the algorithm terminates, since space consumption increases and error decreases as we refine the cut.

4 Multiresolution Representation

Our algorithm can be applied to any nested multiresolution representation, i.e., a representation that satisfies the multiresolution analysis criteria presented by Rodler [10]. In particular, the multiresolution representation must subdivide the entire data space, with nodes at each higher level in the representation spatially bounding their child nodes. Furthermore, the accuracy of the representation must not decrease as the cut is refined. Therefore, the error of each child of a node must be smaller than or equal to the error at that node.

There are many representations that satisfy these requirements. Some commonly used ones are octrees and wavelet-based representations. For simplicity, our algorithm was implemented using an octree representation.While octrees are attractive due to their simplicity, several studies have shown that wavelet-based representations are efficient as well. In particular, when using wavelets, it is possible to compress the original data, to alleviate some of the difficulties in large dataset rendering. Haar wavelets are the most commonly used wavelets. Park and Ihm [9] attained both compression and increased performance by using that representation. Indeed, Rodler [10] presented several more complicated wavelet transforms. However, the Haar wavelet transform is the most appropriate method for most multiresolution techniques, due to its simplicity and efficiency.

5 Rendering

Rendering of volumetric data is a well studied topic. Very large datasets, however, pose additional challenges. Levoy [7] implemented a scheme for interactive raycasting. However, his method involves massively parallel rendering. Still, the optimal performance for raycasting a 512^3 dataset with 96 processors was less than two frames per second. Although CPU performance has increased dramatically over recent years, straightforward raycasting of large datasets is not practical for interactive visualization on PCs.

Some of the fastest techniques for rendering volumetric data on a standard PC utilize 3D texture hardware. LaMar et al. [5] showed how to improve image quality by using object-aligned slices. Westermann [11] showed that it is possible to use texture hardware for multiresolution representations. Guthe and Straßer [2] attained about ten-frames-per-second performance for a 40 GB dataset, using their error estimation technique combined with texture hardware. Therefore, once errors are calculated and a cut is determined, any of the previous techniques could be used to render the data.

6 Results and Discussion

Our algorithm provides an improved method for estimating screen-space error associated with levels in a multiresolution data representation. Our method balances storage overhead, processing time, and quality of estimation. Using this improved selection strategy, we pick a better cut through the multiresolution representation, meaning that the same-sized cut yields lower error.

In Sect. 3, we showed that the storage overhead for this method is dependent on the size of the histograms used and the size of the dataset. Therefore, we can balance speed of the algorithm with the memory footprint.

Figure 4 shows that, given an error threshold, increasing the histogram size estimates the error of a cut with higher accuracy. In this case, we wish to find the smallest cut such that the error threshold is satisfied. For a "nice" dataset, i.e., a data set with smoothly changing function values, the error approximation improves with size of the histogram in a reasonable manner. However, with sparse noise inserted into the dataset, we see a substantial improvement with increased histogram size. One advantage of our method is that small perturbations in the dataset do not significantly increase the estimated error, in contrast to the algorithm presented in Guthe and Straßer [2] (Table 2).

We performed our analysis for the human skull dataset, which has size 256^3 and integer data values in the range [0, 255]. This dataset has both high-frequency and low-frequency regions, and therefore is suitable for analysis purposes. Results were generated with a 2 GHz Pentium 4 processor with 512 Mb of main memory. The dataset was divided into 32^3 blocks and rendered using a straightforward raycasting method. Recalculation of the error when changing the transfer function required less than 1 ms. The recalculation of the error does not scale with dataset size, it scales only with the size of the blocks and the size of the cut. Therefore, interactive modification of the transfer function is possible.

As expected, increasing the size of the cut improves the accuracy of the final image. As more data is used, the error decreases. However, the real benefit of this

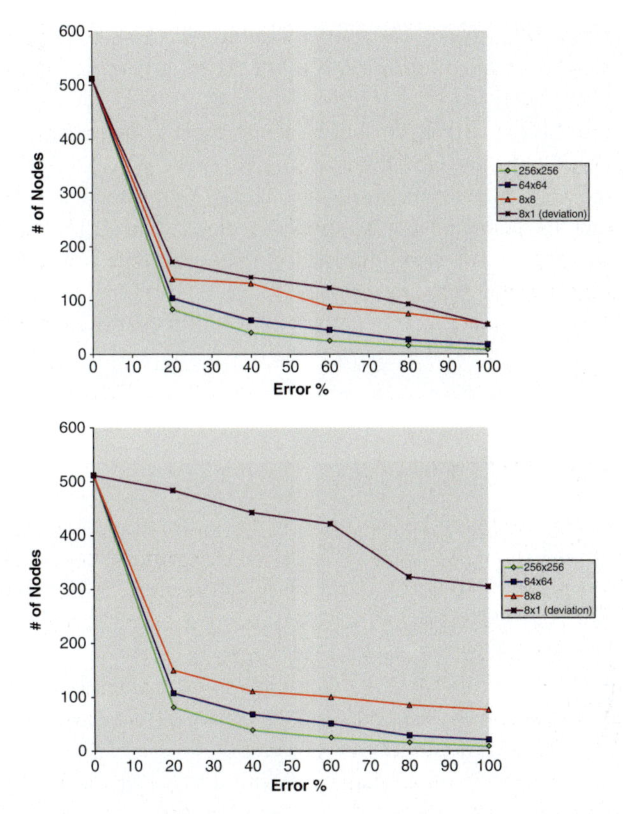

Fig. 4. Increasing error tolerance decreases the size of the cut needed to satisfy the error condition. Error is measured as percentage of the maximum error, which is the error associated with the coarsest resolution. "8 × 1 (deviation)" refers to the algorithm presented by Guthe and Straßer [2]. *Left:* Application to "nice" artificial dataset consisting of distances to points distributed in the domain. *Right:* Application to same dataset with some sparse noise inserted

Table 2. RMS errors associated with image generated by our algorithm for each histogram size. The last column is the RMS error of the image generated by Guthe and Straßer [2] using eight bins

Histogram size	4^2	16^2	64^2	Guthe and Straßer [2]
RMS error	6.20	5.71	4.68	6.29

selection strategy is this: When the cut size is held constant, using a larger histogram reduces the final error by selecting a better cut. Figure 5 shows the inverse relationship between histogram size and screen-space error.

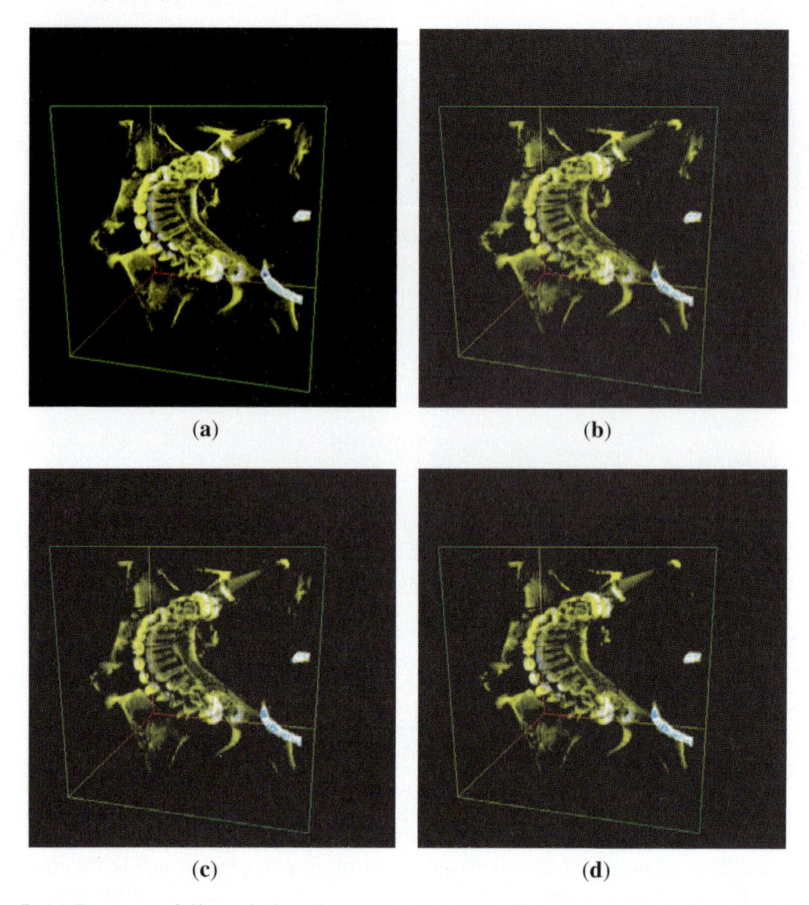

Fig. 5. (**a**) Dataset at full resolution. Images (**b**), (**c**), and (**d**) show a cut of 63 nodes selected by our strategy using histogram sizes of 64^2, 16^2, and 4^2, respectively

7 Conclusions and Future Work

We have presented a cut selection strategy for multiresolution direct volume rendering of large data that supports interactive modification of a transfer function. We have improved previous methods [2, 6], since our method can deal with any scalar-valued dataset, using an improved error estimation scheme. Our histogram approach is memory-efficient and can be used in any multiresolution direct volume rendering method. An important property of our approach is that the size of a histogram determines accuracy of rendering results. Therefore, we can balance computational and memory costs with quality. Although the histograms we showed in the previous section are square, it is possible to attain good results with nonsquare histograms. Our results suggest that it is possible to tune the histogram size automatically for a particular architecture and dataset size.

We plan to extend this algorithm to calculate errors for time-varying volumetric data. Another possible future research area is error calculation for vector fields.

Acknowledgments

This work was supported by the National Science Foundation under contract ACI 9624034 (CAREER Award), through the Large Scientific and Software Data Set Visualization (LSSDSV) program under contract ACI 9982251, through the National Partnership for Advanced Computational Infrastructure (NPACI) and a large Information Technology Research (ITR) grant; and the National Institutes of Health under contract P20 MH60975-06A2, funded by the National Institute of Mental Health and the National Science Foundation. We thank the members of the Visualization and Computer Graphics Research Group at the Institute for Data Analysis and Visualization (IDAV) at the University of California, Davis.

References

1. Imma Boada, Isabel Navazo, and Roberto Scopigno. Multiresolution volume visualization with a texture-based octree. *The Visual Computer*, 17(3):185–197, 2001.
2. Stefan Guthe and Wolfgang Straßer. Advanced techniques for high-quality multi-resolution volume rendering. *Computers and Graphics*, 28(1):51–58, 2004.
3. Stefan Guthe, Michael Wand, Julius Gonser, and Wolfgang Straßer. Interactive rendering of large volume data sets. In *Proceedings of the conference on Visualization '02*, pages 53–60. IEEE Computer Society, 2002.
4. Eric LaMar, Bernd Hamann, and Kenneth I. Joy. Multiresolution techniques for interactive texture-based volume visualization. In *VIS '99: Proceedings of the conference on Visualization '99*, pages 355–361, Los Alamitos, CA, USA, 1999. IEEE Computer Society Press.
5. Eric C. LaMar, Mark A. Duchaineau, Bernd Hamann, and Kenneth I. Joy. Multiresolution techniques for interactive texturing-based rendering of arbitrarily oriented cutting-planes. In W.C. de Leeuw and R. Van Liere, editors, *Proceedings of VisSym 00 The Joint Eurographics and IEEE TCVG Conference on Visualization*, pages 105–114. Springer, Berlin, 2000.
6. Eric C. LaMar, Bernd Hamann, and Kenneth I. Joy. Efficient error calculation for multiresolution texture-based volume visualization. In Gerald Farin, Bernd Hamann, and Hans Hagen, editors, *Hierachical and Geometrical Methods in Scientific Visualization*, pages 51–62. Springer, Heidelberg, 2003.
7. Marc Levoy. Display of surfaces from volume data. *IEEE Computer Graphics and Applications*, 8(3):29–37, 1988.
8. Ky Giang Nguyen and Dietmar Saupe. Rapid high quality compression of volume data for visualization. *Computer Graphics Forum*, 20(3):C49–C56, 2001.
9. Sanghun Park and Insung Ihm. Wavelet-based 3D compression scheme for interactive visualization of very large volume data. *Computer Graphics Forum*, 18(1):3–15, 1999.
10. Flemming Friche Rodler. Wavelet based 3D compression with fast random access for very large volume data. In *Proceedings of the 7th Pacific Conference on Computer Graphics and Applications*, page 108. IEEE Computer Society, 1999.
11. Rüdiger Westermann. A multiresolution framework for volume rendering. In Arie Kaufman and Wolfgang Krüger, editors, *1994 Symposium on Volume Visualization*, pages 51–58, 1994.

Massive Data Visualization: A Survey

Kenneth I. Joy

Institute for Data Analysis and Visualization, University of California, Davis
kijoy@ucdavis.edu

Summary. Today's scientific and engineering problems require a different approach to address the massive data problems in organization, storage, transmission, visualization, exploration, and analysis. Visual techniques for data exploration are now common in many scientific, engineering, and business applications. However, the massive amount of data collected through simulation, collection and logging is inhibiting the use of conventional visualization methods. We need to discover new visualization methods that allow us to explore the massive multidimensional time-varying information streams and turn overwhelming tasks into opportunities for discovery and analysis.

1 Introduction

Scale is the grand challenge of visualization! We face a situation where the problem analyst is continually overwhelmed with massive amounts of information from multiple sources, where the relevant information content exists in a few "nuggets." The visualization challenge is to create new methods that allow the analyst to visually examine this massive, multidimensional, multisource, time-varying information stream and make decisions in a time critical matter.

The definition of "massive" is a moving target, because it changes over time as computational resources and algorithmic methods improve. We will apply the term "massive" to information streams that overwhelm some critical computation or display resource necessary for analysis. The most common application of this term is to information streams that do not fit into local disk or main memory, although many researchers apply the term to data that cannot be displayed on a "single display device."

Various approaches have been used to address the problems of scale. There is a continual effort to construct computer systems that can store and process massive amounts of data, and design display systems that can display data at ever increasing resolution. However, our ability to collect and generate data is increasing at a faster rate than the projected increase in computational and display power, and we wish to

T. Möller et al. (eds.), *Mathematical Foundations of Scientific Visualization, Computer Graphics, and Massive Data Exploration*, Mathematics and Visualization,
DOI: 10.1007/978-3-540-49926-8,

focus on methods that can address the problems of scale independent of computer system and display advances. These methods can be classified as follows:

- Compression and simplification – Reduce the size of the information stream by utilizing data compression methods or by simplifying the data to remove redundant or superfluous items.
- Multiresolution methods – Design multiresolution hierarchies to represent the information at varying resolutions, allowing the user (or the application) to dictate the resolution displayed.
- Memory external methods – Create innovative methods that store the data externally and develop methods that access relevant data subject to the interactivity required by the user.
- Abstraction – Create new visual metaphors that display data in new ways.

Many areas outside of traditional visualization are also critical in meeting the challenge of scale, including an understanding of application areas. Visualization research must reach out to encompass issues from numerical methods, data analysis, software engineering, database methods, networking, image processing, cognitive and perceptual psychology, human–computer interaction, and machine learning to address the problems before us. We must also develop evaluation techniques to "measure" scalability so new tools can be analyzed.

This paper attempts a partial survey of the methods used to address scale in visualization systems. We do not attempt to reproduce existing surveys for compression [13, 51, 70], but focus on simplification, multiresolution techniques, memory-external methods, and visual scalability through abstraction.

2 Driving Problems

Our interest in massive data is driven by the many domains in which people actively collect data. Simulation applications are driven by problems in computational fluid dynamics, engineering analysis, high-energy physics, and microprocessor design. Technical advances are now producing massive time-varying data in medical, biomedical and biological imaging applications. Bioinformatics data collections (e.g., The Human Genome Project) will provide massive multidimensional data to be analyzed. The Sloan Digital Sky Survey is collecting 8 TB per year of astronomical data, and satellites turned towards the Earth collect vast amounts of image data as well. The dramatically decreasing cost and increasing capabilities of sensors have led to an explosion in the collection of real-world data. Dynamic processes, arising in business or telecommunication, generate massive streams of sensor data, web click streams, network traffic logs, or credit card transactions.

Consider the various data types that possibly affect the domain examined by the problem analyst. These could include the following:

- Textual data – Massive textual data from documents, speeches, e-mails, or web pages now influence the problem domain. This data can be truly massive, contain billions of items per day, and much of it must be analyzed in a time-critical manner [34, 76, 77].

- Simulation data – Terascale simulations are now producing tens of terabytes of output for several-day runs on the largest computer systems. As an example, the Gordon–Bell-Prize-winning simulation of a Richtmyer–Meshkov instability in a shock-tube experiment [54], produces isosurfaces of the mixing interface with 460 million unstructured triangles using conventional extraction methods. The pipeline necessary to visualize these massive datasets is described well by Duchaineau et al. [19] (Fig. 1).
- Databases – Many corporate and government entities have constructed huge databases containing a wealth of information. We require new algorithms for the efficient discovery of previously unknown patterns in these large databases.
- Geospatial data – consider the data collected by satellites that image the earth. We now have satellites that can create images at less than 1-m resolution and that can collectively image the land surface of the planet in a very short time. These images must be examined in a time-critical matter.
- Sensor data – the revolution in miniaturization for computer systems has allowed us to produce a myriad of compact sensors. The sensors can collect data about their environment (location, proximity, temperature, light, radiation, etc.), can analyze this data, and can communicate between themselves. Collections of sensors can produce very large streaming sets of data.
- Video and image data – Image and video data are being used more and more to enhance the effectiveness of the security in high-risk operations. Content analysis, combined with massive recording capabilities, is also being used as a powerful tool for improving business processes and customer service. This streaming data paradigm creates new opportunities for visualization applications.

(a)

(b)

Fig. 1. Scientific Simulations can produce massive isosurface data using conventional techniques (courtesy, M. Duchaineau)

Each of these categories can produce massive data streams containing information that is applicable to a specific application domain. The grand challenge in the area of scalability is to develop new tools to distill the relevant nuggets of information from these widely disparate information streams, creating an explorable information space that can be examined by analytic or visual means to influence the decision of the data analyst. We must provide mechanisms that can visualize connections between the relevant information in the information streams, and allow the problem analyst to make decisions based on the totality of the information.

3 How Do We Explore Massive Data?

Nearly all methods to explore large, complex datasets rely on two methods: data transformations and visual transitions. To transform a massive dataset into an "explorable" format, most researchers create a set of simple transformations that can be repeatedly applied to the data. To display meaningful visualization without screen clutter, the researcher must develop alternation visual metaphors and abstractions to illustrate the data. These transformations and abstractions are similar to the "zooms" presented by Bosch et al. [10] and Stolte et al. [71–73] Their framework utilizes multiple zooming panels for both data in visual abstractions. They think of multiscale visualizations as a graph, were each node corresponds to a particular data representation and visual abstractions, and each edge represents a zoom. Zooming in the multiscale visualization is equivalent of traversing this graph.

In the case of massive multidimensional abstract data, consider the (possibly infinite) set of data and visual representations of this information. Together, these represent the nodes of our graph. To explore this information, we need to establish edges between these nodes, and establish methods to find paths through this massive graph. What we wish to find, of course, are the representations that satisfy the basic principles we desire: uncluttered visualizations, maximal information, maximal decision support, etc.

Given an information stream that must be explored, we can view the "visualization process" as a huge, possibly infinite, graph. Each node of this graph contain a representation of the data, and a visual representation method to be used on the data. The edges of the graph represent transitions between two data representations, two visual representations, or two data-representation/visual-representation pairs. In many scientific and engineering applications, the visual representation is given by a representation of three-dimensional space, and only the data transitions must be generated. In applications that work with abstract data, there is no inherited geometry, and visual metaphors and transitions are paramount. We will use this "transformation" and "transition" model throughout this paper.

4 Simplification Methods

Simplification and multiresolution methods for meshes, especially polygonal surface meshes, has dominated the research literature for several years. Most methods attempt to simplify the data through elementary data transformations.

In the case of triangle mesh simplification, most algorithms create a simple data transformation by first removing a contiguous set of triangles from the mesh, creating a hole. The hole is then retriangulated, resulting in a mesh with fewer triangles. These simple data transformations are measured by use of an "error estimation" strategy and ordered by a priority scheme, creating a complex data transformation that (hopefully) preserves the features of the mesh, while producing a simplified mesh that can be examined. These data transformations can be classified into three categories: algorithms that simplify a mesh by removing vertices; algorithms that simplify a mesh by removing edges; and algorithms that simplify a mesh by removing higher-level simplices (Fig. 2).

Schroeder et al. [67] and Renze and Oliver [65] have developed algorithms that simplify a mesh by removing vertices and the triangles containing a vertex. They use a recursive loop-splitting procedure to generate a triangulation of the hole, while Renze and Oliver fill the hole by using an unconstrained Delaunay triangulation algorithm.

Hoppe [36, 37] and Hoppe and Popović [63] describe a progressive-mesh representation, which effectively removes individual edges and their associated triangles. The "edge-collapse" operation forces a retriangulation of the resulting hole. They place edges in a priority queue, ordered by the expected energy cost of its collapse. As edges are collapsed, the priorities of the edges in the neighborhood of the transformation are recomputed and reinserted into the queue. The result is an initial coarse representation of the mesh, and a linear list of edge-collapse operations, each of which can be regenerated to produce finer representations of the mesh. Other edge-collapse algorithms have been described by Xia and Varshney [78], who use the constructed hierarchy for view-dependent simplification and rendering of models, and by Garland and Heckbert [27], who utilize quadratic error metrics for efficient calculation of the hierarchy.

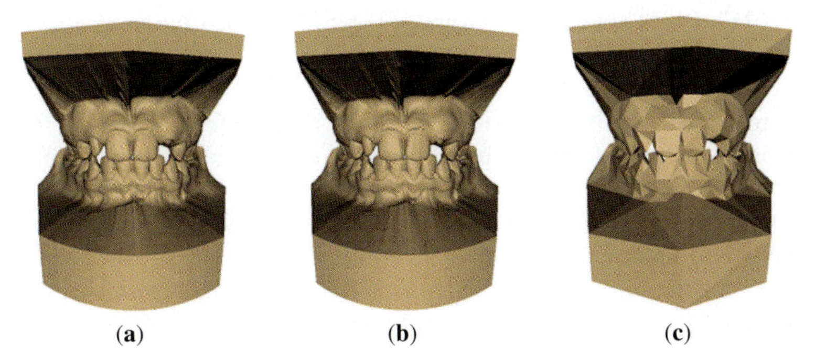

(a) (b) (c)

Fig. 2. Various simplifications of a dataset representing a set of teeth (courtesy, Mike Garland)

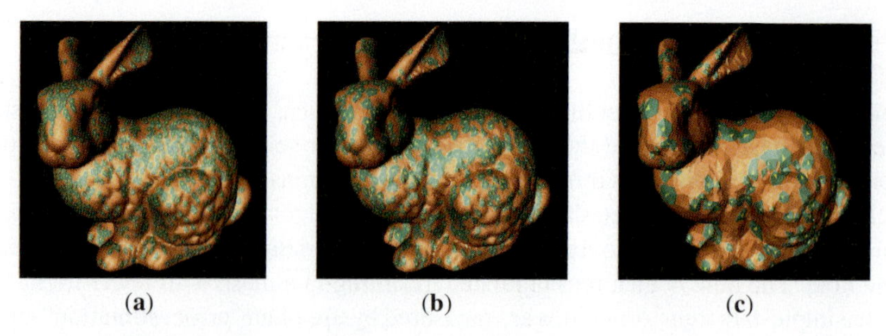

Fig. 3. Simplification by triangle collapse: Individual triangles are identified and collapsed to a point. The collapse point is based upon a best-fit quadratic polynomial to the surface in the neighborhood of the triangle

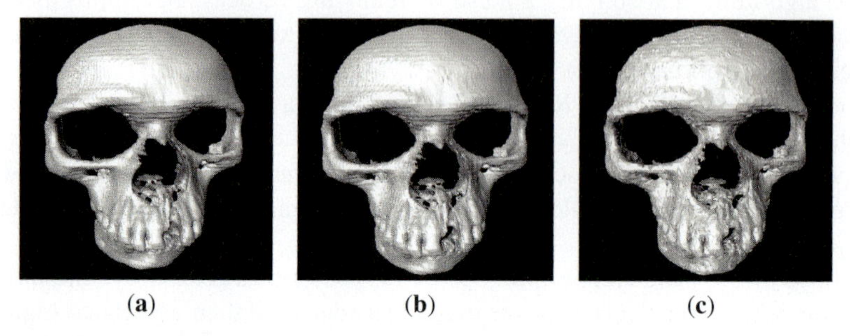

Fig. 4. Tetrahedral collapse algorithm: The method of Trotts et al. [74] is used on a skull dataset. Isosurfaces from the original mesh is shown in (**a**), an intermediate mesh in (**b**), and a coarse mesh in (**c**)

Hamann [32, 33], and Gieng et al. [28, 29] have developed algorithms that simplify triangle meshes by removing triangles. These algorithms order the triangles using weights based partially on the curvature of a surface approximation, partially on the changes in the topology of the mesh due to a triangle collapse, and partially due to a predicted error estimate of a collapse operation. Figure 3 illustrates the results of their method.

These surface algorithms have been lifted to tetrahedral meshes by several researchers [12, 69, 74] Staadt and Gross [69] have extended the progressive mesh algorithm of Hoppe [36] to tetrahedral meshes. Trotts et al. [74] use a strategy similar to the triangle collapse algorithm of Gieng et al. (Fig. 4). Chopra and Meyer [12] identify individual tetrahedra T and remove the immediate neighbors of T. They force a retriangulation of the hole by simulating a collapse of the tetrahedron T to a point.

Cignoni et al. [14] treat the tetrahedral mesh problem by using a top-down, Delaunay-based procedure. The mesh is refined by selecting a data point whose associated function value is poorly approximated by an existing mesh, and the mesh is modified locally to preserve the Delaunay property.

Error approximation for the elementary data transformation steps vary among the algorithms. Vertices to be removed by Schroeder et al. [67] are identified through a distance-to-simplex measure. The data reduction problem is formulated in terms of a global mesh optimization problem by Hoppe [39], ordering the edges according to an energy minimization function. The most frequently used measure is the quadric error metric, introduced by Garland and Heckbert [27].

Error approximation between meshes is addressed by several researchers. Cohen et al. [18] utilize an edge collapse-strategy to simplify polygonal models. They order the edges by using the distances between corresponding points of the mapping. A similar technique by Hoppe [38] is used for error calculations in level-of-detail rendering of terrain modeling. Klein et al. [46] calculate the Hausdorff distance between two meshes.

5 Multiresolution Methods

Several of the simplification methods of the previous section can also be used in a multiresolution manner. For example, the progressive mesh representation of Hoppe [36] can be used to create an initial coarse representation of a mesh, together with a linear list of edge-collapse operations. These can be used to regenerate various resolutions of the mesh by collapsing and expanding appropriate edges.

Many multiresolution geometry schemes exist for terrain rendering. Systems that use a regular grid approach are built on a hierarchy of right triangles [22, 53], while triangulated irregular networks (TINs) [23, 38, 68] work to solve this problem by not restricting triangulations to a regular grid. Both schemes have advantages and disadvantages.

Several researchers have based terrain rendering on regular grids. Lindstrom et al. [49] present a method based upon an adaptive 4–8 mesh, using longest-edge bisection (LEB) as a fundamental operation to refine the mesh. They use a bottom-up vertex-reduction method to reduce the size of the mesh for display purposes. Duchaineau et al. [20] present a system for visualizing terrain also based upon a LEB paradigm. Their system uses a dual-queue management system that splits and merges cells in the hierarchy according to the visual fidelity of the desired image. Lindstrom and Pascucci [50] describe a framework for out-of-core rendering and management of terrain surfaces. They present a view-dependent refinement method along with a scheme for organizing the terrain data to improve coherence and reduce the number of paging events from external storage to main memory. Again, they organize the mesh using a longest-edge bisection strategy, using triangle stripping, view frustum culling and smooth blending of geometry. Pajarola [58] utilizes a restricted quadtree triangulation, similar to an adaptive 4–8 mesh for terrain visualization.

Cignoni et al. [15–17] have demonstrated the ability to display both adaptive geometry and texture of large terrain datasets in real-time. They utilize a quadtree texture hierarchy and a bintree of triangle patches (TINs) for the geometry. The triangle patches are constructed off-line with high-quality simplification and triangle stripping algorithms. Hierarchical view frustum culling and view-dependent texture

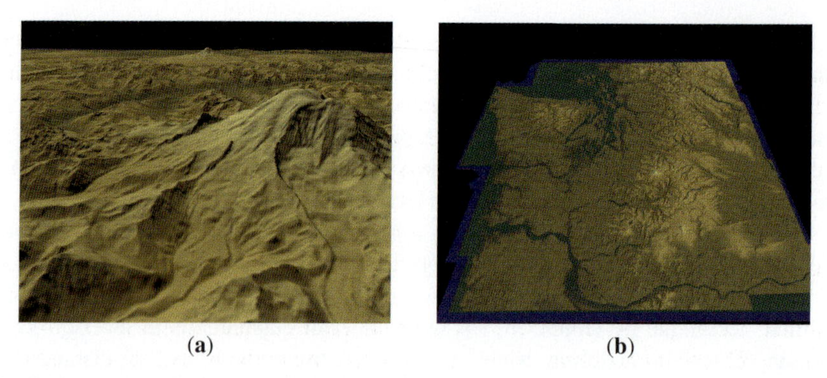

Fig. 5. Hwa et al. [40] display massive texture on terrain using a multiresolution system based on 4–8 textures. These illustrations show two views of a $120k \times 120k$ texture mapped onto terrain data from the State of Washington

and geometry refinement are both performed each frame. Textures are managed as rectangular tiles, resulting in an quadtree representation of the textures. The rendering system traverses the texture quadtree until acceptable error conditions are met, and then selects corresponding patches in the geometry bintree system. Once the texture has been chosen, the geometry is refined until the geometry space error is within tolerance.

Hwa et al. [40] adaptively texture a dynamic terrain mesh using an adaptive 4–8 mesh structure coupled with a diamond data structure and queues for processing both geometry and texture. Results of their technique are showing in Fig. 5. Their method can process massive textures in a multiresolution out-of-core fashion for interactive rendering of textured terrain.

Multiresolution methods can also be constructed around a novel data structure. The refinement of a tetrahedral mesh via longest edge bisection is described in detail in several papers.

Gregorski et al. utilizes a novel longest-edge-bisection mesh to build a multiresolution hierarchy of a volume dataset. They combine coarse-to-fine and fine-to-coarse refinement schemes for this mesh to create an adaptively refinable tetrahedral mesh. This adaptive mesh supports a dual priority queue split/merge algorithm similar to the ROAM system [20] for view-dependent terrain visualization. Sets of tetrahedra that share a common refinement edge are grouped into diamonds, which function as the unit of operation in the mesh hierarchy and simplify the process of refining and coarsening the mesh. At runtime, the split/merge refinement algorithm is used to create a lower resolution dataset that approximates the original dataset to within a given error tolerance. The lower resolution dataset is a set of tetrahedra, possibly from different levels of the hierarchy, that approximate the volume dataset to within this isosurface error tolerance. The isosurface is extracted from the tetrahedra in this lower resolution representation using linear interpolation. The authors have shown this method to be useful for fly-throughs of both static and dynamic three-dimensional datasets.

Abello and Vitter [3] provide an excellent presentation of external memory algorithms and their use in visualization. Abello and Korn [1,2] present a multiresolution method to display massive digraphs. Their method changes many graph nodes into a single node depending on the available screen space to display the graph. They utilize out-of-core methods to access relevant portions of the data.

6 External Memory Methods

External memory algorithms [75], also known as out-of-core algorithms, have been rising to the attention of the computer science community in recent years as they address, systematically, the problem of the nonuniform memory structure of modern computers. This issue is particularly important when dealing with large data-structures that do not fit in main memory since the access time to each memory unit is dependent on its location. New algorithmic techniques and analysis tools have been developed to address this problem in the case of geometric algorithms [3,4,30,52] and visualization [11]. Closely related issues emerge in the area of parallel and distributed computing where remote data transfer can become a primary bottleneck in the computation. In this context, space-filling curves are often used as a tool to determine, very quickly, data distribution layouts that guarantee good geometric locality [31,56,59].

Space filling curves [66] have been also used in the past in a wide variety of applications [5] because of their hierarchical fractal structure as well as for their well known spatial locality properties. The most popular is the Hilbert curve [35] which guarantees the best geometric locality properties [57]. The pseudo-Hilbert scanning order [7,8,45] generalizes the scheme to rectilinear grids that have different number of samples along each coordinate axis. Recently Lawder [47,48] explored the use of different kinds of space-filling curves to develop indexing schemes for data storage layout and fast retrieval in multidimensional databases.

Balmelli et al. [6] use the Z-order (Lebesgue) space-filling curve to navigate efficiently a quad-tree data-structure without using pointers (see Fig. 6). Out-of-core computing [75] addresses specifically the issues of algorithm redesign and data layout restructuring, necessary to enable data access patterns with minimal out-of-core processing performance degradation (Fig. 7). Research in this area is also valuable in parallel and distributed computing, where one has to deal with the similar issue of balancing processing time with data migration time. The solution of the out-of-core processing problem is typically divided into two parts:

1. Algorithm analysis to understand the data access patterns of the problem and "redesigning" the data to maximize locality
2. Storage of the data in secondary memory with a layout consistent with the access patterns of the algorithm, amortizing the cost individual I/O operations over several memory access operations

In the case of hierarchical visualization algorithms for volumetric data, the 3D input hierarchy is traversed to build derived geometric models with adaptive levels

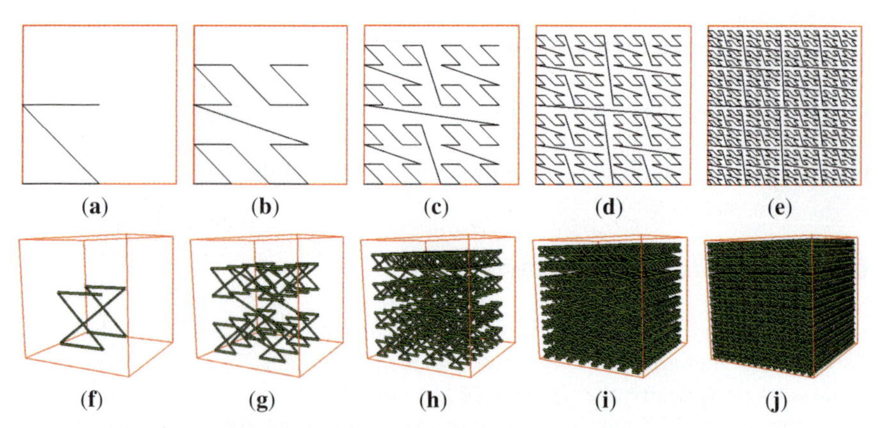

Fig. 6. (**a–e**) The first five levels of resolution of the 2D Lebesgue space-filling curve. (**f–j**) The first five levels of resolution of the 3D Lebesgue space-filling curve

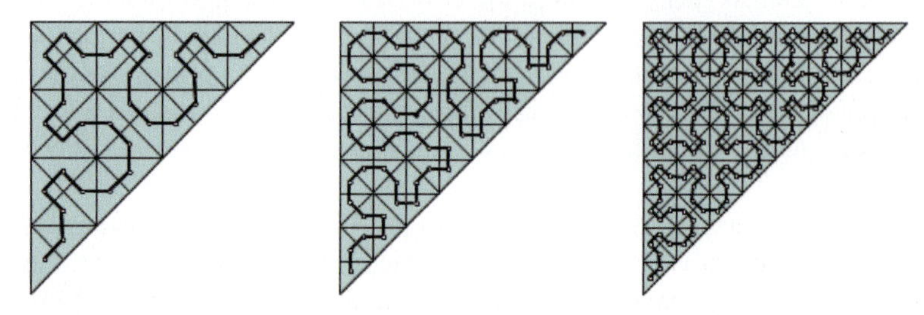

Fig. 7. Hwa et al. utilize an out-of-core storage system of bintree elements by utilizing a method based upon a Sierpinski curve. The Sierpinski curve arises naturally from the indexing of a bintree

of detail. The shape of the output models are then modified dynamically with incremental updates of their level of detail. The parameters that govern this continuous modification of the output geometry are dependent on runtime user interaction, making it impossible to determine, a priori, what levels of detail will be constructed. For example, parameters can be external, such as the viewpoint of the current display window or internal, such as the isovalue of an isocontour or the position of an orthogonal slice. The general structure of the access pattern can be summarized into two main points: (1) the input hierarchy is traversed coarse to fine, level by level so that the data in the same level of resolution is accessed at the same time and (2) within each level of resolution the data is mainly traversed coherently in regions that are in close geometric proximity.

Pascucci and Frank [60] have introduced a new static indexing scheme that induces a data layout satisfying both requirements (1) and (2) for the hierarchical traversal of n-dimensional regular grids. In their method, the order of the data is independent of the out-of-core blocking factor so that its use in different settings (e.g., local disk access or transmission through a network) does not require any large data

Fig. 8. Streaming meshes. Triangles of the mesh are sorted by one coordinate, then input in order. This allows processing of a contiguous portion of the mesh, and out-of-core storage (courtesy, M. Isenburg and P. Lindstrom)

reorganization. Conversion from the standard Z-order indexing to the new indexing scheme can be implemented with a simple sequence of bit-string manipulations. In addition, their algorithm requires no data replication, avoiding any performance penalty for dynamic updates or any inflated storage typical of most hierarchical and out-of-core schemes.

Isenburg et al. [42–44] utilize a streaming mesh paradigm to process massive meshes. They first sort the triangles of a mesh according to a selected coordinate. They can then input triangles up to the capacity of the system's main memory and process these triangles. Triangles that are no longer necessary can be discarded and new triangles can be input in order. They have used this methodology to perform simplification algorithms. The processing order on the dragon model is shown in Fig. 8.

7 Visual Scalability

Visual scalability refers to the capability of data exploration tools to display massive data, in terms of either the number or the dimension of individual data elements, on a display device. The quality of visual displays, the visual metaphors used in the display of information, the techniques used to interact with the visual representations, and the perception capabilities of the human cognitive system, all affect the methods used to present information.

Many visualization problems arise from the complexity and dynamic nature of datasets. The effective display of this information is beyond the capability of current presentation techniques. The sheer size of the data makes it very difficult to effectively generate global views and one must generate novel interaction methods, coupled with innovative visual metaphors to allow the program analyst to explore the data.

Eick and Karr [21] proposed a scalability analysis and concluded that many visual metaphors do not scale effectively, even for moderately sized datasets. Scatterplots for example, one of the most useful graphical techniques for understanding relationships between two variables, can be overwhelmed by a few thousand points.

"Focus+Context" refers to display methods that focus on detail while the less-important data is shown in a smaller representation (Fig. 9). Interesting approaches include fish-eye views [26,61] and Magic Lens filters [9]. The table lens [64] applies fish-eye techniques to table-oriented data. Plaisant has proposed the SpaceTree [62], a tree-browser which adds dynamic rescaling of branches of the tree to best fit the available screen space. Munzner et al. have illustrated similar concepts in TreeJuxtaposer [55]. Hierarchical Parallel Coordinates [24,25], a multiresolution version of Parallel Coordinates [41], uses multiple views at different levels of detail for representation of large-scale data.

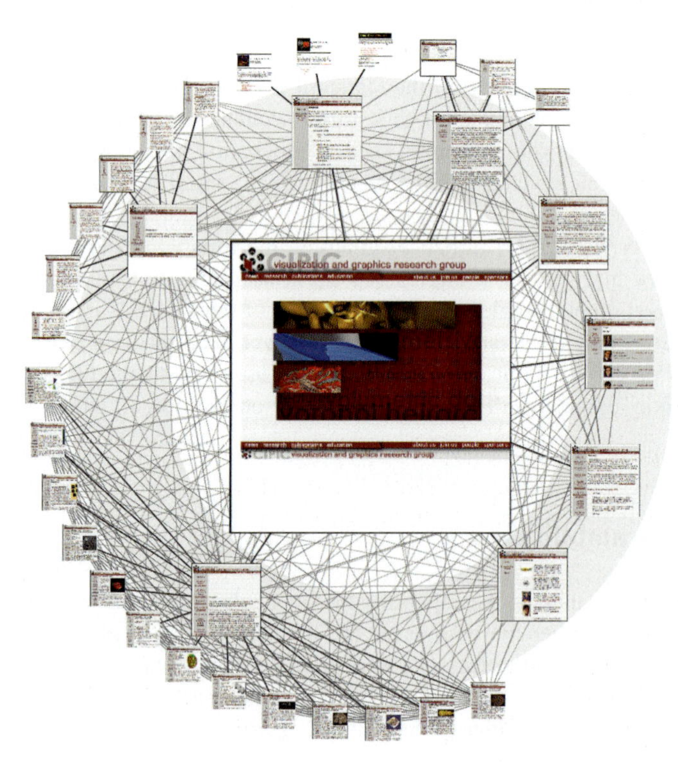

Fig. 9. Focus+Context views (courtesy, T.J. Jankun-Kelly)

8 Conclusions

The research in massive data exploration is only in its infancy. The grand challenge problem, to explore dynamic multidimensional massive data streams, is still largely unexplored. We must extend the state-of-the-art in visual and data transformations to be able to explore these complex information streams. We must develop new interaction methods to explore massive data in a time-critical matter. We must develop new techniques for information fusion that can integrate the relevant nuggets of information from multisource multidimensional information streams. We must develop new methods to address the complexity of information, and create a seamless integration of computational and visual techniques to create a proper environment for analysis. We also must augment our methods to consider visual limits, human perception limits, and information content limits.

There is much to do.

Acknowledgments

This work was supported by the National Science Foundation under contracts ACR 9882251 and ACR 0222909, and by Lawrence Livermore National Laboratory under contract B523818. Thanks to Valerio Pascucci and Mark Duchaineau for their valuable assistance.

References

1. ABELLO, J., AND KORN, J. Visualizing massive multi-digraphs. In *INFOVIS 2000: Proceedings of the IEEE Symposium on Information Vizualization 2000* (2000), IEEE Computer Society, p. 39.
2. ABELLO, J., AND KORN, J. Mgv: A system for visualizing massive multidigraphs. *IEEE Trans. Vis. Comput. Graph. 8*, 1 (2002), 21–38.
3. ABELLO, J., AND VITTER, J. S., Eds. External memory algorithms and visualization. DIMACS Series in Discrete Mathematics and Theoretical Computer Science. American Mathematical Society Press, Providence, RI, 1999.
4. ARGE, L., AND MILTERSEN, P. B. On showing lower bounds for external-memory computational geometry problems. In *External Memory Algorithms and Visualization*, J. Abello and J. S. Vitter, Eds., DIMACS Series in Discrete Mathematics and Theoretical Computer Science. American Mathematical Society Press, Providence, RI, 1999.
5. ASANO, T., RANJAN, D., ROOS, T., AND WELZL, E. Space filling curves and their use in the design of geometric data structures. Lecture Notes in Computer Science, vol. 911. Springer, Berlin, 1995, pp. 36–44.
6. BALMELLI, L., KOVAνCEVIĆ, J., AND VETTERLI, M. Quadtree for embedded surface visualization: Constraints and efficient data structures. In *IEEE International Conference on Image Processing (ICIP)* (Kobe Japan, October 1999), pp. 487–491.
7. BANDOU, Y., AND KAMATA, S.-I. An address generator for a 3-dimensional pseudo-hilbert scan in a cuboid region. In *International Conference on Image Processing, ICIP99* (1999), vol. I.

8. BANDOU, Y., AND KAMATA, S.-I. An address generator for an n-dimensional pseudo-hilbert scan in a hyper-rectangular parallelepiped region. In *International Conference on Image Processing, ICIP 2000* (2000), pp. 707–714.
9. BIER, E., STONE, M., AND PIER, K. Enhanced illustration using magic lens filters. *IEEE Comput. Graph. Appl. 17*, 6 (1997), 62–70.
10. BOSCH, R., STOLTE, C., TANG, D., GERTH, J., ROSENBLUM, M., AND HANRAHAN, P. Rivet: a flexible environment for computer systems visualization. *SIGGRAPH Comput. Graph. 34*, 1 (2000), 68–73.
11. CHIANG, Y., AND SILVA, C. T. I/O optimal isosurface extraction. In *IEEE Visualization 1997* (Nov. 1997), R. Yagel and H. Hagen, Eds., IEEE, pp. 293–300.
12. CHOPRA, P., AND MEYER, J. Tetfusion: An algorithm for rapid tetrahedral mesh simplification. In *VIS 2002: Proceedings of the conference on Visualization 2002* (Washington, DC, USA, 2002), IEEE Computer Society.
13. CHUI, C. K. An introduction to wavelets. Wavelet Analysis and its Applications, vol. 1. Academic, New York, 1992.
14. CIGNONI, P., DE FLORIANI, L., MONTONI, C., PUPPO, E., AND SCOPIGNO, R. Multiresolution modeling and visualization of volume data based on simplicial complexes. In *1994 Symposium on Volume Visualization* (Oct. 1994), A. Kaufman and W. Krueger, Eds., ACM SIGGRAPH, pp. 19–26.
15. CIGNONI, P., GANOVELLI, F., GOBBETTI, E., MARTON, F., PONCHIO, F., AND SCOPIGNO, R. BDAM: Batched dynamic adaptive meshes for high performance terrain visualization. In *Proceedings of the 24th Annual Conference of the European Association for Computer Graphics (EG-03)* (Oxford, UK, Sept. 1–6 2003), P. Brunet and D. Fellner, Eds., vol. 22, 3 of *Computer Graphics forum*, IEEE Computer Society/Blackwell, Malden, pp. 505–514.
16. CIGNONI, P., GANOVELLI, F., GOBBETTI, E., MARTON, F., PONCHIO, F., AND SCOPIGNO, R. Interactive out-of-core visualization of very large landscapes on commodity graphics platforms. In *ICVS 2003*, Lecture Notes in Computer Science. Springer, New York, pp. 21–29.
17. CIGNONI, P., GANOVELLI, F., GOBBETTI, E., MARTON, F., PONCHIO, F., AND SCOPIGNO, R. Planet–sized batched dynamic adaptive meshes (P-BDAM). In *Proceedings IEEE Visualization* (Conference held in Seattle, WA, USA, Oct. 2003), IEEE Computer Society, IEEE Computer Society Press, pp. 147–155.
18. COHEN, J., MANOCHA, D., AND OLANO, M. Simplifying polygonal models using successive mappings. In *IEEE Visualization 1997* (Nov. 1997), R. Yagel and H. Hagen, Eds., IEEE, pp. 395–402.
19. DUCHAINEAU, M. A., PORUMBESCU, S., BERTRAM, M., HAMANN, B., AND JOY, K. I. Dataflow and remapping for wavelet compression and view-dependent optimization of billion-trangle isosurfaces. In *Approximation and Geometrical Methods for Scientific Visualization*, G. Farin, H. Hagen, and B. Hamann, Eds., Springer, Heidelberg, 2002, pp. 1–17.
20. DUCHAINEAU, M. A., WOLINSKY, M., SIGETI, D. E., MILLER, M. C., ALDRICH, C., AND MINEEV-WEINSTEIN, M. B. ROAMing terrain: Real-time optimally adapting meshes. In *IEEE Visualization 1997* (Nov. 1997), R. Yagel and H. Hagen, Eds., IEEE Computer Society Press, Los Alamitos, CA, pp. 81–88.
21. EICK, S., AND KARR, A. Visual scalability. *Int. J. Comput. Graph. Stat. 11*, 1 (Mar. 2002), 22–43.
22. EVANS, W., KIRKPATRICK, D., AND TOWNSEND, G. Right triangular irregular networks. Tech. Rep. TR97-09, The Department of Computer Science, University of Arizona, May 30 1997.

23. FOWLER, R. J., AND LITTLE, J. J. Automatic extraction of irregular network digital terrain models. *Computer Graphics (SIGGRAPH 1979 Proc.) 13*, 2 (Aug. 1979), pp. 199–207.
24. FUA, Y.-H., WARD, M. O., AND RUNDENSTEINER, E. A. Hierarchical parallel coordinates for exploration of large datasets. In *IEEE Visualization 1999 (Vis 1999)* (Washington, Brussels, Tokyo, Oct. 1999), IEEE, pp. 43–50.
25. FUA, Y.-H., WARD, M. O., AND RUNDENSTEINER, E. A. Structure-based brushes: A mechanism for navigating hierarchically organized data and information spaces. *IEEE Trans. Vis. Comput. Graph. 6*, 2 (2000), 150–159.
26. FURNAS, G. W. Generalized fisheye views. In *CHI 1986: Proceedings of the SIGCHI conference on Human factors in computing systems* (New York, USA, 1986), ACM, pp. 16–23.
27. GARLAND, M., AND HECKBERT, P. S. Surface simplification using quadric error metrics. In *SIGGRAPH 1997: Proceedings of the 24th annual conference on Computer graphics and interactive techniques* (1997), ACM/Addison-Wesley, pp. 209–216.
28. GIENG, T. S., HAMANN, B., JOY, K. I., SCHUSSMAN, G. L., AND TROTTS, I. J. Smooth hierarchical surface triangulations. In *Proceedings of Visualization 1997* (Oct. 1997), H. Hagen and R. Yagel, Eds., IEEE Computer Society Press, Los Alamitos, CA, pp. 379–386.
29. GIENG, T. S., HAMANN, B., JOY, K. I., SCHUSSMAN, G. L., AND TROTTS, I. J. Constructing hierarchies for triangle meshes. *IEEE Trans. Vis. Comput. Graph. 4*, 2 (1998), 145–161.
30. GOODRICH, M. T., TSAY, J.-J., VENGROFF, D. E., AND VITTER, J. S. External-memory computational geometry. In *Proceedings of the 34th Annual IEEE Symposium on Foundations of Computer Science (FOCS 1993)* (Palo Alto, CA, November 1993).
31. GRIEBEL, M., AND ZUMBUSCH, G. W. Parallel multigrid in an adaptive pde solver based on hashing and space-filling curves. *Parallel Comput. 25* (1999), 827–843.
32. HAMANN, B. A data reduction scheme for triangulated surfaces. *Comput. Aided Geometric Des. 11* (1994), 197–214.
33. HAMANN, B., AND CHEN, J.-L. Data point selection for piecewise linear curve approximation. *Computer-Aided Geometric Design 11*, 3 (June 1994), 289–301.
34. HAVRE, S., HETZLER, B., AND NOWELL, L. Themeriver: Visualizing theme changes over time. In *INFOVIS 2000: Proceedings of the IEEE Symposium on Information Vizualization 2000* (Washington, DC, USA, 2000), IEEE Computer Society, p. 115.
35. HILBERT, D. Über die stetige abbildung einer linie auf ein flachenstück. *Mathematische Annalen 38* (1891), 459–460.
36. HOPPE, H. Progressive meshes. In *SIGGRAPH 1996 Proc.* (Aug. 1996), pp. 99–108.
37. HOPPE, H. View-dependent refinement of progressive meshes. In *SIGGRAPH 1997 Proc.* (Aug. 1997).
38. HOPPE, H. Smooth view-dependent level-of-detail control and its application to terrain rendering. In *Proceedings of IEEE Visualization 1998* (Jan. 1998), IEEE, Piscataway, NJ, 1998, pp. 35–42.
39. HOPPE, H., DEROSE, T., DUCHAMP, T., MCDONALD, J., AND STUETZLE, W. Mesh optimization. In *Computer Graphics (SIGGRAPH 1993 Proceedings)* (Aug. 1993), J. T. Kajiya, Ed., vol. 27 (4), pp. 19–26.
40. HWA, L. M., DUCHAINEAU, M. A., AND JOY, K. I. Adaptive 4-8 texture hierarchies. In *VIS 2004: Proceedings of the conference on Visualization 2004* (Washington, DC, USA, 2004), IEEE Computer Society, pp. 219–226.

41. INSELBERG, A., AND DIMSDALE, B. Parallel coordinates for visualizing multi-dimensional geometry. In *Computer Graphics 1987 (Proceedings of CG International 1987)*, T. L. Kunii, Ed., Springer, Berlin, pp. 25–44.
42. ISENBURG, M., AND LINDSTROM, P. Streaming meshes. In *Proceedings of IEEE Visualization 2005* (2005), IEEE Computer Society Press, pp. 231–238.
43. ISENBURG, M., LINDSTROM, P., GUMHOLD, S., AND SNOEYINK, J. Large mesh simplification using processing sequences. In *Proceedings of IEEE Visualization 2003* (2003), IEEE Computer Society Press, pp. 465–472.
44. ISENBURG, M., LINDSTROM, P., AND SNOEYINK, J. Streaming compression of triangle meshes. In *Proceedings of Eurographics Symposium on Geometry Processing* (2005), pp. 111–118.
45. KAMATA, S.-I., AND BANDOU, Y. An address generator of a pseudo-hilbert scan in a rectangle region. In *International Conference on Image Processing, ICIP97* (1997), vol. I, pp. 707–714.
46. KLEIN, R., LIEBICH, G., AND STRASSER, W. Mesh reduction with error control. In *Proceedings of IEEE Visualization 1996* (Oct. 1996), IEEE Computer Society Press, Los Alamitos, CA, pp. 311–318.
47. LAWDER, J. K. *The Application of Space-filling Curves to the Storage and Retrieval of Multi-Dimensional Data.* PhD thesis, School of Computer Science and Information Systems, Birkbeck College, University of London, 2000.
48. LAWDER, J. K., AND KING, P. J. H. Using space-filling curves for multi-dimensional indexing. In *proceedings of the 17th British National Conference on Databases (BNCOD 17)* (July 2000), B. Lings and K. Jeffery, Eds., vol. 1832 of *Lecture Notes in Computer Science*. Springer, Berlin, pp. 20–35.
49. LINDSTROM, P., KOLLER, D., RIBARSKY, W., HUGHES, L. F., FAUST, N., AND TURNER, G. Real-Time, continuous level of detail rendering of height fields. In *SIGGRAPH 96 Conference Proceedings* (New Orleans, LA, 04–09 Aug. 1996), H. Rushmeier, Ed., Annual Conference Series, ACM SIGGRAPH, Addison Wesley, pp. 109–118.
50. LINDSTROM, P., AND PASCUCCI, V. Visualization of large terrains made easy. In *Proceedings Visualization 2001* (2001), T. Ertl, K. Joy, and A. Varshney, Eds., IEEE Computer Society Technical Committee on Visualization and Graphics Executive Committee, pp. 363–370.
51. MALLAT, S. *A wavelet tour of signal processing*. Academic, San Diego, 1998.
52. MATIAS, Y., SEGAL, E., AND VITTER, J. S. Efficient bundle sorting. In *SODA 2000: Proceedings of the eleventh annual ACM-SIAM symposium on Discrete algorithms* (2000), Society for Industrial and Applied Mathematics, pp. 839–848.
53. MIRANTE, A., AND WEINGARTEN, N. The radial sweep algorithm for constructing triangulated irregular networks. *IEEE Comput. Graph. Appl. 2*, 3 (May 1982), 11–13, 15–21.
54. MIRIN, A. A., COHEN, R. H., CURTIS, B. C., DANNEVIK, W. P., DIMITS, A. M., DUCHAINEAU, M. A., ELIASON, D. E., SCHIKORE, D. R., ANDERSON, S. E., PORTER, D. H., WOODWARD, P. R., SHIEH, L. J., AND WHITE, S. W. Very high resolution simulation of compressible turbulence on the IBM-SP system. *Supercomputing 99 Conference* (Nov. 1999).
55. MUNZNER, T., GUIMBRETIRE, F., TASIRAN, S., ZHANG, L., AND ZHOU, Y. Treejuxtaposer: scalable tree comparison using focus+context with guaranteed visibility. *ACM Trans. Graph. 22*, 3 (2003), 453–462.
56. NIEDERMEIER, R., REINHARDT, K., AND SANDERS, P. Towards optimal locality in meshindexings, 1997.

57. NIEDERMEIER, R., AND SANDERS, P. On the manhattan-distance between points on space-filling mesh-indexings. Technical Report iratr-1996-18, Universität Karlsruhe, Informatik für Ingenieure und Naturwissenschaftler, 1996.
58. PAJAROLA, R. Large scale terrain visualization using the restricted quadtree triangulation. In *Proceedings Visualization 98* (Los Alamitos, CA, 1998), IEEE, Computer Society Press, pp. 19–26, 515. Extended version available as technical report ftp://ftp.inf.ethz.ch/pub/publications/tech-reports/2xx/292.ps.
59. PARASHAR, M., BROWNE, J., EDWARDS, C., AND KLIMKOWSKI, K. A common data management infrastructure for adaptive algorithms for pde solutions. In *SuperComputing 97* (1997).
60. PASCUCCI, V., AND FRANK, R. J. Global static indexing for real-time exploration of very large regular grids. In *Supercomputing 2001: Proceedings of the 2001 ACM/IEEE conference on Supercomputing (CDROM)* (New York, NY, USA, 2001), ACM, p. 2.
61. PLAISANT, C., CARR, D., AND SHNEIDERMAN, B. Image-browser taxonomy and guidelines for designers. *IEEE Softw. 12*, 2 (1995), 21–32.
62. PLAISANT, C., GROSJEAN, J., AND BEDERSON, B. B. Spacetree: Supporting exploration in large node link tree, design evolution and empirical evaluation. In *INFOVIS 2002: Proceedings of the IEEE Symposium on Information Visualization (InfoVis 2002)* (Washington, DC, USA, 2002), IEEE Computer Society, p. 57.
63. POPOVIĆ, J., AND HOPPE, H. Progressive simplicial complexes. In *SIGGRAPH 97 Conference Proceedings* (Aug. 1997), T. Whitted, Ed., Annual Conference Series, ACM SIGGRAPH, Addison Wesley, pp. 217–224.
64. RAO, R., AND CARD, S. K. The table lens: merging graphical and symbolic representations in an interactive focus+context visualization for tabular information. In *CHI 1994: Proceedings of the SIGCHI conference on Human factors in computing systems* (New York, NY, USA, 1994), ACM, pp. 318–322.
65. RENZE, K. J., AND OLIVER, J. H. Generalized unstructured decimation. *IEEE Comput. Graph. Appl. 16*, 6 (Nov. 1996), 24–32.
66. SAGAN, H. *Space-filling curves*. Springer, New York, 1994.
67. SCHROEDER, W. J., ZARGE, J. A., AND LORENSEN, W. E. Decimation of triangle meshes. *Computer Graphics (SIGGRAPH 1992 Proc.) 26*, 2 (July 1992), 65–70.
68. SILVA, C. T., MITCHELL, J. S. B., AND KAUFMAN, A. E. Automatic generation of triangular irregular networks using greedy cuts. In *Proceedings of IEEE Visualization* (1995), IEEE Computer Society, IEEE Computer Society Press, Los Alamitos, CA, pp. 201–208.
69. STAADT, O. G., AND GROSS, M. H. Progressive tetrahedralizations. In *Proceedings of Visualization 98* (Oct. 1998), D. Ebert, H. Hagen, and H. Rushmeier, Eds., IEEE Computer Society Press, Los Alamitos, CA, pp. 397–402.
70. STOLLNITZ, E. J., DEROSE, T. D., AND SALESIN, D. H. *Wavelets for computer graphics: theory and applications*. Morgann Kaufmann, San Francisco, CA, 1996.
71. STOLTE, C., TANG, D., AND HANRAHAN, P. Multiscale visualization using data cubes. In *INFOVIS 2002: Proceedings of the IEEE Symposium on Information Visualization (InfoVis 2002)* (2002), IEEE Computer Society, p. 7.
72. STOLTE, C., TANG, D., AND HANRAHAN, P. Polaris: A system for query, analysis, and visualization of multidimensional relational databases. *IEEE Trans. Vis. Comput. Graph. 8*, 1 (2002), 52–65.
73. STOLTE, C., TANG, D., AND HANRAHAN, P. Query, analysis, and visualization of hierarchically structured data using polaris. In *KDD 2002: Proceedings of the eighth ACM SIGKDD international conference on Knowledge discovery and data mining* (2002), ACM, pp. 112–122.

74. Trotts, I. J., Hamann, B., Joy, K. I., and Wiley, D. F. Simplification of tetrahedral meshes. In *Proceedings of Visualization 98* (Oct. 1998), D. Ebert, H. Hagen, and H. Rushmeier, Eds., IEEE Computer Society Press, Los Alamitos, CA, pp. 287–296.
75. Vitter, J. S. External memory algorithms and data structures: dealing with massive data. *ACM Comput. Surv. 33*, 2 (2001), 209–271.
76. Wise, J. A., Thomas, J. J., Pennock, K., Lantrip, D., Pottier, M., Schur, A., and Crow, V. *Visualizing the non-visual: spatial analysis and interaction with information for text documents*. Morgan Kaufmann, San Francisco, CA, USA, 1999, pp. 442–450.
77. Wong, P. C., and Thomas, J. Visual analytics. *IEEE Comput. Graph. Appl. 24*, 5 (2004), 20–21.
78. Xia, J. C., and Varshney, A. Dynamic view-dependent simplification for polygonal models. In *Proceedings of IEEE Visualization 1996* (1996), IEEE Computer Society Press, pp. 327–334.

Compression and Occlusion Culling for Fast Isosurface Extraction from Massive Datasets

Benjamin Gregorski[1], Joshua Senecal[1], Mark Duchaineau[1],
and Kenneth I. Joy[2]

[1] Center for Applied Scientific Computing, Lawrence Livermore National Laboratory
7000 East Avenue, Livermore, CA 94551, USA
bfgregorski@ucdavis.edu, senecal1@llnl.gov, duchaine@llnl.gov
[2] Institute for Data Analysis and Visualization, Department of Computer Science, University of California Davis, One Shields Avenue, Davis, CA 95616, USA
kijoy@ucdavis.edu

Summary. We present two algorithms for data compression and occlusion culling that improve interactive, adaptive isosurface extraction from large volume datasets. Our algorithm, based on hierarchical tetrahedral meshes defined by longest edge bisection, allows arbitrary isosurfaces to be adaptively extracted at interactive rates from losslessly compressed volumes where the region of interest, determined at runtime by user interaction, is decompressed *on-the-fly*. For interactive applications, we exploit frame-to-frame coherence between consecutive views to simplify the mesh structure in occluded regions and eliminate occluded triangles significantly reducing the complexity of the visualized surface and the underlying multiresolution volume representation. We extend the use of hardware accelerated occlusion queries to adaptive isosurface extraction applications where the surface geometry and topology change with the level-of-detail and view-point and the user can select an arbitrary isovalue for visualization.

1 Introduction

Isosurface extraction from rectilinear datasets is a well-known problem and significant research has been done to improve the quality of the extracted isosurface, see [23, 28]. Unfortunately, the complexity of these isosurfaces varies considerably and somewhat independently of the original volume's size. In Duchaineau et al. [3], isosurfaces extracted from a single timestep of a large fluid dynamics dataset contained 460 million triangles and exceeded the original volume's storage requirements. Additionally, the visual complexity of the isosurface was very high with as many as 50 occluded surfaces per screen space pixel. The size and complexity of this surface makes it virtually impossible to interactively visualize by extracting and rendering the entire triangle mesh. Furthermore, most volume datasets contain several isosurfaces of interest and extracting the triangle mesh for each isosurface and then rendering it is simply not practical.

T. Möller et al. (eds.), *Mathematical Foundations of Scientific Visualization, Computer Graphics, and Massive Data Exploration*, Mathematics and Visualization,
DOI: 10.1007/978-3-540-49926-8, © 2009 Springer-Verlag Berlin Heidelberg

One solution to this visualization problem is to directly extract the isosurfaces from a volume in an *on-demand* fashion using multiresolution volumetric techniques that allow visualization quality and interactivity to be balanced efficiently. Gregorski et al. [10] developed a view-dependent isosurface extraction algorithm based on longest-edge bisection (LEB) of tetrahedral meshes that allows the extraction of arbitrary isosurfaces. These LEB meshes have been used extensively in scientific visualization for adaptively reconstructing ultrasound data [30], performing isocontouring with controlled topology simplification [8], rendering multiple transparent and opaque isosurfaces [7], and visualizing large polygonal models [2]. One can effectively deal with large datasets through the use of a hierarchical data layout based on space filling curves [10, 20, 29]. This method, however, provides only a partial solution because it does not address the problems of storing these large volumes and effectively rendering extremely complex isosurfaces with high depth complexity. In this paper we address these issues using a fast and efficient data compression algorithm (Sect. 4) that follows the tetrahedral mesh hierarchy defined by LEB and an occlusion culling algorithm (Sect. 5) that uses the diamonds of the mesh hierarchy and hardware occlusion queries to determine and simplify occluded regions.

Our compression algorithm follows the LEB hierarchy and replaces the cubical blocks of an octree, frequently used in compression algorithms, with the diamond shapes developed in Gregorski et al. [10] and used by Linsen et al. [21] for hierarchical representation of large datasets. These meshes have the advantage that the number of elements is increased by a factor of 2 with each refinement rather than the factor of 8 associated with octree refinement. A top-down traversal of the mesh hierarchy uses the data values and gradient components at the vertices of a refinement edge to predict the data at the split vertex. In general, the magnitudes of these predicted values are smaller than the original magnitudes meaning that the resulting data has reduced entropy and is easier to compress. This prediction scheme follows the data access pattern dictated by LEB refinement. Combined with the z-order data layout scheme of Pascucci et al. [29], it ensures that reconstruction masks are of minimal size and that data needed for reconstruction is accessed in a coherent manner. Furthermore, it ensures that the decompression can be confined to local regions of the volume which is essential for visualization of massive datasets. The delta values used for runtime reconstruction are compressed in chunks and stored in z-order. Hierarchy levels are created from subsampled versions of the dataset. Thus, as finer (higher resolution) levels of the volume are reconstructed, values at coarser levels do not need to be updated (as they must be when wavelet lifting schemes are used).

Using the diamonds of the mesh hierarchy to determine occluded regions allows occlusion culling to be utilized regardless of the isosurface being visualized. This is extremely important for isosurface visualization applications where a large number of very different isosurfaces are visualized. In these applications, the precomputation of visible sets or bounding-object hierarchies used for occlusion culling in static environments is not practical due to the large number of different surfaces that can be visualized. Since diamonds are also the unit of refinement, it is straightforward to determine how the mesh must be refined or simplified as regions become visible or occluded. As the diamond hierarchy does not have a nested parent–child relationship that allows for front-to-back rendering, we use frame-to-frame coherence between successive view-points to create an initial set of occluders against which hardware occlusion tests are performed. As the user moves the camera around an isosurface, the visible and occluded regions of the volume are determined, and the mesh is dynamically refined and coarsened accordingly.

Section 3 reviews the mesh refinement, our hierarchical data layout algorithm, and the precomputed data needed to drive the refinement process. Sections 4 and 5 describe the data compression and occlusion culling algorithms and results are given in Sect. 6.

2 Related Work

Visualization of massive isosurfaces is still difficult because of the large storage requirements of the volumes from which they are extracted. Widely available compression libraries such as gzip and bzip are computationally too expensive to be used for these interactive applications. Traditional hierarchical compression algorithms based on wavelet decompositions update data values as the level of detail changes forcing the isosurface to be reextracted from a larger number of elements, thus increasing the cost of the visualization. In order to be effective for interactive applications, we need a decompression algorithm that is fast, follows the hierarchy given by the mesh refinement to leverage memory coherence, and is localized to the region of interest.

Isosurfaces extracted from massive volumes contain millions of triangles and almost always have a depth complexity far greater than one. When viewing an isosurface at a high resolution, extracting and rendering a large number of occluded triangles consumes a large amount of processing, memory, and rendering resources. CPU resources are wasted because the mesh is refined in the occluded regions, increasing the size of the runtime data structures and the amount of data loaded from disk. GPU resources are wasted because a large number of invisible triangles are stored and rendered.

Occlusion culling prevents these invisible triangles from being extracted and rendered. This allows valuable system resources to be allocated to the visible regions of the surface. By preventing additional, invisible triangles from being extracted, occlusion culling saves a large amount of vertex processor work. It can also save fragment processor work because query geometry is rendered without texturing or complex shading whereas the occluded surface geometry is generally rendered with a more complex shader. Occlusion culling's main drawback is that it does not work well for all polygonal models and isosurfaces because they can have regions with low occlusion such as surfaces with small, thin features. In these situations, the cost of performing the occlusion queries can outweigh the cost of rendering and storing the additional geometry and cause an overall slowdown in performance.

Compression techniques for interactive visualization have been utilized in the context of volume rendering by Guthe et al. [12], where wavelets and mpeg style encoding are used to compress time-varying volumes, and by Lum et al. [24] where quantization of DCT coefficients is used. Westermann et al. [34] use octrees to perform adaptive isosurface extraction. The cracks in the isosurface that occur in the transition between octree levels are fixed with triangle fans. Guthe and Strasser [13] use wavelet based compression to adaptively render volume datasets using texture hardware. In their method, an octree decomposition breaks the volume into a hierarchy of 2^{3k} blocks and linear spline wavelets are used to transform the data. Arithmetic and Huffman coding are used to compress the wavelet coefficients. Schneider and Westermann [16] use a hierarchical vector quantization scheme for compressing static and time-varying volumes. The dataset is divided into blocks of 4^3 elements and each block is encoded hierarchically. Their algorithm can reconstruct the data on the graphics card removing the overhead of CPU decompression.

Software based occlusion methods precompute a visibility database to help determine occluded areas. El-Sana et al. [5] divide a dataset against a grid and determine a solidity value for each cell. A cell's visibility is based on the solidity values of the cells that intersect the line segment between the view point and the cell's center. The hierarchical occlusion map developed by Zhang et al. [36] is used to select occluders from a database. Occluders are rendered as white polygons on a black background. An estimated depth buffer is constructed to allow occluded regions to be detected. Modern graphics hardware has the ability to perform occlusion queries and to report the number of pixels affected by some geometry. This feature has been used extensively for occlusion culling in large polygonal environments, see [1, 9, 14, 15, 35]. Because hardware occlusion queries have rasterization and read back overhead, Staneker et al. [33] introduced *Occupancy Maps*, an efficient, software-based occlusion test to reduce the number of occlusion queries performed.

Occlusion culling methods have also been extended to isosurface extraction from volume datasets and to volume rendering applications. Livnat and Hansen [22] traverse an octree decomposition of the dataset front-to-back and use a hierarchical visibility test based on coverage masks to determine occluded regions. As occluded regions are determined during the traversal, blocks of the volume are culled and no isosurface is extracted from them. In Zhang et al. [37], the dataset is decomposed into a set of blocks, and ray casting is used to select a group of blocks that are occluders which are then rendered to create a mask of covered screen pixels. The remaining unoccluded blocks, as determined by this mask, are rendered. Isosurfaces can also be visualized directly from volume datasets using direct volume rendering techniques such as those by Lum et al. [25] and [26]. The advantage of these algorithms is that they do not require the overhead of multiresolution data structures and space for storing explicit isosurface geometry. Their disadvantage is that they are fill-limited and their performance is directly related to the size of the rendered image and available texture memory. We extract isosurfaces directly from the tetrahedral mesh and render explicit geometry. This can provide a better balance between the geometry and fill sections of the graphics pipeline. Recent work by Li et al. [19] and Gao et al. [6] has extended occlusion culling techniques to volume rendering applications where transparent and opaque surfaces are rendered.

3 Mesh Refinement and Preprocessing

In this section we review LEB mesh refinement [10] and its relationship with the z-order space filling curve [29] used for data layout. The mesh refinement begins with the root diamond: a cube divided into six tetrahedra around its major diagonal. Refinement occurs around cube diagonals, face diagonals, and edges of the mesh. Figure 1 shows the three types of polyhedral shapes or diamonds that occur around the diagonals and edges of the mesh. The basic unit of refinement is the diamond. All tetrahedra in a diamond are refined at the same time by inserting a new vertex called the *split vertex* at the midpoint of the diamond's *split edge* which is the edge shared by all tetrahedra in the diamond. See [10] for a complete description of this procedure.

Split and *merge* operations are used to refine and coarsen the mesh. When a diamond is split, the split vertex is inserted and each tetrahedron in the diamond is split into two child tetrahedra. Splitting and merging diamonds ensures that the mesh is always crack-free. The split/merge process is implemented with two error-based priority queues [4, 10], the split queue and the merge queue, which contain all diamonds that can be split and merged, respectively. At the start of each frame, error values are recomputed for all diamonds in the split and merge queues that contain the isosurface. Given an error threshold E, diamonds in the split queue with an error greater than E are refined, and diamonds in the merge queue with an error less than E are coarsened. Diamonds whose error is equivalent to E are never merged and are only split when it is necessary to maintain the continuity of the mesh. Thus, it is not possible to have a situation where a diamond is continuously split and merged.[1]

Figure 2 illustrates the connection between LEB refinement and the z-order space filling curve. Each (i, j, k) index of a point P in the dataset corresponds to a $z-order$ index on the one-dimensional z-order curve. Conversion between (i, j, k) and $z-order$ indices is straightforward and given in Pascucci et al. [29]. Note that each data point, except for the points at the vertices of the root diamond, is the split vertex of a diamond. The order of the points introduced by each level of mesh refinement is essentially the same as that given by the z-order curve. The only difference is that the mesh refinement operates on $(2^k + 1)^3$ grids, and the z-order curve works on

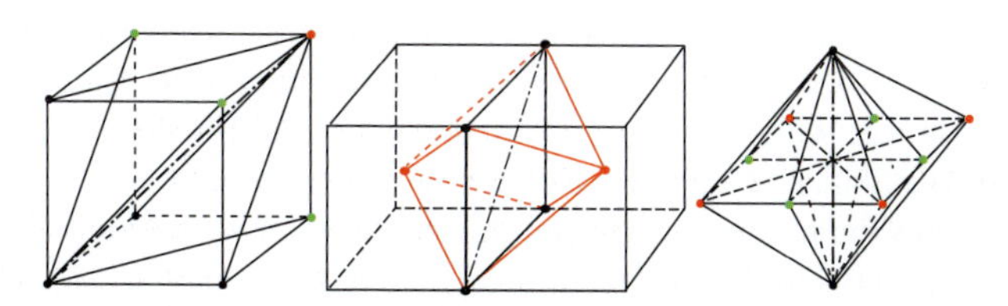

Fig. 1. *From left to right: diamond shapes* for phases 2, 1, and 0 of the mesh refinement. The split or refinement edge is the *bold*, circle-segment *dashed line*. The diamonds occur around cube diagonals, face diagonals, and edges of the mesh

[1] As with similar error based refinement algorithms, this assumes that there are no ambiguities in the floating point comparisons.

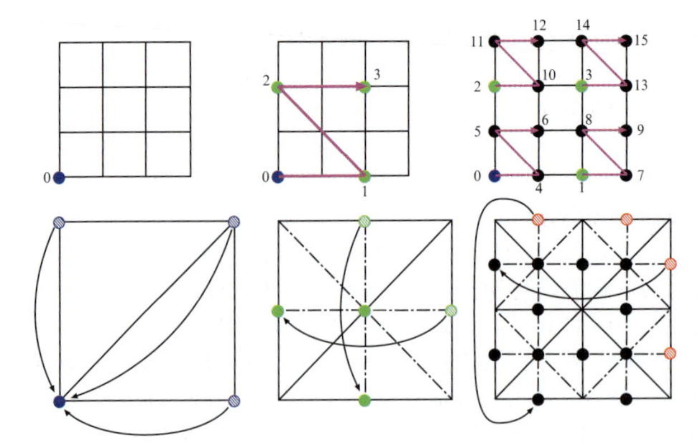

Fig. 2. Two-dimensional example of hierarchical z-order and mesh refinement for a 4×4 dataset. *Top, from left to right:* levels 2 (*blue*), 1 (*green*), and 0 (*black*) of the hierarchical z-order curve showing the points introduced at each level and their order on the one-dimensional curve. *Bottom*, *left* to *right:* levels 2, 1, and 0 of the LEB mesh. The *solid circles* indicate the data points introduced by the LEB refinement at each hierarchy level. The *dashed circles and arrows* show how mesh points with at least one index equal to (2^k) are wrapped around the dataset to where the index equals 0

$(2^k)^3$ grids. This is not a problem because the z-order curve tiles the dataset in all spatial dimensions causing an index I to map to $I \bmod 2^k$. Thus index 2^k on the mesh boundary maps to index 0. The correspondence between the mesh hierarchy and the z-order curve gives excellent disk and memory coherence making them well suited for adaptive, out-of-core visualization of large datasets.

The following information is computed in a preprocessing step:

1. The isosurface error, minimum data value, and maximum data value of the region enclosed by each diamond in the mesh.
2. The normalized gradient vector at each data point. The gradient is used as a texture coordinate to perform diffuse lighting and to highlight the regions orthogonal to the viewing direction.

For a tetrahedron T, the isosurface error is the maximum difference between the isosurface generated using the scalar values at T's vertices and the true isosurface passing through T. A diamond D's error, e_{iso}, is the max of its tetrahedras' errors. At runtime, a sphere with radius e_{iso}, centered at D's split vertex, is projected onto the view screen and its size in pixels is the view-dependent error.

The storage of the minimum, maximum, and error values is reduced by grouping sets of tetrahedra into chunks of 64 tetrahedra called subtrees, see [11]. This reduces the overall error of a tetrahedron, the granularity of the mesh refinement, and the size of the runtime data structures. All tetrahedra in a subtree are contoured at the same time. For a $1{,}024^3$ byte dataset, the size of these values is reduced by a factor of 64 from 3 to 50 MB, and they are stored uncompressed. This reduced storage comes at the cost of an increased number of tetrahedra that do not contain the isosurface being contoured. The data compression algorithm described in Sect. 4 is used for the data values and gradient components. Gradients are precomputed because it is expensive to compute gradients via differencing at runtime due to adjacent data points being far away on the z-order curve.

4 Data Compression

Hierarchical prediction algorithms, such as the *JPEG 2000* algorithm, $\sqrt{2}$ meshes [21], and wavelet transforms based on octrees [13] traverse a dataset in a fine-to-coarse manner building successive, filtered approximations of the original data. One problem with wavelet based techniques is that the function values at vertices change as the vertex moves between levels in the hierarchy. When a vertex V is added, function values at the vertices needed to reconstruct V's value must be updated before V's value is computed. While this may be satisfactory for volume rendering applications where overlapping grids can be used for hardware accelerated texture mapping, for isosurface applications, to prevent cracks, the isosurface must be re-extracted from the elements that use updated values. As regions of the mesh are coarsened, function values must be updated to the appropriate level of the hierarchy. This increases the amount of work that must be done in each frame and makes it more difficult to maintain consistent frame rates as the level-of-detail is changed. The alternative is subsampled hierarchies. Here the values do not change because the coarser resolutions are created by selecting a reduced set of individual data points from the finer resolution without considering neighboring values. We use hierarchies based on subsampling and exploit their static properties so that the error bound can be quickly tightened and the isosurface can be extracted from regions of interest at a high resolution.

Our data compression algorithm is divided into three phases. First, the original volume is traversed along the LEB hierarchy, and the function value and gradient components at a diamond's split vertex are predicted using difference from linear along a diamond's split edge. Gradients are quantized component-wise with 8 bit per component in order to have a meaningful delta for data prediction. Next, these deltas are passed to an encoder which builds a set of codes used for compression. Finally, the stream of deltas is arranged in hierarchical z-order, divided into pages, and compressed.

For multiresolution visualization, it is important to use hierarchical prediction that follows the mesh refinement. This ensures that the data needed for reconstruction can be located efficiently at runtime. Decompressing a region at a high resolution occurs in a small, localized space around the region and in similarly localized spaces at coarser levels. In this paper we consider selective, local refinement applications, such as view-dependent refinement, where the needed regions change at the user's discretion and must be located immediately without having to reconstruct the whole volume. Thus, the encoded data stream must not only be progressive with respect to the overall quality of the dataset but also must allow random access to spatially coherent blocks located across the levels of the hierarchy.

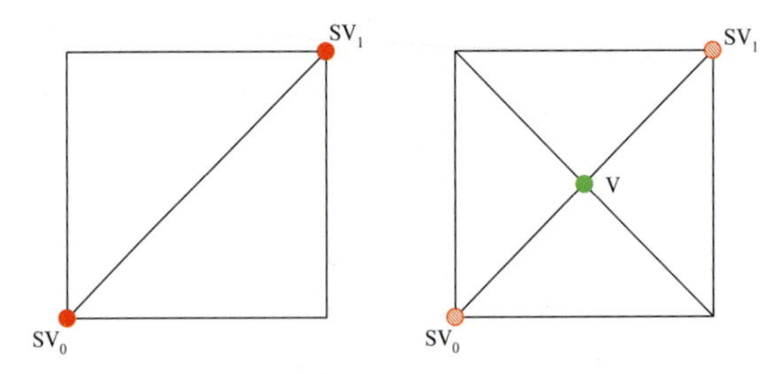

Fig. 3. The value at a vertex V, which is also the split vertex of a diamond, is predicted as the average of the values at its split edge vertices SV_0 and SV_1

Embedded coders and bit-plane coders such as zero-tree coding [32] and SPIHT [31] provide good compression and excellent bit-rates and allow for progressive transmission, but they are not well suited for selective, local refinements where only a subregion of the original data is needed. This is because they encode the most important wavelet coefficients first and the least important ones last; the goal being to represent the whole dataset as accurately as possible with the fewest number of bits. However, in view-dependent and region-of-interest refinement, the needed coefficients change and are not known until runtime. Thus, we cannot use an algorithm that requires a globally fixed ordering and must instead use an algorithm that follows the mesh hierarchy's local refinement. Data prediction using data values along the split edges accomplishes this while maintaining small reconstruction masks necessary for good performance.

Our data prediction algorithm, based on difference from linear prediction along a diamond's split edge, is illustrated in Fig. 3. If D_{SV_0} and D_{SV_1} are the data values and gradient components at the vertices of the split edge, then we define the predicted value to be

$$D_p = \frac{(D_{SV_0} + D_{SV_1})}{2}, \tag{1}$$

and the encoded delta value is

$$\delta = D_V - D_p. \tag{2}$$

This simple predictor handles boundary conditions easily because the diamonds on the faces of the volume are always phase one and zero diamonds, and the diamonds on the edges (except for the corners which are not predicted) are phase zero diamonds, see Fig. 1. Figure 4 shows histograms of the data before and after prediction.

After data prediction, the original data has been converted to a representation that is easier to compress since the transformed data's coefficient magnitudes are clustered toward smaller values than those of the original data. Thus, there are a larger runs of zeros in the most significant bits of the coefficients indicating that the transformed data has more redundancy. For example, in an 8-bit transform coefficient, such as 00000101, the high-order bits 000001 are more compressible than the

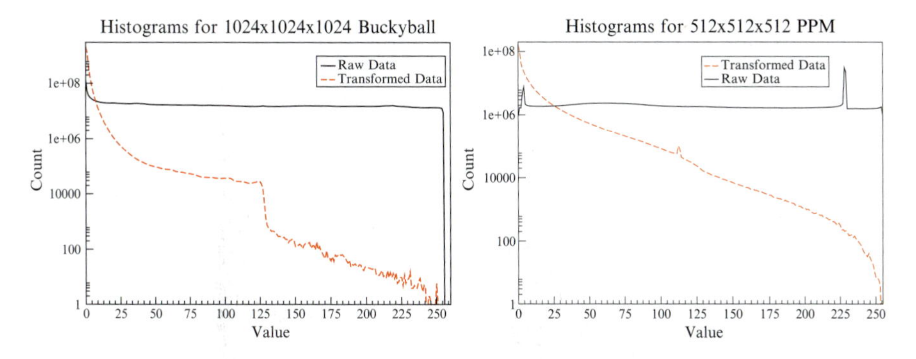

Fig. 4. Histograms of the magnitudes of raw and transformed values (data and quantized gradient components) for the Buckball and PPM datasets (8-bit) on a logarithmic scale

remaining bits 01.[2] Histograms showing the distribution of the data values before and after data prediction are shown in Fig. 4.

To exploit this property of the transformed data, we use *lead-1 encoding*. If the transformed data values require 8 bit for representation (signed data), there are 128 different coefficient magnitudes and 256 possible values. Instead of encoding all possible coefficient magnitudes, we encode the position of the magnitude's leading 1. The remaining bits of the magnitude and the sign bit are passed through uncompressed. For 8-bit transform coefficients, the number of symbols that need to be represented is reduced to 9 from 256, i.e., the eight possible positions of the leading 1 and the zero coefficient vs. the 256 possible values that can be represented with 8 bit. The codes for the leading 1 position are generated using the Huffman algorithm. The compressed output consists of the codeword followed by the sign bit if the magnitude is nonzero and the remaining low-order bits after the leading one. The encoding process uses one value of context to generate the codes. Given the sequence of transformed values generated by the prediction process, the encoder creates a table of leading-1 counts based upon the previous and current lead-1 positions. This 2D table of counts is used by the encoder to build the Huffman codes for the leading 1 positions. An encoding and decoding table is required for each previous leading 1 position.

Lead-1 encoding and Huffman codes were chosen for several reasons. First, encoding and decoding can be done with direct table lookups which is very fast. Second, the number of symbols that need to be encoded is greatly reduced, minimizing the size of the encoding and decoding tables. Lead-1 encoding reduces the number of symbols that need to be encoded thus reducing the size of the codes and the table size. An entry in the decoding table stores the length of the codeword that indexes to that entry and the leading 1 position indicated by the code. This information is packed into a single byte which ensures that the decoding tables can fit in cache memory. Lastly, when compared with other types of coders such as arithmetic and

[2] Only lossless compression is considered since our test datasets have already been quantized before entering the visualization pipeline.

Table 1. Compressed and uncompressed sizes of datasets in megabytes. The Data and Grad columns show the compression ratios for just the data and gradients respectively. As expected, the original data values compress much better than the gradient components

Dataset	Uncomp	Comp	Data+Grad	Data	Grad
Bucky $1{,}024^3$	4,096	1,334	3.071	5.51	2.6
PPM 512^3	512	284.9	1.797	2.43	1.65
Engine 256^3	64	24.6	2.6	4.35	2.22
XMasTree 512^3	640	378.1	1.707	4.06	1.42

arithmetic-type encoders, lead-1 encoding offers an acceptable compression ratio with superior speed and throughput.

The data being compressed is the data value and quantized gradient components at each vertex. The transformed coefficients are stored in hierarchical z-order. To allow for local decompression, the transformed coefficients are divided into pages of 4×2^p elements, and each page is encoded separately. A lookup table is used at runtime to find the disk offset for loading a page. Given an index $P_{(i,j,k)}$ for a point P, the data value and gradient at P are reconstructed as follows:

1. Convert the (i, j, k) index $P_{(i,j,k)}$ to its z-order index P_z. Using P_z, locate the disk page DP that contains P, and decompress DP to get the delta values associated with DP's data points.
2. Since the deltas are stored in z-order, we know the z-order index P_z for all points in DP. For each point in DP, compute its (i, j, k) index $P_{(i,j,k)}$ from P_z.
3. Given that $P_{(i,j,k)}$ is the split vertex of a diamond, compute the (i, j, k) indices of the vertices necessary to reconstruct the value at P.
4. Fetch the surrounding values needed for reconstruction and compute the original data value. This may require recursive decompression as necessary to obtain all of the surrounding values.

The data values d_i needed to reconstruct the data at a point $P_{(i,j,k)}$ all have z-order indices z_i such that $z_i < P_z$. This means that they come from coarser levels of the mesh and from earlier positions (relative to the file offset) in the data file than P. Since z-order stores the data points first by level-of-detail and then by spatial proximity within each level, accessing the data at the split edge vertices needed for reconstruction exploits the storage order's coherence properties. Compression results for test datasets are given in Table 1.

5 Occlusion Culling

Since a dataset typically contains a large number of interesting isosurfaces whose topology and geometry can vary greatly with level-of-detail and isovalue, our occlusion culling algorithm is based upon the mesh hierarchy and not the extracted isosurface geometry. Occlusion queries on modern graphics hardware allow one to render geometry and determine the number of samples that pass the stencil and depth tests. There is a noticeable delay between the time when an occlusion query is issued and the time when the results are available since the query geometry must be rendered and the query results read back from the graphics card. Staneker et al. [33] show that the query latency time is related to the number of rasterized pixels of the query geometry. One of the keys to using hardware occlusion queries efficiently is to fill this time gap with useful computations. The cost and efficient use of occlusion queries is also detailed in the specifications for *GL_NV_occlusion_query* and *GL_ARB_occlusion_query*. To minimize this delay, we use a two-pass algorithm that first initiates occlusion queries for the diamonds in the split and merge queues that contain the isosurface and then reads the query results and computes the diamonds' error values.

Using diamonds instead of tetrahedra for occlusion culling allows us to quickly eliminate regions of refinement with a single test. Since a diamond's children are not contained within its convex hull, the occlusion queries cannot be performed top-down.[3] Furthermore, since we maintain a hierarchy of diamonds but not of tetrahedra, we cannot render the isosurface front-to-back to generate the depth buffer for the occlusion queries. Instead we exploit the isosurface's coherence between successive views to generate a set of occluders from which a depth buffer is created.

Our occlusion culling algorithm works as follows:

1. At frame 0, the mesh contains only the root diamond which is assumed to be visible.
2. To exploit frame-to-frame coherence, at the start of a frame f_i, we render the isosurface geometry from frame f_{i-1} using f_i's viewing parameters. The geometry from f_{i-1} provides a good initial guess for the set of occluders and gives the correct depth buffer for them. The occlusion queries in frame f_i are performed against this depth buffer.
3. An occlusion query is then initiated for all diamonds in the split and merge queues that are in the view frustum, contain the isosurface, and were occluded in frame f_{i-1}. This is done by rendering the diamond's bounding box.[4] Queries are performed on occluded diamonds every frame. Visible diamonds are allowed to remain visible for several frames before occlusion tests are performed.
4. After the queries have been issued, the results are read back, and occluded diamonds are given an error of zero. Occluded diamonds are never split and they are the first ones to be merged.
5. The triangles extracted from the tetrahedra added to the mesh in frame f_i are rendered to fill the holes caused by rendering f_{i-1}'s isosurface at f_i's viewpoint.

Our method allows visible regions to remain visible for several frames. This greatly reduces the number of occlusion tests performed per frame and randomizes

[3] A tetrahedron's children are contained in its convex hull, but because diamonds are groups of tetrahedra around a split edge, not all of a diamond's tetrahedra are contained in the convex hull of a parent diamond.

[4] The bounding boxes are rendered with backface culling on because only the front three quads or six triangles need to be drawn.

when the queries are performed. Empirical testing indicates that allowing diamonds to remain visible for 5–7 frames provides a good balance between the number of occlusion queries and invisible diamonds marked as visible. Our occlusion culling method is conservative because diamonds marked as visible can become occluded for a few frames before an occlusion test is performed and because we render bounding boxes to determine a diamonds's visibility. Thus, some occluded triangles will be extracted and rendered.

When the isovalue is changed, the currently extracted isosurface can no longer be used as a valid set of occluders. When the isovalue is changed by the user, the mesh is reinitialized with the root diamond. All cached triangles, outstanding occlusion queries, and other data structures are cleared, and the new isosurface is extracted starting from this base configuration. The depth buffer given by the new isosurface can now be used for occlusion queries, and the refinement continues as described above in Sect. 5. As new error values are computed and new occluded regions are discovered, the mesh structure refines around the visible region of the new isosurface and coarsens everywhere else.

6 Results

Our test machine is a 2 GHz Pentium with 1 GB of main memory and a GeForce4 Ti 4600 graphics card. We have tested our algorithms on several large volume datasets. The buckyball dataset is a synthetic dataset made from Gaussian functions. The $1{,}024^3$ dataset was made by a $2 \times 2 \times 2$ tiling of a 512^3 dataset. The PPM (Piecewise Parabolic Method) dataset is a 512^3 chunk from timestep 273 of the Richtmyer–Meshkov simulation dataset from Lawrence Livermore National Laboratory [27]. The engine dataset is originally $256 \times 256 \times 128$ and has been resampled to 256^3 with the cell aspect ratios appropriately adjusted at runtime. For the Christmas Tree CT dataset [17], we are using the high resolution version of the near isotropic and rotated dataset. Its initial dimensions are $512 \times 512 \times 499$. The region containing the base plate has been removed, and the volume has been padded to make it 512^3. As stated by Lee et al. [18], the extension and padding of the dataset comes with a small increase in entropy. Because it contains a large number of small, thin features, this dataset is not a good candidate for occlusion culling and was only used for data compression tests. All isosurfaces were extracted from the compressed volumes.

Each dataset includes the actual data and three bytes for a quantized gradient, one byte for each component. The four values are predicted independently, and one encoding table is used for compression. Tests using separate encoding tables for each component or separate tables for the data and gradients did not yield significant improvements in compression. Compression results are shown in Table 1. For several test datasets, we have lossless compression ratios between 1.7:1 to 3.1:1 for the data and gradient components which is comparable to the lossless compression rates for large volume datasets given in [13]. The amount of main memory available for uncompressed data was limited to the lesser of 512 MB or $(1/2)$ of the uncompressed dataset size. As described in Sect. 4, the dataset is compressed in chunks of 4×2^p

elements to allow for local decompression. The chunk of main memory allocated for decompressed data is divided into pages of 4×2^p elements and managed with a demand paging algorithm that uses an LRU replacement strategy. Decompressed data is cached in main memory, and compressed data is read from disk and decompressed as needed.

To test the performance of our runtime decompression and reconstruction algorithms, we recorded measurements over several interaction sessions in which the viewing position, isovalue, and error level were all changed at some point. Decompression takes n input bytes, where n is the size of the compressed page, and produces 4×2^p deltas; the time to load data from disk is not considered. Over these test runs the decompression speed ranged from 10–110 MB s^{-1} with the average decompression speed between 50–70 MB s^{-1}. The *reconstruction rate* is measured as decompression time plus the time to compute the actual data and gradients using the deltas. This includes the time to invert the z-order indices, compute the vertices needed for reconstruction and fetch their associated data and gradients. It also includes the time to load data from disk since this significantly affects the performance of out-of-core algorithms. The reconstruction speed is dictated by the time to load data from disk and the cost of fetching the needed data for the split edge vertices. Reconstructing a single point requires two memory lookups or a total of 2^{p+1} accesses to reconstruct one page. For example, the buckyball dataset was compressed with $p = 12$. It contains 2^{18} pages, each uncompressed disk page contains $4 * 2^{12}$ or 2^{14} bytes, and 2^{13} lookups are required to reconstruct one page. Table 2 summarizes the performance results of our data reconstruction algorithm. Performance is given in megabytes per second, data points per second, and disk pages per second to show the overall raw performance of the algorithm as well as its performance with respect to metrics more pertinent to our application where data points are reconstructed in fixed sized chunks.

To further evaluate our data compression and reconstruction algorithms, we extracted isosurfaces using occlusion culling from the same viewpoint using compressed and uncompressed volumes. The results, shown in Table 3, indicate that isosurface extraction from compressed volumes achieves nearly the same performance as isosurface extraction from uncompressed volumes while using a fraction of the disk storage. This shows that our data compression and reconstruction algorithms can be effectively used for multiresolution visualization applications. Furthermore, since

Table 2. Reconstruction performance measurements for Buckyball and PPM datasets in megabytes per second (MB s^{-1}), data points per second (DP s^{-1}), and pages per second (Pages s^{-1}). The differences between the minimum and maximum performance values are caused by memory coherence and the speed of external storage

Dataset	p	Minimum			Maximum			Average		
		MB s^{-1}	DP s^{-1}	Pages s^{-1}	MB s^{-1}	DP s^{-1}	Pages s^{-1}	MB s^{-1}	DP s^{-1}	Pages s^{-1}
Bucky	12	0.66	174K	42	11.9	3.12M	761	10.1	2.65M	646
PPM	12	1.38	326K	79	12.7	3.32M	810	5.8	1.52M	371
PPM	11	0.25	66K	32	12.4	3.24M	1582	6.7	1.75M	854

Table 3. Comparison of contouring using occlusion culling from uncompressed and compressed volumes. Total time for isosurface extraction and the number of split/merge iterations are shown. The number of split/merge iterations differs between compressed and uncompressed volumes because the a fixed amount of time is allocated per frame for the split/merge process and the number of operations performed depends upon a large number of factors

Dataset	Uncompressed		Compressed		Disk pages loaded	Compressed data loaded (MB)
	Time (s)	Iters	Time (s)	Iters		
PPM (Fig. 8)	20	87	22	80	3,153	19.85
Buckyball (Fig. 6)	20.9	93	23.4	94	2,369	9.4

our algorithm progressively refines the isosurface over time, the user is provided with immediate visual feedback in the form of a coarse isosurface that is continuously being refined and can be interacted with as the visualization quality is incrementally improved.

The time to perform an occlusion query includes the time to issue the query, render the query geometry, and read back the results. As described earlier, to use occlusion queries efficiently, the time the graphics card spends rendering the query geometry must be filled with useful computations. To measure the time needed for rendering the query geometry, we tested two occlusion query algorithms. The first (Occ 1) issues an occlusion query and immediately reads back the results, and the second (Occ 2), a two pass algorithm, issues all queries first so that when the results are read back the query geometry has finished rendering and the results are immediately available. The average time to issue an occlusion query was $(3–4) \times 10^{-6}$ s. The average time to request the occlusion query results from the graphics system takes on average $(7–8) \times 10^{-6}$ and $(1–2) \times 10^{-6}$ s for Occ 1 and Occ 2 respectively. Since Occ 1 requests the query results immediately after issuing the occlusion query, the difference between these two results indicates that rendering the query geometry takes about 6×10^{-6} s. To analyze the performance of these two algorithms, we extracted high resolution isosurfaces from the Engine, PPM and Buckyball datasets using both algorithms and rotated them around a predefined axis to ensure that occluded regions became visible as often as possible. Issuing multiple occlusion queries before reading the results improved overall performance by 25–40%. The results are given in Table 4. As expected, reading the query results directly after issuing the query (Occ 1) requires significantly more time. The actual time to issue the occlusion queries is the same in both algorithms, however, it occupies a larger percentage of the overall frame time in Occ 2 because reading the query results takes significantly less time in the Occ 2 algorithm than in the Occ 1 algorithm.

To test the performance of our occlusion culling algorithm we extracted isosurfaces from the same viewpoint with and without occlusion culling. The isosurfaces were extracted at low error bounds, about $\frac{1}{2}$ a pixel in screen space tolerance, to generate high resolution surfaces with a large number of triangles. At lower resolutions with fewer triangles, occlusion culling's benefits diminish, and the extra overhead can decrease overall performance. This is because occlusion culling renders a small

Table 4. Comparison of different occlusion query algorithms using the Engine (Error = 0.5, Isovalue = 90), PPM (Error = 0.5, Isovalue = 227), and Bucky (Error = 0.5, Isovalue = 130) datasets. The *Occ issue* and *Occ query* columns show the percentage of the frame spent issuing the queries and requesting the query results

Dataset	Triangles		Diamonds		Frames per second			Occ issue (%)		Occ query (%)	
	Occ	No occ	Occ	No occ	Occ 1	Occ 2	No Occ	Occ 1	Occ 2	Occ1	Occ 2
Engine	1.2M	2.67M	48K	89K	3.0–4.0	4.0–5.0	3.0–4.0	18	27	35	8.5
PPM	1.6M	4.05M	78K	102K	2.5–3.0	3.5–4.5	1.0–1.5	15	20	28	5.5
Bucky	1.75M	4.08M	95K	132K	2.2–2.7	3.3–3.9	0.6–1.3	11	14	18	4.7

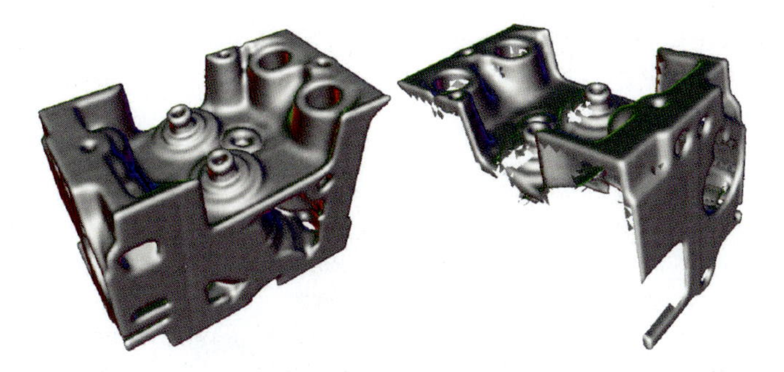

Fig. 5. High resolution isosurface from the engine dataset, Isovalue = 80, Error = 0.5. Occlusion culling results are given in Table 5

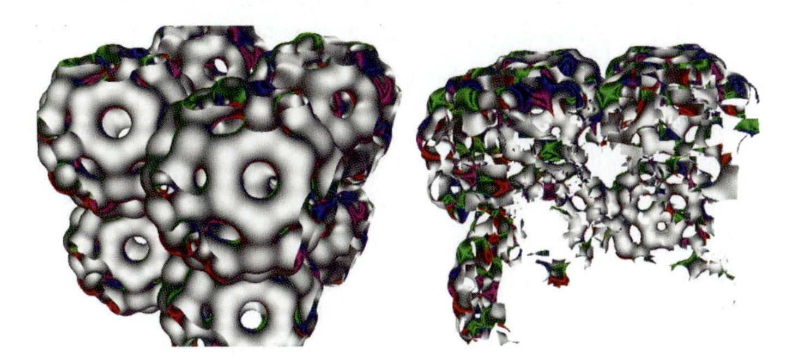

Fig. 6. High resolution rendering of the $1{,}024^3$ synthetic buckyball dataset, Isovalue = 80, Error = 0.5. Occlusion culling results are given in Table 5

amount of geometry to try and eliminate a large amount of geometry. Thus the amount of geometry eliminated must be large enough to justify the overhead of the occlusion test. At lower resolutions with fewer triangles, each occlusion test eliminates less geometry. Occlusion culling results for the images shown in Figs. 5–8 are summarized in Table 5. These figures show the extracted isosurface from the initial viewpoint used for view-dependent rendering and occlusion culling and a second

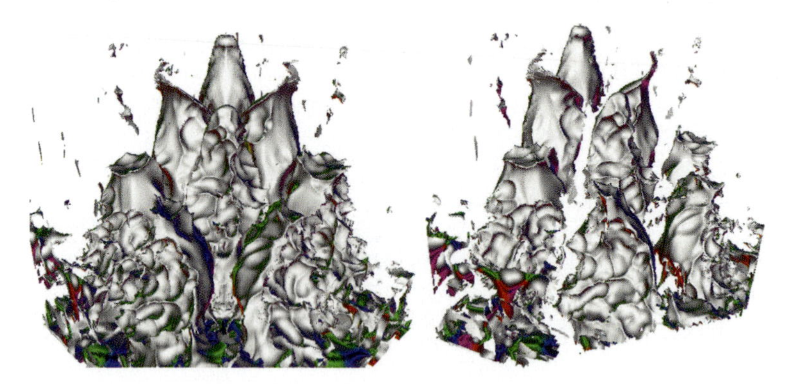

Fig. 7. Isosurface from the PPM dataset: Isovalue = 228, Error = 0.50. Occlusion culling results are given in Table 5

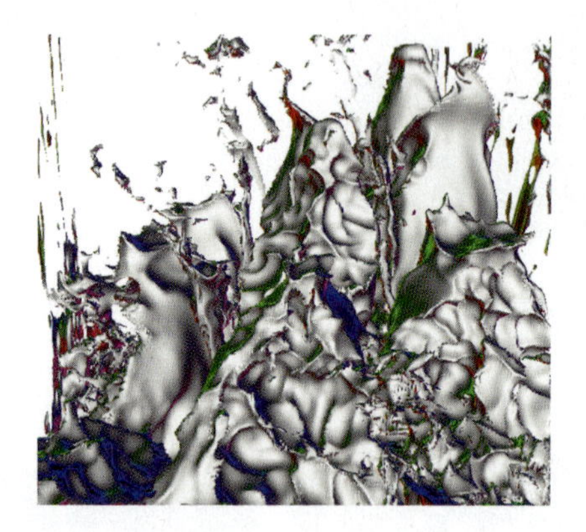

Fig. 8. PPM isosurface extracted using occlusion culling: Isovalue = 225, Error = 0.55. Results are given in Tables 3 and 5

Table 5. Occlusion culling performance for isosurfaces in Figs. 5–8. *Top and bottom rows* show values without and with culling. Results are the average of several runs using the same error and isosurface value

Dataset	Error	Triangles	Diamonds	Occlusion tests per frame	FPS
Engine (256^3)	0.50	2.61M	87K	–	4.4
	0.50	867K (33%)	47K (54%)	8–9K	5.8
PPM (Fig. 7)	0.50	4.4M	215K	–	2.5
	0.50	1.38M (31%)	148K (69%)	15–19K	4.6
PPM (Fig. 8)	0.55	3.98M	170K	–	3.0
	0.55	1.22M (31%)	122K(72%)	13–18K	5.0
Buckyball ($1{,}024^3$)	0.50	2.99M	156K	–	3.3
	0.50	1.1M, (37%)	100K (64%)	15–18K	4.2

viewpoint that shows the regions of the isosurface removed by occlusion culling. For our test datasets, occlusion culling reduced the number of extracted triangles by 50–70% and the number of diamonds in the mesh by 30–50%. The extracted isosurface geometry, vertices and normals, and triangle indices are stored in vertex buffer objects (VBOs). Since a large number of triangles are kept between frames, this allows for fast, asynchronous rendering.

Depending on the viewpoint, the engine has a depth complexity between 2 and 6, and the buckyball has a depth complexity between 2 and 8. For these datasets, occlusion culling greatly reduces the number of triangles, diamonds, and overall system resource usage. The reduction in mesh size reduces the number of error calculations that needs to be performed, the number of data pages that needs to be decompressed, and the amount of isosurface geometry that must be extracted and stored. The benefits of occlusion culling are limited by the time to perform the queries, i.e., setting up and rasterizing the bounding box geometry, and the readback speed from the graphics card. In future graphics systems the performance of occlusion queries should improve greatly. Compared to the engine and buckyball datasets, isosurfaces from the PPM dataset are much more complex. They generally contain a larger number of triangles when compared to isosurfaces from similarly sized volumes at the same error bound, and they have high depth complexity and a large number components and small features where occlusion culling can perform poorly. As described in Sect. 2, occlusion culling works best for surfaces that contain larger occluded regions where the cost of performing the occlusion query is outweighed by the cost of refining the mesh structure and extracting the isosurface.

7 Conclusions and Future Work

We have presented new data compression and occlusion culling algorithms for fast isosurface extraction from massive datasets. Our fast and efficient compression algorithm reduces the storage requirements of these massive volumes and our occlusion culling algorithm reduces the number of triangles in the extracted isosurfaces. The synergistic combination of these two algorithms allows large isosurfaces consisting of millions of triangles to be visualized at interactive rates from massive, compressed datasets.

We utilize a hierarchical tetrahedral mesh structure based on longest-edge bisection to form an adaptive volume representation. The adaptive representation is the basis for a hierarchical prediction technique used to compress the volume data and its precomputed, quantized gradients. A fast coder based on lead-1 encoding and Huffman codes achieves 1.7:1 to 3.1:1 compression with 2.4:1 to 5:1 compression for the data and 1.65:1 to 2.6:1 compression for the gradients. Occlusion culling, performed using the diamond representation of the mesh, allows unseen portions of the volume to be culled and simplified, thus reducing the number of rendered triangles and runtime computation. Frame-to-frame coherence is utilized to create a good initial set of occluders from which an approximate depth buffer is created. Hardware occlusion queries are performed against this set of occluders and the mesh is quickly refined and coarsened in the visible and occluded regions.

Our future work is focused on developing new predictors, better methods for quantizing and compressing the gradients, and new shading techniques. Since the accuracy of the predictor can greatly improve the compression performance, new predictors that can further reduce the entropy without a large computation cost are desired. While our technique provides good lossless compression for the data, the compression of the quantized gradients is not nearly as good. New techniques for predicting the gradients for compression and quantizing the gradients for accurate prediction are needed to improve the compression performance. The compression of the gradients is necessary because they are needed for shading. New shading techniques, based on other surface and volume properties such as curvature and second derivatives, that do not require gradients or can use lower quality gradients are needed to improve the quality of the visualization and reduce the gradient's storage overhead.

Acknowledgments

The Christmas Tree dataset was obtained from Department of Radiology, University of Vienna and the Institute of Computer Graphics and Algorithms, Vienna University of Technology. This work was performed under the auspices of the US Department of Energy by University of California Lawrence Livermore National Laboratory under contract No. W-7405-Eng-48. This work was supported by the National Science Foundation under contracts ACR 9982251 and ACR 0222909, through the National Partnership for Advanced Computing Infrastructure (NPACI) and by Lawrence Livermore National Laboratory under contract B523818. We thank the members of the Visualization and Graphics Research Group at the Center for Image Processing and Integrated Computing (CIPIC) at the University of California, Davis. Joshua Senecal's work was supported in part by a United States Department of Education Government Assistance in Areas of National Need (DOE–GAANN) grant #P200A980307.

The engine dataset was obtained from `http://www.volvis.org`. The buckyball dataset was obtained from Oliver Kreylos at UC Davis.

References

1. Dirk Bartz, Dirk Staneker, Wolfgang Strasser, Briam Cripe, Tom Gaskins, Kristanni Orton, Michael Carter, Andreas Johannsen, and Jeff Trom. Jupiter: A toolkit for interactive large model visualization. In *Proceedings of IEEE Symposium on Parallel and Large Data Visualization and Graphics*, 2001.
2. Paolo Cignoni, Fabio Hanovelli, Enrico Gobbetti, Fabio Marton, Federico Ponchio, and Roberto Scopigno. Adaptive tetrapuzzles: efficient out-of-core construction and visualization of gigantic multiresolution polygonal models. In *SIGGRAPH 2004 Conference Proceedings*, 2004.

3. Mark A. Duchaineau, Serban Porumbescu, Martin Bertram, Bernd Hamann, and Kenneth I. Joy. Dataflow and re-mapping for wavelet compression and view-dependent optimization of billion-triangle isosurfaces. In G. Farin, H. Hagen, and B. Hamann, editors, *Hierarchical Approximation and Geometrical Methods for Scientific Visualization*. Springer, Berlin, 2002.
4. Mark A. Duchaineau, Murray Wolinsky, David E. Sigeti, Mark C. Miller, Charles Aldrich, and Mark B. Mineev-Weinstein. ROAMing terrain: real-time optimally adapting meshes. In *Proceedings of IEEE Visualization 1997*, pages 81–88, 1997.
5. J. El-Sana, N. Sokolovsky, and C. Silva. Integrating occlusion culling with view-dependent rendering. In *Proc. of IEEE Visualization*, 2001.
6. Jinzhu Gao, Jian Huang, Han-Wei Shen, and James Arthur Kohl. Visibility culling using plenoptic opacity functions for large volume visualization. In *Proceedings of IEEE Visualization 2003*, pages 341–348, 2003.
7. Thomas Gerstner. Fast Multiresolution Extraction Of Multiple Transparent Isosurfaces. In *Data Visualization 2001 Proceedings of VisSim 2001*, Annual Conference Series. Springer, 2001.
8. Thomas Gerstner and Renato Pajarola. Topology preserving and controlled topology simplifying multiresolution isosurface extraction. In *Proceedings of IEEE Visualization 2000*, pages 259–266. IEEE Computer Society Press, 2000.
9. Naga K. Govindaraju, Brandon Lloyd, Sung-Eui Yoon, Avneesh Sud, and Dinesh Manocha. Interactive shadow generation in complex environments. In *Proceedings of SIGGRAPH 2003*, pages 501–510, 2003.
10. Benjamin Gregorski, Mark A. Duchaineau, Peter Lindstrom, Valerio Pascucci, and Kenneth I. Joy. Interactive view-dependent rendering of large isosurfaces. In *Proceedings of the IEEE Visualization 2002*, 2002.
11. Benjamin Gregorski, Joshua Senecal, Mark Duchaineau, and Kenneth I. Joy. Adaptive extraction of time-varying isosurfaces. In *IEEE Transactions on Visualization and Computer Graphics, available as LLNL UCRL UCRL-JP-200087*, 2004.
12. Stefan Guthe and Wolfgang Staser. Real-time decompression and visualization of animated volume data. In *Proceedings of IEEE Visualization 2001*, pages 349–358, 2001.
13. Stefan Guthe, Michael Wand, Julius Gonser, and Wolfgang Staser. Interactive rendering of large volume data sets. In *Proceedings of IEEE Visualization 2002*, pages 53–60, 2002.
14. Haeyoung Ha. Out-of-core interactive display of large meshes using an oriented bounding box-based hardware depth query. Master's thesis, University of California, Davis, September 2003. Available as Department of Computer Science Technical Report CSE-2003-25.
15. K. Hillesland, B. Salomon, A. Lastra, and D. Manocha. Fast and simple occlusion culling using hardware-based depth queries. Technical Report UNC-CH-TR02-039, Computer Science Department, University of North Carolina at Chapel Hill, Chapel Hill, North Carolina, 2002.
16. Schneider J. and Rudiger Westermann. Compression domain volume rendering. In *Proceedings of IEEE Visualization 2003*, pages 293–300, 2003.
17. Armin Kanitsar, Thomas Theußl, Lukas Mroz, Milos Srámek, Anna Vilanova Bartrolí, Balázs Csébfalvi, Jirí Hladùvka, Dominik Fleischmann, Michael Knapp, Rainer Wegenkittl, Petr Felkel, Stefan Guthe, Werner Purgathofer, and Meister Eduard Gröller. Christmas tree case study: computed tomography as a tool for mastering complex real world objects with applications in computer graphics. In *Proceedings of the 13th IEEE Visualization 2002 Conference (VIS-02)*, pages 489–492, Piscataway, NJ, October 27–November 1 2002. IEEE Computer Society.

18. Haeyoung Lee, Mathieu Desbrun, and Peter Schroder. Progressive encoding of complex isosurfaces. In *Proceedings of SIGGRAPH 2003*, pages 471–476, 2003.
19. Wei Li, Klaus Mueller, and Ari Kaufman. Empty space skippping and occlusion clippinf for texture-based volume rendering. In *Proceedings of IEEE Visualization 2003*, pages 317–326, 2003.
20. Peter Lindstrom and Valerio Pascucci. Visualization of large terrains made easy. In *Proceedings of IEEE Visualization 2001*, pages 363–370. IEEE Computer Society Press, 2001.
21. Lars Linsen, Jevan T. Gray, Valerio Pascucci, Mark A. Duchaineau, Bernd Hamann, and Kenneth I. Joy. Hierarchical large-scale volume representation with '3rd-root-of-2' subdivision and trivariate b-spline wavelets. In *Geometric Modeling for Scientific Visualization*. Springer, Heidelberg, 2003.
22. Y. Livnat and C. Hansen. View Dependent Isosurface Extraction. In *Proceedings of IEEE Visualization 1998*, pages 175–180. IEEE Computer Society Press, 1998.
23. Adriano Loes and Ken Brodlie. Improving the robustness and accuracy of the marching cubes algorithm for isosurfacing. *IEEE Transactions on Visualization and Computer Graphics*, volume 9(1), January–March 2003.
24. Eric Lum, Kwan-Liu Ma, and John Clyne. Texture hardware assisted rendering of time-varying volume data. In *Proceedings of IEEE Visualization 2001*, pages 263–270, 2001.
25. Eric B. Lum and Kwan-Liu Ma. Rendering isosurfaces directly from 3D textures. Technical Report CSE-2003-10, Computer Science Department, University of California Davis, 2003.
26. Eric B. Lum, Brett Wilson, and Kwan-Liu Ma. High quality lighting and efficient pre-integration for volume rendering. In *Proceedings of IEEE TVCG Symposium on Visualization*, 2004.
27. Arthur A. Mirin, Ron H. Cohen, Bruce C. Curtis, William P. Dannevik, Andris M. Dimits, Mark A. Duchaineau, D. E. Eliason, Daniel R. Schikore, S. E. Anderson, D. H. Porter, , and Paul R. Woodward. Very high resolution simulation of compressible turbulence on the IBM-SP system. In *Proceedings of SuperComputing 1999*. (Also available as Lawrence Livermore National Laboratory technical report UCRL-MI-134237), 1999.
28. Gregory M. Nielson. Mc*: star functions for marching cubes. In *Proceedings of IEEE Visualization*, pages 59–66, 2003.
29. Valerio Pascucci. Multi-resolution indexing for out-of-core adaptive traversal of regular grids. In G. Farin, H. Hagen, and B. Hamann, editors, *Proceedings of the NSF/DoE Lake Tahoe Workshop on Hierarchical Approximation and Geometric Methods for Scientific Visualization*. Springer, Berlin, 2002. (Available as LLNL technical report UCRL-JC-140581).
30. Tom Roxborough and Gregory M. Nielson. Tetrahedron based, least squares, progressive volume models with application to freehand ultrasound data. In *Proceedings Visualization 2000*, pages 93–100. IEEE Computer Society Press, 2000.
31. Amir Said and William A. Pearlman. A new fast and efficient image codec based on set partitioning in hierarchical trees. In *IEEE Transactions on Circuits and Systems for Video Technology*, volume 6, pages 243–250, June 1996.
32. J. M. Shapiro. Embedded image coding using zerotrees of wavelets coeefficients. *IEEE Transactions on Signal Processing*, volume 41, pages 3445–3462, December 1993.
33. D. Staneker, D. Bartz, and M. Meissner. Improving occlusion query efficiency with occupancy maps. In *Proc. of Symposium on Parallel and Large Data Visualization and Graphics*, pages 111–118, 2003.

34. R. Westermann, L. Kobbelt, and T. Ertl. Real-time exploration of regular volume data by adaptive reconstruction of isosurfaces. *The Visual Computer*, volume 15, pages 100–111, 1999.
35. Sung-Eui Yoon, Brian Salomon, and Dinesh Manocha. Interactive view-dependent rendering with conservative occlusion culling in complex environments. In *Proceedings of IEEE Visualization 2003*, 2003.
36. H. Zhang, D. Manocha, T. Hudson, and K. Hoff III. Visibility culling using hierarchical occlusion maps. In *Proceedings of SIGGRAPH*, pages 77–88, August 1997.
37. Xiaoyu Zhang, Chandrajit Bajaj, and Vijaya Ramachandran. Parallel and out-of-core view-dependent isocontour visualization. In *Proceedings of the Joint Eurographics - IEEE TCVG Symposium on Visualizatation (VisSym-02)*, Vienna, Austria, May 27–29 2002. Springer.

Volume Visualization of Multiple Alignment of Large Genomic DNA

Nameeta Shah[1,2], Scott E. Dillard[1], Gunther H. Weber[1,2], and Bernd Hamann[1,2]

[1] Institute for Data Analysis and Visualization (IDAV), Department of Computer Science, University of California, Davis, One Shields Avenue, Davis, CA 95616-8562, USA
{nyshah,sedilla,ghweber,bhamann}@ucdavis.edu

[2] Visualization Group, National Energy Research Scientific Computing Center (NERSC), Lawrence Berkeley National Laboratory, One Cyclotron Road, Berkeley, CA 94720, USA

Summary. Genomes of hundreds of species have been sequenced to date, and many more are being sequenced. As more and more sequence data sets become available, and as the challenge of comparing these massive "billion basepair DNA sequences" becomes substantial, so does the need for more powerful tools supporting the exploration of these data sets. Similarity score data used to compare aligned DNA sequences is inherently one-dimensional. One-dimensional (1D) representations of these data sets do not effectively utilize screen real estate. As a result, tools using 1D representations are incapable of providing informatory overview for extremely large data sets. We present a technique to arrange 1D data in 3D space to allow us to apply state-of-the-art interactive volume visualization techniques for data exploration. We demonstrate our technique using multi-millions-basepair-long aligned DNA sequence data and compare it with traditional 1D line plots. The results show that our technique is superior in providing an overview of entire data sets. Our technique, coupled with 1D line plots, results in effective multi-resolution visualization of very large aligned sequence data sets.

1 Introduction

The human genome consists of about three billion basepairs, of which only a small percentage is well-understood. In order to decipher the rest of the genome, and to understand general principles of genome structure and function, biologists look for overrepresented patterns in the genomes. Another approach to understanding genetic code is through comparison of genomes, or parts of genomes, of different species. Although many techniques for visualization of DNA sequences have been developed and various tools exist for visualization of alignment data, there exists a need for visualization techniques that can handle very large data sets.

T. Möller et al. (eds.), *Mathematical Foundations of Scientific Visualization, Computer Graphics, and Massive Data Exploration*, Mathematics and Visualization,
DOI: 10.1007/978-3-540-49926-8,

```
Human      AATTCCGATGGGAACTACTGGATC-CGG
Chimp      AATTCCAATGGGAA--ACTGGATCCCGG
Mouse      AAATCCG---GAAACCACTGG----AGG
Rat        A--TCCG---GAAACCACTGG----AGG

All        6316663111626631666661110266
           (6-100%, 3-50%, 2-33%, 1-17%)

Primates   1111110111111100111111110111
           (1-100%, 0-0%)

Rodents    1001111000111111111110000111
           (1-100%, 0-0%)
```

Fig. 1. Multiple alignment of four species: human, chimp, mouse, and rat, and sum-of-pairs similarity scores for three different plots. (A sum-of-pair similarity score is computed by adding one for every basepair consisting of the same base, considering all possible distinct species pair)

1.1 Multiple Alignment

Currently, biologists compare genomes of many species at different evolutionary distances by examining multiple alignments [7]. A multiple alignment is a set of sequences in a "rectangular arrangement," where each row consists of one sequence padded by gaps, such that the columns highlight similarity/conservation between positions (`http://www.cryst.bbk.ac.uk/BCD/bcdgloss.html`). Figure 1 shows an example of a multiple alignment of four sequences, where the characters A, T, C and G represent the bases adenine, thymine, cytosine and guanine, respectively. Below the alignment we show similarity scores for the multiple alignment. The similarity score for each column shows the level of conservation among sequences, considering all possible pairwise comparisons of characters (six in the case of four species). Different schemes are used to calculate similarity scores, including *entropy, sum-of-pairs, weighted sum-of-pairs, parsimony, etc.* (`http://lepo.it.da.ut.ee/~mremm/kurs/multali.htm`). We assume that similarity scores are provided as input for our visualization purposes.

1.2 Related Work

Comparison of biological sequences is a very important aspect of genome research. Fractal-based visualization using chaos automata and iterated function systems [1] and space-filling curves [4, 12] have been used for identifying patterns, similarities and dissimilarities in biological sequences. PATTVision is a 3D visualization tool that uses texture mapping for viewing patterns in multiple sequences [23]. Arc diagrams have been used to visualize shared patterns among sequences [22]. These methods are effective for relatively small-sized sequences consisting of up to few thousand basepairs. Alignment is one of the most extensively used techniques for comparing DNA sequences. With larger sequences being aligned, text-based alignment viewers are inadequate, and alignment visualization using line-based glyphs in 3D space have been developed as a response [5, 6]. Currently, several tools for the visualization of alignment data are publicly available. One highly popular and successful tool is VISTA [16]. VISTA represents the level of conservation between species as a curve calculated by sliding a window of predefined size over the given alignment and computing the average similarity score over the window. VISTA shows pairwise similarity scores. Phylo-VISTA [19] extends the VISTA concept to the visualization of multiple (more than two) alignments. Other commonly used tools like MultiPipMaker [18] and SynPlot [9] also use 1D line or dot plots. Various genome browsers [11, 14, 26] also use textual display and 1D line plots to show alignment scores. SequenceJuxtaposer [21] uses "focus+context" techniques to provide interactive multiresolution navigation of sequences up to two million basepairs in total.

1.3 Motivation

With advancements in sequencing technology, increases in computational power, and the development of better computational methods, it is now possible to align several-million-basepair-long sequences. It is clear that the need exists, or will exist in the very near future, to develop new visualization techniques to support the interactive, visual exploration for such extremely large sequence data. The basic principle or the *Visual Information Seeking Mantra* is [20]:

Overview first, zoom and filter, then details-on-demand.

As the size of a typical alignment reaches several million basepairs, all existing techniques for visualization of alignment data fail to provide a good overview of an entire data set that will highlight regions of interest for further focus. Our work is driven by the need for visually presenting large 1D data sets in their entirety such that interesting features of the data for more detailed exploration are clearly visible.

Earlier work by Wong et al. [27] used space-filling *Hilbert curves* to transform sequential data into 2D space. This transformation allows one to display one million basepairs using a $1{,}000 \times 1{,}000$ pixel image. Application of digital image processing filters to such images reveals interesting patterns in the data. This work motivated us to represent multiple alignment data in 3D space, embedded in a fixed volume by using 3D space-filling curves. This approach allows us to apply various volume visualization techniques to render the data. We use hardware-accelerated volume rendering with maximal intensity projection [10] for visualization. In general, choosing a transfer function for volume rendering is not a trivial task. In our application, however, we specify a transfer function based on parameters relevant from the perspective of the driving biological problem.

2 Our Approach

With current multiple alignment algorithms, million-basepair alignments are becoming increasingly common. Techniques are required to examine such large data sets that contain more data points than there are screen pixels. "Focus+context" approaches have been effectively utilized in such cases where the most important data is given more screen space and is displayed in more detail than the rest of the data, which is still displayed at lower-resolution to provide context. This method is not very helpful when a user might need to examine the complete data set to locate interesting features in the data. Our goal is to make better use of the screen real estate to provide more screen area per data element.

2.1 From 1D Sequential Data to 3D Volume Data

A naïve approach for arranging sequential data in 3D space is using a scanline traversal in 2D planes and stacking these planes in perpendicular direction. In a 64^3 volume grid, for example, such an arrangement will place the 0th position (mapped to (0, 0, 0) in 3D space) and 4, 096th position (mapped to (0, 0, 1) in 3D space) next to each other. When two positions that are distant in the 1D sequence happen to be adjacent in a 3D arrangement, interpretation of data/visuals is difficult. To mitigate this problem, a 3D arrangement that maximizes spatial coherence should be used. Work by Keim et al. [13] and Voorhies [25] has shown that a Hilbert curve-based mapping is among the most coherent space-filling-curve-based approaches. Coherence is defined as the amount by which neighboring pixels (voxels in our case) are at sequential positions on the curve [27]. Figure 4 shows a 3D Hilbert curve. We map a score at position i in a multiple alignment to a position in 3D space using the algorithm described by Max [15].

2.2 Volume-Based Visualization

Once the 1D alignment data is transformed to volume data, we apply volume rendering to the data. In the following, we describe the details of volume-based visualization.

Color Channels

Consider a multiple alignment of sequences of four species: human, chimp, mouse and rat, see Fig. 1. Biologists are interested in:

- Conserved features in all four species
- Primate-specific features
- Rodent-specific features

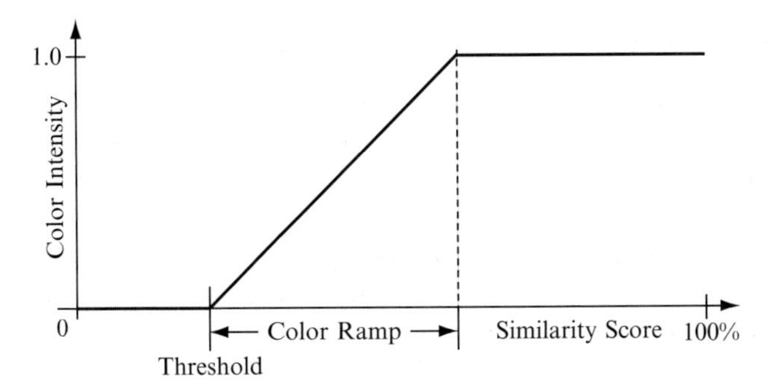

Fig. 2. Transfer function

These features can be identified by considering the following three different similarity plots and comparing them:

- A similarity plot for all four species
- A similarity plot for human and chimp (primates)
- A similarity plot for mouse and rat (rodents)

In this scenario, we can take advantage of three color channels, red, green and blue, to visualize three similarity plots: We map the first similarity plot to the red channel, the second plot to the green channel and the third plot to the blue channel. This approach allows a user to compare common and distinct features of all plots. Different colors highlight different features of the data. For example, blue represents a primate-specific feature absent in rodents, green represents a rodent-specific feature absent in primates, and cyan represents a feature specific to both primates and rodents. Further, white (the combination of all colors) represents a region conserved among all species.

Transfer Function

A separate transfer function is associated with each color channel. We use a linear transfer function like the one shown in Fig. 2. We use a user-defined similarity score threshold, below which everything is rendered transparently. The slope of the function is adjustable.

Maximal Intensity Projection

The Hilbert curve transforms a 1D sequence into a 3D volume that is just as large in Z direction as in X and Y directions. When this volume is projected on to 2D screen space, there is a substantial overlap of data. If the data is rendered fully opaquely, then the frontmost layer will be the only visible output, occluding the rest of the data. A back-to-front rendering allows for alpha blending and transparency, so that uninteresting areas of the data are drawn transparently and do not occlude anything. The user can freely rotate the data, in order to avoid arrangements in which interesting features occlude each other. But it is possible that a feature in the data may be completely surrounded by opaque voxels, so that rotation alone will not reveal it (Fig. 7c). We employ a maximal intensity projection that maps the intensity of a voxel to its screen depth. This depth is then used as input to the Z-buffer algorithm, common to most graphics hardware available today. This method guarantees that highly conserved regions in the DNA are always displayed before less conserved regions. Data is still occluded, but it is assumed that the user is more interested in regions of high conservation rather than regions of less conservation (Fig. 7d). In situations where this assumption is false, the user may invert the data for display, which is effectively a minimal intensity projection.

2.3 Annotations

For analysis of multiple alignment data, biologists often need additional biological information, such as information concerning a known gene model. For example, it is desirable to show the start and end coordinates of a gene, exons (the protein coding part of a gene), etc. This information can be provided in the form of annotations next to genome sequences. Thus, in addition to displaying similarity plots of a multiple alignment, displaying annotations is a major aspect of genomic data visualization. We draw annotations as "pipes" following the 3D Hilbert curve (Fig. 4). As the size of an annotation grows larger so does the number of pipes we draw. This method may clutter the display and slow down interaction. We handle this problem by using a multiresolution Hilbert curve. Figure 3 shows a Hilbert curve drawn in 2D

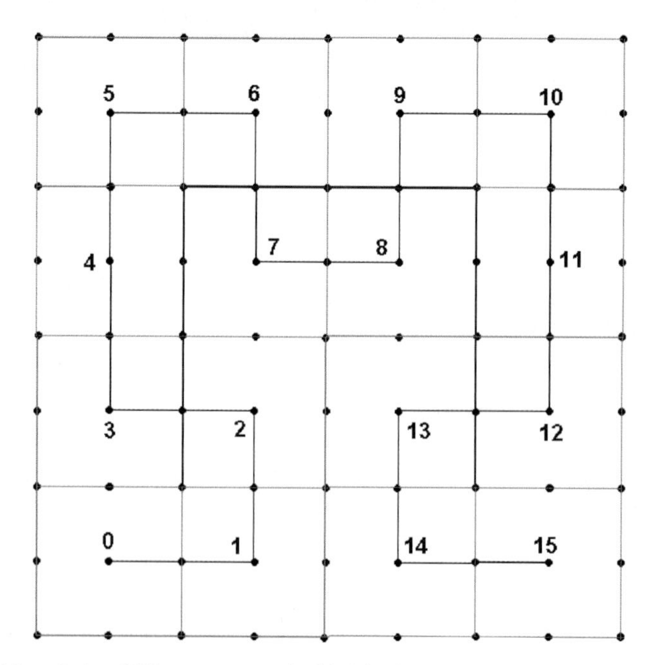

Fig. 3. Multiresolution Hilbert curve embedded in 2D space. The *black curve* is a high-resolution Hilbert curve; a lower-resolution version is shown in *red*

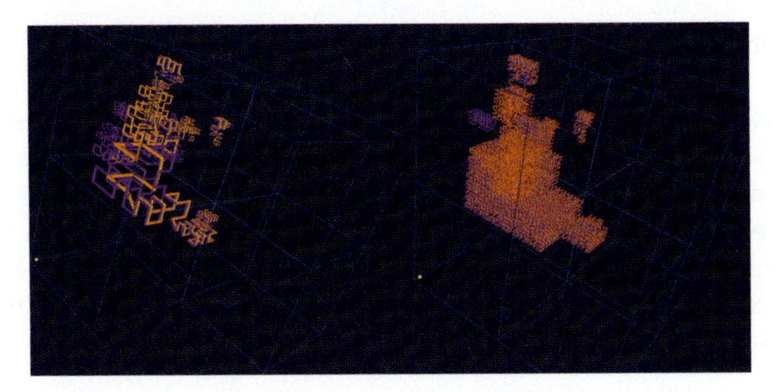

Fig. 4. Annotations. *Left:* annotations drawn using multiresolution Hilbert curve; *right:* annotations drawn using high-resolution Hilbert curve. *Purple and orange pipes* represent two different sets of annotations

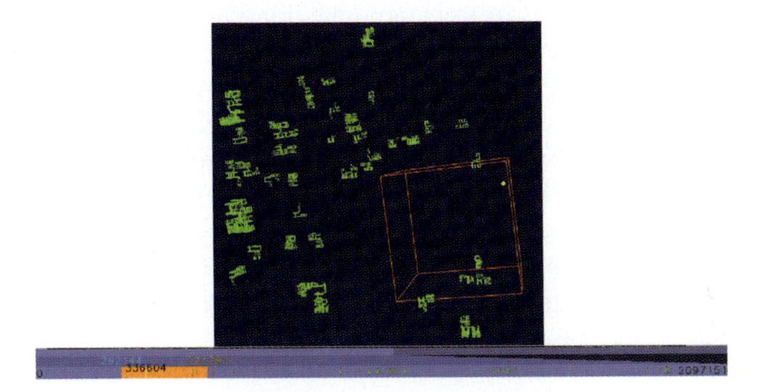

Fig. 5. Effective use of screen real estate. Three annotations of sizes 63, 54, and 63 separated by 565 and 11,260 basepairs, respectively, on 2,097,151-basepair-long sequence are distinctly visible in 3D space (*in orange cube*). But the first two merge in a 1D representation (*in orange rectangle*)

space using two resolutions. The black curve is a high-resolution curve with 15 line segments; the red curve is a lower-resolution version with just three segments. The image on the left-hand side in Fig. 4 shows annotations drawn using a multiresolution Hilbert curve approach. The image on the right-hand side uses the same annotations drawn as a high-resolution Hilbert curve. We draw lower-resolution curves as pipes with large diameters to show that they cover a greater volume. We plot overlapping annotations by drawing multiple Hilbert curves with a slight offset (Fig. 4). As a result of our volume-based visualization approach annotations can also be displayed at a higher resolution. Consider a 2,097,151-basepair-long genome sequence with three annotations of sizes 63, 54, and 63 basepairs and separated by 565 and 11,260 basepairs, respectively. The first two annotations, when displayed using a traditional 1D plot at 1,000-pixel resolution, would occupy the same pixel position on the screen, and the two annotations would be indistinguishable. Using our 3D visualization method all three annotations can be seen distinctly (Fig. 5).

3 Implementation

The adaptation of volume rendering techniques to modern commodity graphics hardware has made it easy and affordable to achieve interactive framerates. The massive parallelism of a modern GPU is well suited to evaluating the relatively simple transfer function millions of times per second, while the SIMD architecture allows for three channels of data to be processed as quickly as one. Using an NVidia 5900 FX graphics processor, our implementation displays 15 frames s^{-1} for a 64^3 data set ($\sim$260,000 basepairs), where each frame consists of 800×800 pixels.

3.1 Volume Renderer

We render images using view-perpendicular slices with a back-to-front ordering approach. This approach allows us to incorporate transparency into the transfer function. We use a fragment program to evaluate the transfer function [2]. This program compares the value at each voxel with the three thresholds (for the three color channels.) If a value is above the threshold, that color channel is activated for that voxel. If no value is above the threshold, the voxel is transparent rather than black. The color ramp (see Fig. 2) can be increased to yield a smooth transition between states. In this case, values near the threshold will be only slightly visible while values much greater than the threshold will clearly stand out. The maximal intensity projection is evaluated within the fragment program. A specific color channel, or the sum of all three, is mapped to the Z-buffer.

3.2 User Interface

We provide a user interface for our volume renderer that allows a biologist to explore intuitively large alignment data. We describe the specifics in the following.

Volume Rendering Controls

A user can choose thresholds for three color channels, which will automatically set the transfer function. Any of the three channels, or the sum of all three, can be chosen for the maximal intensity projection. A box filter of variable size can be applied to the similarity score data for smoothing.

Annotation Controls

Multiple sets of annotations can be loaded and displayed with our prototype system. User-defined colors are associated with each set of annotations. In addition, the diameters of the pipes can be adjusted.

Navigation Controls

A Hilbert curve is *fractal* in nature [17]. As a result, a data string embedded in a 3D volume can be organized using an octree-like structure. For navigation purposes, a currently displayed sequence portion, which always has the shape of a cube, is divided into octants. A user can then select an octant for zooming in. It is also possible to shift the selected octant by half of its size in both direction. To provide a context for the currently displayed sequence portion, the bounding box of the volume corresponding to the complete sequence can be displayed.

A 3D representation of inherently 1D data poses problems for navigation in 3D space. A user must know the 1D position of the underlying (sequence) data, but one can lose one's orientation when navigating in 3D space. We tackle this problem by using a 1D representation of the sequence shown next to the volume display. This representation allows a biologist to keep track of the position within the sequence in a more traditional fashion. This additional display consists of two bars, see Fig. 6, where the upper bar corresponds to the complete sequence and the lower bar corresponds to the displayed portion of a sequence. A highlighted region in the upper bar represents the displayed portion of the complete sequence and is connected to the lower bar, indicating the correspondence of the displayed sequence portion to the complete sequence.

Annotations are also displayed using rectangles on the lower bar. Selecting an octant for zooming in can be accomplished by clicking on the corresponding octant in the volumetric image, or by clicking on the corresponding part of the 1D bar representing the displayed portion of the sequence. Zooming out leads a user to the next lower level of detail. The 1D sequence representation can also be used to move a marker through the 3D display volume. This marker is symbolized by a vertical line in the 1D sequence display. Dragging this line by using a mouse moves the marker through the volume along the Hilbert curve. We show the similarity score for all

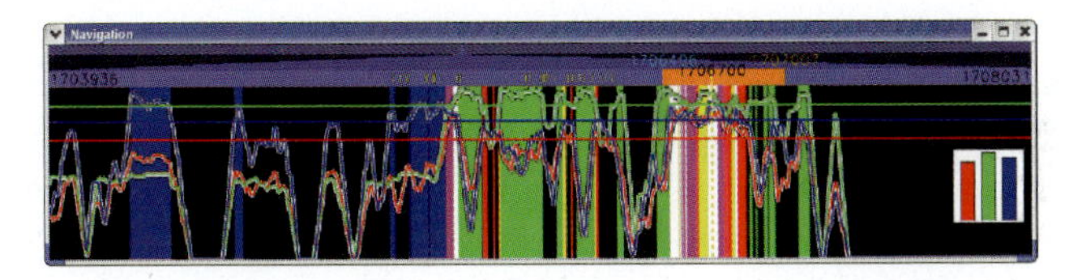

Fig. 6. Control for navigation on original 1D data sequence. The *upper bar* corresponds to the complete sequence, and the *lower bar* corresponds to the displayed portion of a sequence. A highlighted region in the *upper bar* represents the displayed portion of the complete sequence and is connected to the *lower bar*, showing the correspondence of the displayed sequence portion to the complete sequence. The highlighted region in the *lower bar* denotes the selected octant. *Yellow rectangles* are the annotations. 1D plots for all three channels are drawn as *red, blue and green curves*. The background color is the same as the corresponding color of the voxel in the volume. The similarity score for all three channels is shown at the marker position using a bar graph as shown in the *square on the right-hand side*

three channels at the marker position using a bar graph as shown in the square on the right-hand side of Fig. 6. We also show 1D plots for all three channels and we use the color in the volume as background for the line plot.

4 Results

We have applied our method to multiple alignment datasets that were created using MLAGAN [3]. We have used these two test datasets:

1. Stem Cell Leukemia (SCL) dataset.
   The SCL dataset is a multiple alignment DNA sequence data set of sequences from five species: human, mouse, chicken, pufferfish and zebrafish. All sequences contain the SCL gene. The alignment consists of 150,000 basepairs. These sequences were aligned in order to discover regulatory elements of the SCL gene. Regulatory elements are short DNA sequences consisting between 6 and 12 basepairs. They are generally found in a region in front of a gene called *promoter*. The underlying assumption is that they are conserved in evolutionary distant species because regulatory elements are functionally important.
2. Cystic Fibrosis Transmembrane Conductance (CFTR) dataset.
   The CFTR dataset is a multiple alignment DNA data set of sequences from 12 species: human, chimp, baboon, cat, dog, cow, pig, mouse, rat, chicken, fugu-fish and zebrafish. The sequences are from the region containing a gene coding the CFTR regulator, and nine other genes. The alignment is four million base-pairs long. This alignment can help biologists with the discovery of regulatory elements as well as their identification of subclass-specific features.

We compare volume-based visualizations of these datasets with 1D similarity plots and SequenceJuxtaposer. Most of the currently available tools including genome browsers like UCSF, Ensembl and NCBI use line plots and yield similar results to the 1D plots we compare our results with. We are not aware of any other tools that can handle well alignment data as large as four million basepairs.

Figure 7 shows the visualization of the SCL dataset. We visualize three similarity plots: the similarity plot for all five species, which is mapped to the red channel; the similarity plot for human–mouse–chicken, which is mapped to the green channel; and the similarity plot for the two fish species, which is mapped to the blue channel. The resolution of the displayed volume is 64^3 (Fig. 7c,d). Yellow pipes show exons (protein-coding parts of a gene) for the human sequence. In Phylo-VISTA plots, exons are shown as purple bars below the plot. Two bars exist: an upper one showing exons of the human sequence and a bottom one showing exons of the mouse sequence (Fig. 7b). A box filter with a width of 50 was used to smooth data for all three similarity scores. A threshold of 25% was used for all plots. We also generated a visualization of the SCL dataset using SequenceJuxtaposer [21] (Fig. 7a).

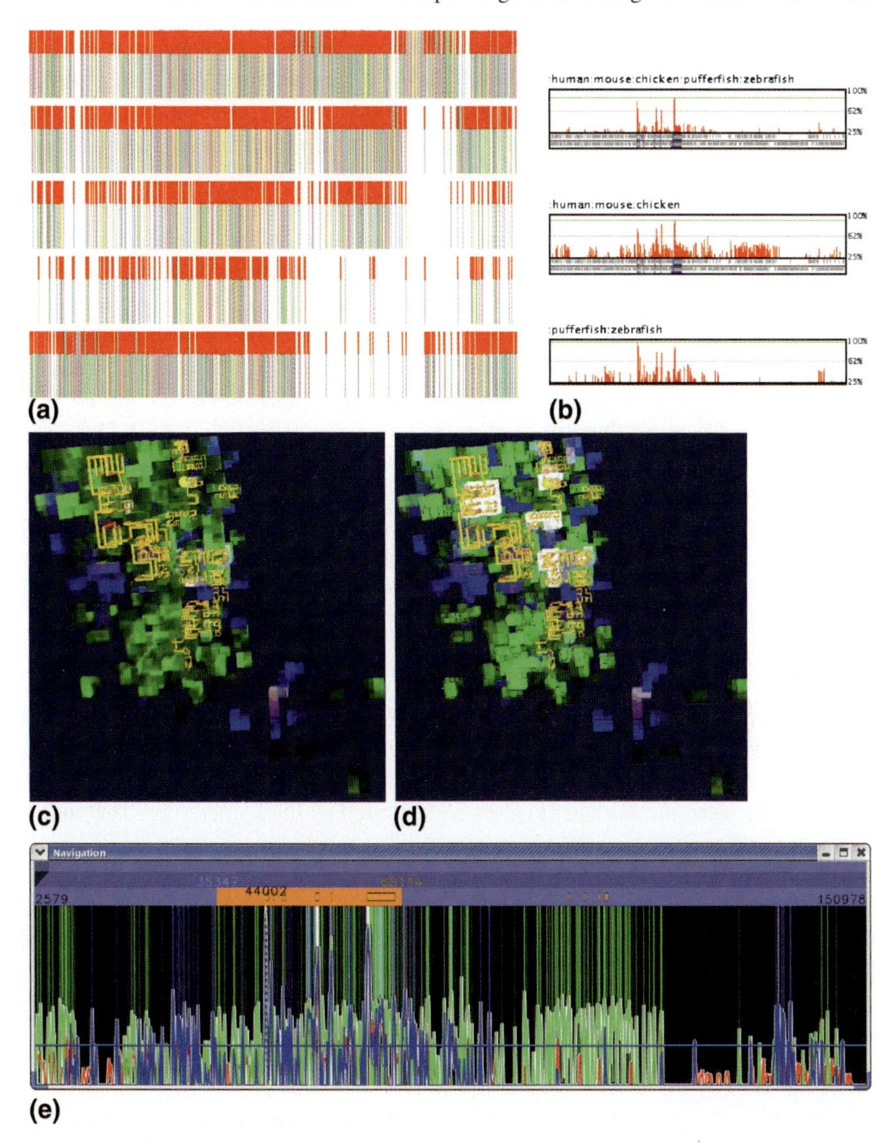

Fig. 7. Using volume rendering to discover a regulatory region for the SCL gene: (**a**) *Red lines* show differences among the five sequences. This image was generated using SequenceJuxtaposer. (**b**) 1D plots created with Phylo-VISTA. (**c**) The volume-rendered image with *yellow pipes* showing annotations. The 3D representation occludes some of the interesting features of the data. (**d**) The *white regions* in the volume-rendered image correspond to regions that are highly conserved in all sequences. The volume-rendered representation using maximal intensity projection allows users to detect these regions instantaneously, without the need to compare multiple plots. (**e**) 1D line representation obtained by overlaying all three similarity plots. *White lines* show conserved regions

Red lines in Fig. 7a show differences among the five sequences. It is difficult to find regions of high similarity looking at this picture. White spots in the volume-rendered image (Fig. 7d) indicate regions of high conservation in all three similarity plots. These spots are seen as peaks in all three corresponding Phylo-VISTA plots (Fig. 7b). Green spots are conserved regions in the human–mouse–chicken plot that are absent in the fish plot. Similarly, blue spots are fish-specific conserved regions. The conserved regions seen in these images contain the regulatory elements of the SCL gene [8]. In order to compare three 1D plots a user has to inspect them by eye and determine whether there are peaks in different plots at the same position. This analysis approach may create problems, especially when one pixel represents the similarity score for more than one column in a multiple alignment. In the case of the volume-rendered image, a user needs to consider only the color to compare all three plots at once. Of course, the same color scheme can be applied to 1D plots, and this is show in Fig. 7e. One can see that the four white peaks are barely visible in this plot whereas the corresponding four white spots in the volume-rendered image are strikingly visible. One of the issues in volume rendering is occlusion as can be seen in Fig. 7c, where the high similarity regions are hidden inside the volume. We handle this problem by using maximal intensity projection, as a result of which the high-similarity regions show through (Fig. 7d).

Figure 8 shows a visualization of the CFTR dataset. This figure shows only one plot for the similarity among all 12 species, mapped to the red channel. The resolution of the volume is 256^3. The sequence data fills 25% of the volume. White pipes show genes, and purple pipes indicate the exons of the human sequence. The data was smoothed using a box filter of width 50. The threshold was 25%. The correspondence between the 3D and 1D plots is illustrated by thick gray lines. The middle image shows the entire dataset. Exons and their conservation are much more clearly and distinctly visible in the 3D plot than in the 1D plot. The red spots in the first half of the dataset indicate conservation among all species. The top and the bottom images show zoomed-in views from different viewpoints of the same circled part of the middle image. The circled part in the top image indicates conservation of the first exon of a gene and the promoter region. The image to its right shows a zoomed-in view. The conserved region seen in this right-most top image contains regulatory elements of CAV2 gene [24]. Similarly, the bottom images show the conservation of regulatory elements of the CAPZA2 gene [24].

Figures 9 and 10 show the visualization of the CFTR dataset. Three different similarity plots are used in the volume visualization. The similarity plot for primates (human, chimp, baboon), artiodactyls (cow, pig) and carnivores (cat, dog) are mapped to the red channel. The green channel is used to display the similarity plot for primates, and the blue channel for showing the similarity plot of carnivores and artiodactyls. Figure 9a shows the entire dataset. We use different thresholds for different channels: 70% for red, 90% for green and 80% for blue. The 3D visualization reveals many more features when compared to the 1D plot. Larger primate-specific (green) and artiodactyls-carnivores-specific (blue) regions can be identified from both 3D and 1D plots. In the 3D plot, we can see a white spot in an otherwise blue–green region (shown with the red arrow). This feature of the dataset is not visible in the 1D plot

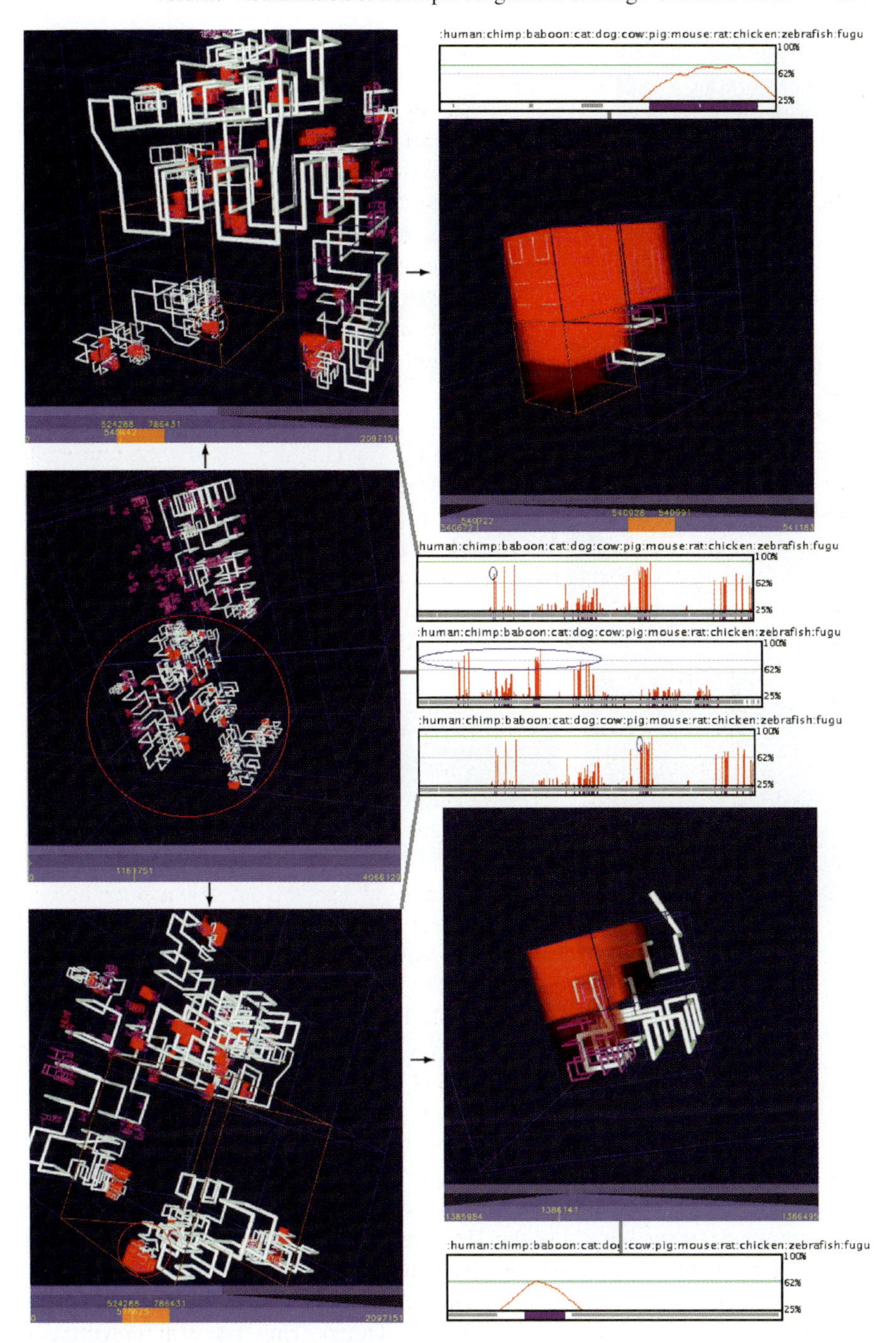

Fig. 8. Visualizing the CFTR data set: The *red spots* indicate similarities among all species in a region corresponding to the CAPZA2 gene. In the volume-rendered image, exons and their conservation are much more clearly and distinctly visible than in the 1D plot

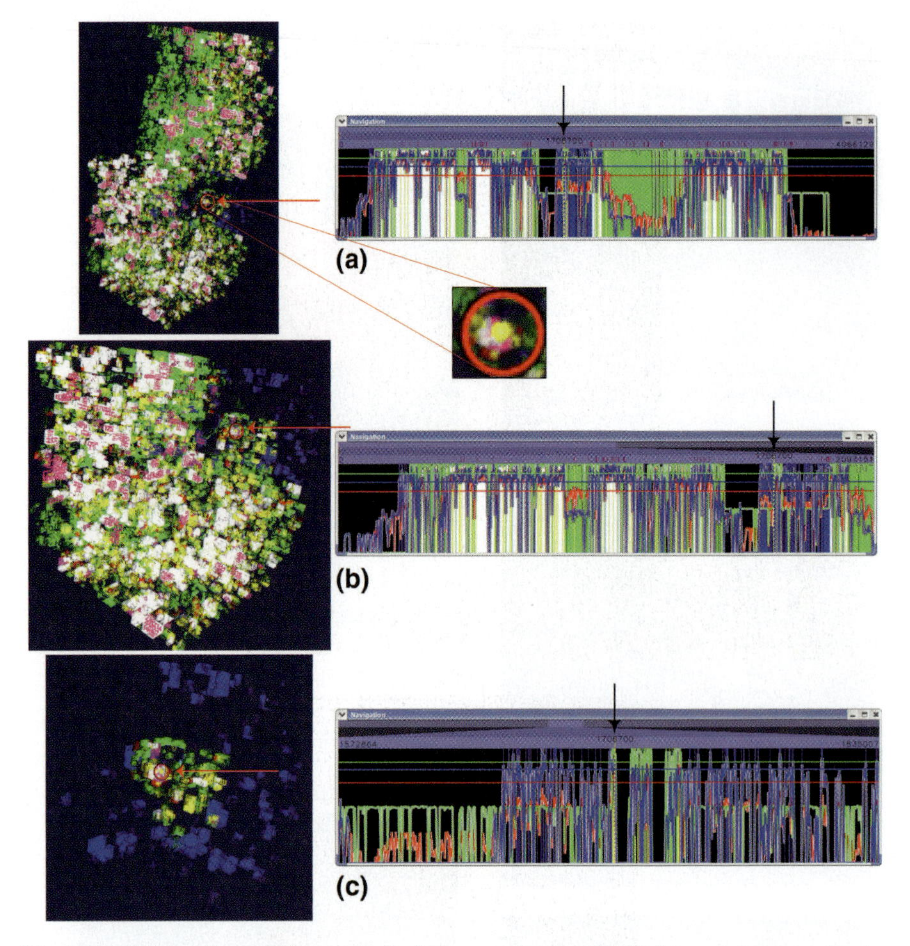

Fig. 9. Visualizing three different similarity scores for the CFTR data set. Larger primate-specific (*green*) and artiodactyls-carnivores-specific (*blue*) regions can be seen in both 3D and 1D plots. (**a**) A *white spot* in an otherwise blue–green region (indicated by the *red arrow*). The *black arrow* indicates the corresponding region in 1D plot with the feature not visible. (**b**, **c**) Zoomed-in views with the feature in the 3D plots being invisible in 1D plots. (Figure continued on next page)

(indicated by the black arrow). As we zoom-in further (Fig. 9b,c), the spot remains visible in the 3D plots but not in the 1D plots. In Fig. 10d, we start seeing a white line in the 1D plot, which becomes more distinctly visible when we zoom-in further (Fig. 10e). This conserved region is in a noncoding region that is far away from any gene but can potentially be a distant regulatory element. Thus, our 3D representation allowed us to detect an interesting feature immediately that would have been missed by looking at just 1D plots.

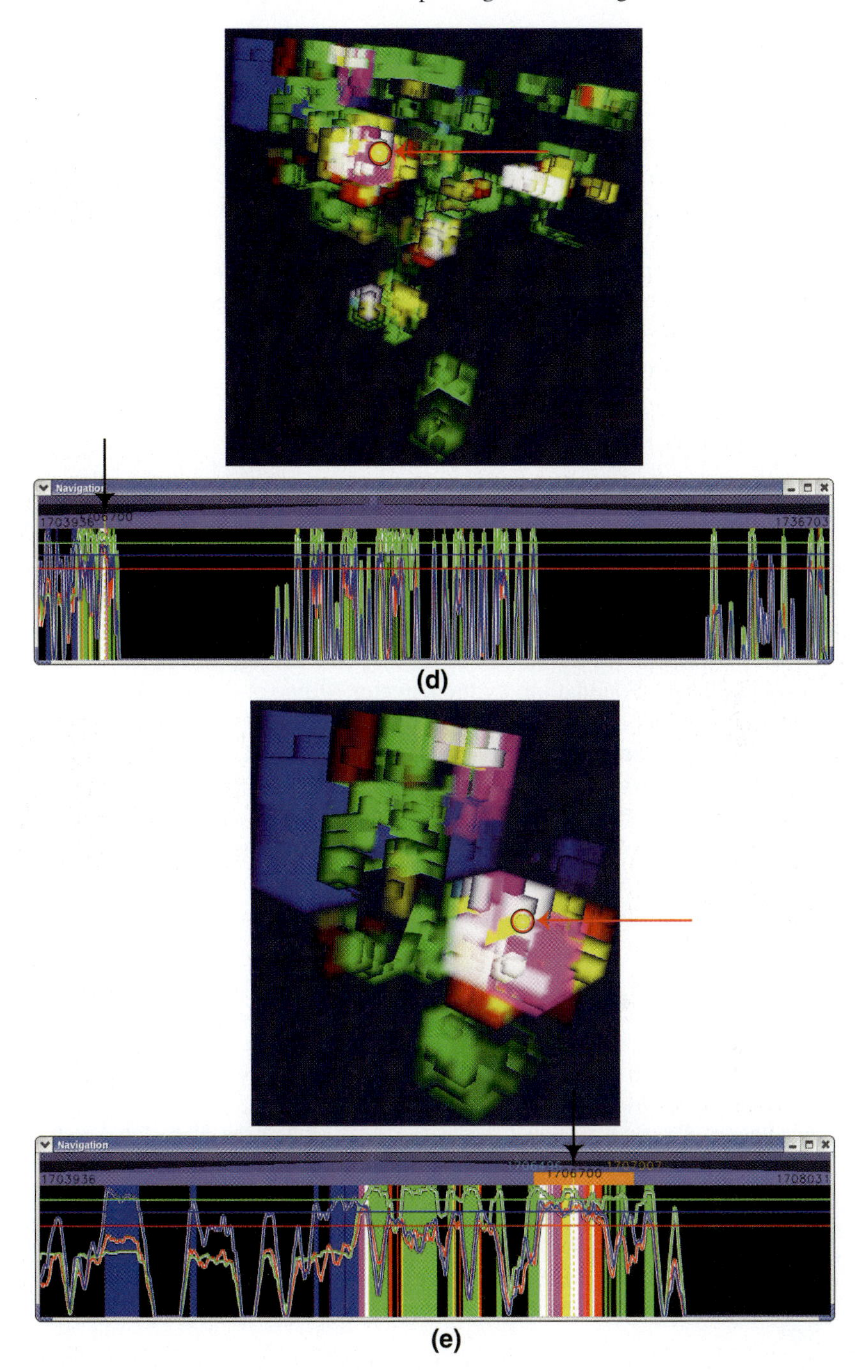

Fig. 10. Zoomed-in views of three different similarity scores for the CFTR data set. (**d**) A feature visible in the 3D plot is visible as a thin, white line in the corresponding 1D plot. (**e**) A feature visible in the 3D plot is clearly visible in the 1D plot after zooming-in

5 Conclusions

We have presented a volume-based visualization technique for analyzing multiple alignment data. Our results demonstrate that 3D representations and visualizations of genome data are quite effective and utilize 3D display space efficiently. As a result, we can convey information more compactly, especially for billion-basepair sequence data.

Although developed for a particular biological application, our method can be applied to other kinds of massive sequential 1D data sets. Other volume-based visualization techniques, like isosurfacing or plane slicing, etc. could also be used when appropriate for a given application.

Acknowledgments

This work was supported by the National Science Foundation under contracts ACI 9624034 (CAREER Award), through the Large Scientific and Software Data Set Visualization (LSSDSV) program under contract ACI 9982251, through the National Partnership for Advanced Computational Infrastructure (NPACI) and a large Information Technology Research (ITR) grant; the National Institutes of Health under contract P20 MH60975-06A2, funded by the National Institute of Mental Health and the National Science Foundation; by the Director, Office of Science, USA. Department of Energy under contract DE-AC03-76SF00098; and the Lawrence Berkeley National Laboratory through a Laboratory Directed Research Development (LDRD) project. We thank Chris Co and Oliver Kreylos for their helpful suggestions. We also thank the members of the Visualization and Graphics Research Group at the Institute for Data Analysis and Visualization (IDAV) at the University of California, Davis, and the members of the Genome Sciences Department and the NERSC Visualization Group of the Lawrence Berkeley National Laboratory for their support.

References

1. Dan Ashlock and Jim Golden. *Ch.11 Evolutionary Computation and Fractal Visualization of Sequence Data*. Morgan Kaufmann, 2002.
2. Bob Beretta, Pat Brown, Matt Craighead, Cass Everitt, Evan Hart, Jon Leech, Bill Licea-Kane, Bimal Poddar, Jeremy Sandmel, Jon Paul Schelter, Avinash Seetharamaiah, and Nick Triantos. GL_ARB_fragment_program Specification. Online OpenGL Extension Registry, August 2003.
3. Michael Brudno, Chuong Do, Gregory Cooper, Michael F. Kim, Eugene Davydov, Eric D. Green, Arend Sidow, and Serafim Batzoglou. Lagan and multi-lagan: efficient tools for large-scale multiple alignment of genomic DNA. *Genome Research*, 13(4):721–731, 2003.
4. Hsuan T. Chang, Neng-Wen Lo, Wei C. Lu, and Chung J. Kuo. Visualization and comparison of DNA sequences by use of three-dimensional trajectories. In *Proceedings of the First Asia-Pacific Bioinformatics Conference on Bioinformatics 2003*, pages 81–85, Adelaide, Australia, 2003.

5. Ed Huai-hsin Chi, Phillip Barry, Elizabeth Shoop, John Carlis, Ernest Retzel, and John Riedl. Visualization of biological sequence similarity search results. In Gregory M. Nielson and Deborah Silver, editors, *Proceedings of IEEE Visualization 1995*, IEEE Visualization, Annual Conference Series, pages 44–51, Atlanta, USA, 1995. IEEE, IEEE Computer Society Press.
6. Ed Huai-hsin Chi, John Riedl, Elizabeth Shoop, John V. Carlis, Ernest Retzel, and Phillip Barry. Flexible information visualization of multivariate data from biological sequence similarity searches. In Roni Yagel and Gregory M. Nielson, editors, *Proceedings of IEEE Visualization 1996*, IEEE Visualization, Annual Conference Series, pages 133–140, San Francisco, USA, 1996. IEEE, IEEE Computer Society Press.
7. Kelly A. Frazer, Laura Elnitski, Deanna M. Church, Inna Dubchak, and Ross C. Hardison. Cross-species sequence comparisons: A review of methods and available resources. *Genome Research*, 13(1):1–12, 2003.
8. Berthold Göttgens, Linda M. Barton, Michael A. Chapman, Angus M. Sinclair, Bjarne Knudsen, Darren Grafham, James G.R. Gilbert, Jane Rogers, David R. Bentley, and Anthony R. Green. Transcriptional regulation of the stem cell leukemia gene (scl) comparative analysis of five vertebrate scl loci. *Genome Research*, 12(5):749–759, 2002.
9. Berthold Göttgens, James G R. Gilbert, Linda M. Barton, Darren Grafham, Jane Rogers, David R. Bentley, and Anthony R. Green. Long-range comparison of human and mouse scl loci: Localized regions of sensitivity to restriction endonucleases correspond precisely with peaks of conserved noncoding sequences. *Genome Research*, 11(1):87–97, 2001.
10. W. Heidrich, M. McCool, and J. Stevens. Interactive maximum projection volume rendering. In Gregory M. Nielson and Deborah Silver, editors, *Proceedings of IEEE Visualization 1995*, IEEE Visualization, Annual Conference Series, pages 11–18, Atlanta, USA, 1995. IEEE, IEEE Computer Society Press.
11. T. Hubbard, D. Barker, E. Birney1, G. Cameron, Y. Chen, L. Clark, T. Cox, J. Cuff, V. Curwen, T. Down, R. Durbin, E. Eyras, J. Gilbert, M. Hammond, L. Huminiecki, A. Kasprzyk, H. Lehvaslaiho, P. Lijnzaad, C.Melsopp, E. Mongin, R. Pettett, M. Pocock, S. Potter, A. Rust, E. Schmidt, S. Searle, G. Slater, J. Smith, W. Spooner, A. Stabenau, J. Stalker, E.Stupka, A. Ureta-Vidal, I. Vastrik, and M. Clamp. The ensembl genome database project. *Nucleic Acids Research*, 30(1):38–41, 2002.
12. H. Joel Jeffrey. Chaos game representation of gene structure. *Nucleic Acids Research*, 18(8):2163–2170, 1990.
13. Daniel A. Keim, Mihael Ankerst, and Hans-Peter Kriegel. Recursive pattern: A technique for visualizing very large amounts of data. In Gregory M. Nielson and Deborah Silver, editors, *Proceedings of IEEE Visualization 1995*, IEEE Visualization, Annual Conference Series, pages 279–288, Atlanta, Georgia, 1995. IEEE, IEEE Computer Society.
14. W. James Kent, Charles W. Sugnet, Terrence S. Furey, Krishna M. Roskin, Tom H. Pringle, Alan M. Zahler, and David Haussler. The human genome browser at UCSC. *Genome Research*, 12(6):996–1006, 2002.
15. Nelson L. Max. Visualizing Hilbert curves. IEEE visualization 1998. In David S. Ebert, Holly Rushmeier, and Hans Hagen, editors, *Proceedings of IEEE Visualization 1998*, IEEE Visualization, Annual Conference Series, pages 447–450, North Carolina, USA, 1998. IEEE, IEEE Computer Society Press.
16. C. Mayor, M. Brudno, J. R. Schwartz, A. Poliakov, E. M. Rubin, Kelly A. Frazer, Lior S. Pachter, and Inna Dubchak. VISTA: visualizing global DNA sequence alignments of arbitrary length. *Bioinformatics*, 16(11):1046–1047, 2000.
17. Hans Sagan. *Space-Filling Curves*. Springer-Verlag, 1994.
18. S. Schwartz, L. Elnitski, M. Li, M. Weirauch, C. Riemer, A. Smit, E. D. Green, R. C. Hardison, W. Miller, and NISC Comparative Sequencing Program. Multipipmaker and

supporting tools: Alignments and analysis of multiple genomic DNA sequences. *Nucleic Acids Research*, 31(13):3518–3524, 2003.
19. Nameeta Shah, Olivier Couronne, Len A. Pennacchio, M. Brudno, Serafim Batzoglou, E. W. Bethel, E. M. Rubin, Bernd Hamann, and Inna Dubchak. Phylo-vista: interactive visualization of multiple DNA sequence alignments. *Bioinformatics*, 20(5):636–643, 2004.
20. Ben Shneiderman. The eyes have it: A task by data type taxonomy for information visualizations. In *Proceedings of IEEE Symposium on Visual Languages 1996*, pages 336–343. IEEE Computer Society, 1996.
21. James Slack, Kristian Hildebrand, Tamara Munzner, and Katherine St. John. Sequence-juxtaposer: Fluid navigation for large-scale sequence comparison in context. In Robert Giegerich and Jens Stoye, editors, *German Conference on Bioinformatics*, volume 53 of *LNI*. GI, 2004.
22. Rhazes Spell, Rachael Brady, and Fred Dietrich. BARD: A visualization tool for biological sequence analysis. In Tamara Munzner and Stephen North, editors, *IEEE Symposium on Information Visualization, 2003*, pages 219–226. IEEE Computer Society, 2003.
23. Praveen Thiagarajan and Guang Gao. Visualizing biosequence data using texture mapping. In Pak Chung Wong and Keith Andrews, editors, *IEEE Symposium on Information Visualization, 2002*, pages 103–109. IEEE Computer Society Press, 2002.
24. J. W. Thomas, J. W. Touchman, R. W. Blakesley, G. G. Bouffard, S. M. Beckstrom-Sternberg, E. H. Margulies, M. Blanchette, A. C. Siepel, P. J. Thomas, J. C. Mcdowell, B. Maskeri, N. F. Hansen, M. S. Schwartz, R. J. Weber, W. J. Kent, D. Karolchik, T. C. Bruen, R. Bevan, D. J. Cutler, S. Schwartz, L. Elnitski, J. R. Idol, A. B. Prasad, S.-Q Lee-Lin, V. V. B. Maduro, T. J. Summers, M. E. Portnoy, N. L. Dietrich, N. Akhter, K. Ayele, B. Benjamin, K. Cariaga, C. P. Brinkley, S. Y. Brooks, S. Granite, X. Guan, J. Gupta, P. Haghighi, S.-L Ho, M. C. Huang, E. Karlins, P. L. Laric, R. Legaspi, M. J. Lim, Q. L. Maduro, C. A. Masiello, S. D. Mastrian, J. C. Mccloskey, R. Pearson, S. Stantripop, E. E. Tiongson, J. T. Tran, C. Tsurgeon, J. L. Vogt, M. A. Walker, K. D. Wetherby, L. S. Wiggins, A. C. Young, L.-H Zhang, K. Osoegawa, B. Zhu, B. Zhao, C. L. Shu, P. J. De Jong, C. E. Lawrence, A. F. Smit, A. Chakravarti, D. Haussler, P. Green, W. Miller, and E. D. Green. Comparative analyses of multi-species sequences from targeted genomic regions. *Nature*, 424(14):788–793, 2003.
25. Douglas Voorhies. Space-filling curves and a measure of coherence. In James Arvo, editor, *Graphics Gems II*, Graphics Gems, pages 26–30. Academic Press, 1991.
26. David L. Wheeler, Colombe Chappey, Alex E. Lash, Detlef D. Leipe, Thomas L. Madden, Gregory D. Schuler, Tatiana A. Tatusova, and Barbara A. Rapp. Database resources of the national center for biotechnology information. *Nucleic Acids Research*, 28(1):10–14, 2000.
27. Pak Chung Wong, Kwong Kwok Wong, Harlan Foote, and Jim Thomas. Global visualization and alignment of whole bacterial genomes. *IEEE Transactions on Visualization and Computer Graphics*, 9(3):361–377, 2003.

Model-Based Visualization: Computing Perceptually Optimal Visualizations

Jarke J. van Wijk

Department of Mathematics and Computer Science, Technische Universiteit Eindhoven, The Netherlands
vanwijk@win.tue.nl

Summary. Visualization is often more an art than a science. Here a model-based approach to visualization is promoted in order to make steps forward in this respect. The main ingredients of this approach are the derivation of a quantitative model of the perception of (certain aspects of) a visualization, the definition of an optimal visualization, followed by its computation. Some examples of this approach are presented, with optimal zooming and panning as main example. Advantages and problems are discussed. In conclusion, we think this approach to be challenging and promising, although we acknowledge that it is not an easy route.

1 Introduction

The development of effective novel visualization methods and techniques is hard. It requires creativity, knowledge, experience, and effort. Traditionally, the focus was on the creative part, the development of a new idea itself. In the early days of visualization, many problems were not addressed yet, and each first solution was welcome. However, now that the field is getting more mature, the demands are getting higher. Just a new idea, followed by parameter tuning until nice images result is not enough, comparisons and validation are needed to convince the community that the new idea is an improvement indeed.

Validation afterwards is always important to achieve higher quality visualization methods. In this paper we want to attract attention to another approach that can lead to a higher quality: *Model-based Visualization*.

The visualization pipeline consists of many steps. One step is crucial: The step from the image on the screen to the perception and cognition of the user. If this step fails, everything else is useless. Hence, a good understanding of the relation between visual stimuli and their perception is vital to obtain effective visualization methods [9]. We propose to take this further. If we have a quantitative model of the relation between stimulus and perception, and if we know what should be perceived, we might be able to calculate the corresponding optimal stimulus.

In the following we elaborate on this. In the next section we present the steps to be followed, followed by some examples where this approach has been used. One

T. Möller et al. (eds.), *Mathematical Foundations of Scientific Visualization, Computer Graphics, and Massive Data Exploration*, Mathematics and Visualization,
DOI: 10.1007/978-3-540-49926-8,

of these is the calculation of optimal zoom-pan trajectories, which was the main inspiration for the ideas presented here. In the last section we discuss the advantages and disadvantages of the approach, and suggest possible applications.

2 Approach

The approach consists of the following steps:

1. Define a model $V(d, v)$ of the visualization. Given a data-set d and a setting of visualization parameters v, the result is a visualization $V(d, v)$. A visualization here is something that can be perceived: an image, an animation, or just some aspect of it, such as a color or line width.
2. Define which aim is pursued in a perceptual sense, i.e., state what qualities an optimal visualization should have.
3. Define a model $P(V(d, v), p)$ for the perception of a visualization, where p refers to free parameters in the model.
4. Define a quantity $Q(P(V(d, v), p), d)$ to be optimized as a function of v.
5. Compute v such that Q is optimal.
6. Find good values for p.

We defined these steps based on a generalization of earlier work we have done. The results we obtained, as well as the positive feedback on our work, made us wonder if the approach taken could be used for other problems as well. We were obviously not the first to use this approach, the same pattern can be found in other cases.

2.1 Zooming and Panning

The problem we have attacked seemed simple, but turned out to be non-trivial [7,8]. Here we give a short overview, in the original papers much more detail can be found. Given a large 2D information space, the user must be enabled to focus on different areas with different scales. Suppose we want to offer the viewer a smooth animation from one view to another view, such that he can understand the relation between the two views. One way to solve this problem is to use some heuristics, invent possible trajectories of the virtual camera, and validate these afterwards. Our challenge was to find an optimal solution.

The first step was the precise definition of the visualization problem. In this case, a view can be defined by a center point $\mathbf{c} = (c_x, c_y)$ and a width w. When two views are given, defined by $(\mathbf{c}_i, w_i)$, with $i = 0, 1$, an animation from the first to the second can be defined by functions $\mathbf{c}(s)$ and $w(s)$, $s \in [0, S]$, where the parameter s is along a path from the first to the second view, and S denotes the final value. The functions $\mathbf{c}(s)$ and $w(s)$ denote the path of a virtual camera and the width shown along the path. An animation can now be defined by setting

$$s = Vt, \ t \in [0, S/V],$$

where V denotes the constant animation speed, and t wall clock time, for instance in seconds. Assuming a straight path between the start and end view, this can be reduced to:

$$\mathbf{c}(s) = \mathbf{c}_0 + \frac{\mathbf{c}_1 - \mathbf{c}_0}{\|\mathbf{c}_1 - \mathbf{c}_0\|}\, u(s), \;\; u \in [u_0, u_1],$$

$$\text{with } u_0 = 0 \text{ and } u_1 = \|\mathbf{c}_1 - \mathbf{c}_0\|.$$

where the parameter $u(s)$ denotes panning along a straight line. Functions $u(s)$ and $w(s)$, $s \in [0, S]$, must be found such that at least

$$\begin{aligned} u(0) &= u_0, \;\; w(0) = w_0, \\ u(S) &= u_1, \;\; w(S) = w_1. \end{aligned}$$

Secondly, we defined what properties an optimal animation should have. We claim this can be summarized in two words: The optimal path should be *smooth* and *efficient*. Smoothness is a constraint. The path should be at least continuous in the first order, in the sense that no sudden steps are made or abrupt changes in direction occur. Also, when the camera moves along the path, the viewer should get the impression of a uniform and constant motion of the projected image on the screen. Efficiency is the aspect to be optimized: We want to go from the first view to the second as quick as possible, without detours.

Thirdly, we defined a relation between the motion of the virtual camera and the perception of the motion of the image. The average velocity of the moving image is given by

$$V_{\text{RMS}}^2 = V^2 \left(\frac{1}{w^2}\, \dot{u}^2 + \frac{1}{6w^2}\, \dot{w}^2 \right).$$

We generalize this in the form of a metric on (u, w) space:

$$ds^2 = \frac{\rho^2}{w^2}\, du^2 + \frac{1}{\rho^2 w^2}\, dw^2.$$

This metric gives the distance ds traveled, when u and w are changed with du and dw. The parameter ρ represents a trade-off between zooming and panning. A high value indicates that zooming has little impact, a low value indicates that panning has less impact.

Fourth, we made our perceptual aim explicit. Both smoothness and efficiency can be achieved by using the shortest path in (u, w) space, i.e., the geodesic. Each step along the path should have the same length, where length is measured using the metric. Also, we want to make the overall motion as short as possible, hence we use the geodesic.

The fifth step is to calculate the optimal solution. In this particular case, we could find an analytic solution for the optimal path for arbitrary parameters using

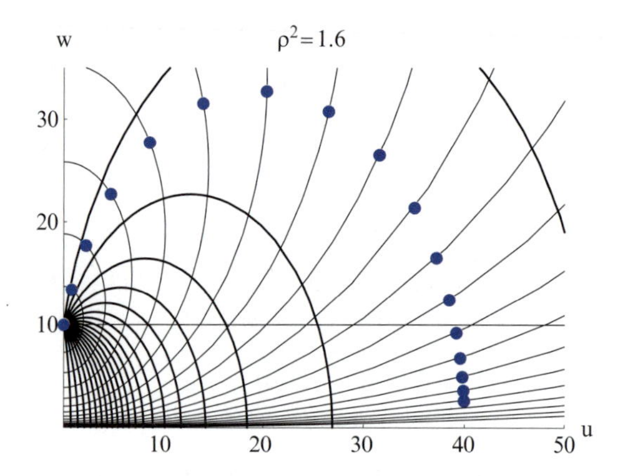

Fig. 1. Paths and iso-distance curves in (u, w) space

differential geometry. The path of the camera $(u(s), w(s)), s \in [0, S], u_0 \neq u_1$ is given by

$$
\begin{aligned}
u(s) &= \frac{w_0}{\rho^2} \cosh r_0 \tanh(\rho s + r_0) - \frac{w_0}{\rho^2} \sinh r_0 + u_0, \\
w(s) &= w_0 \cosh r_0 / \cosh(\rho s + r_0), \\
S &= (r_1 - r_0)/\rho, \\
r_i &= \ln(-b_i + \sqrt{b_i^2 + 1}), \;\; i = 0, 1, \;\; \text{and} \\
b_i &= \frac{w_1^2 - w_0^2 + (-1)^i \rho^4 (u_1 - u_0)^2}{2 w_i \rho^2 (u_1 - u_0)}, \;\; i = 0, 1.
\end{aligned}
$$

Figure 1 shows sets of geodesic paths, starting from $u = 0$ and $w = 10$ in different directions, for a certain value of ρ. Furthermore, in each plot a set of contours is shown as thin lines. Each contour represents a set of points at an equal distance from the start point. Both the paths and the contours are parts of ellipses, where ρ determines their shapes.

The last step is to select good values for the free perceptual parameters. In this case, we had introduced two: The overall velocity V and the trade-off parameter ρ. We have done a first user experiment to obtain insight in preferred values for V and ρ. Figure 2 shows a scatterplot of the results. The variation in individual preferences was considerable. The average values ($V = 0.90$, and $\rho = 1.42$) can be used as default values, for an optimal result per user the settings of these parameters can be made customizable.

2.2 Other Examples

The overall approach sketched can be found also in other research. One area is the use of color. Color has been studied intensively, and much knowledge is available on

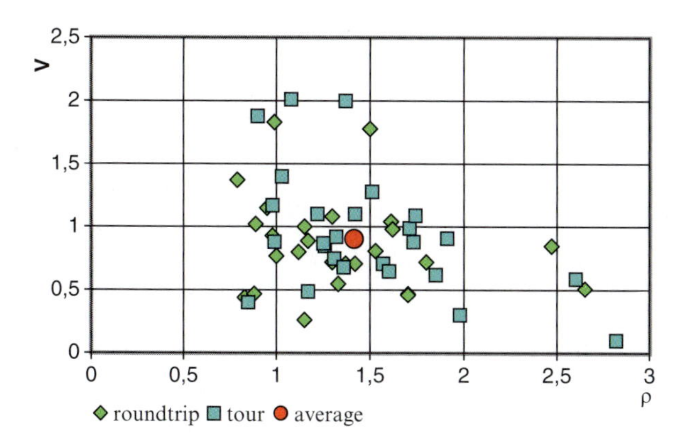

Fig. 2. Scatterplot of V and ρ values found

its perception. Color scales are often used for the representation of single-parameter distributions. Levkowitz and Herman [3] have studied how optimal color scales can be calculated, based on a set of requirements and knowledge on color perception. They required that a color scale preserves the order of the values, the difference in perceived colors should match the difference between the corresponding values, and the scale should be continuous. Furthermore, given a start and end color, they defined an optimal scale as a scale with a maximum number of distinct perceived colors along the scale, i.e., the scale should be as long as possible in a perceptual sense. Next, the constraints, criteria, and perceptual aspects are modeled mathematically, followed by a numerical optimization. This led to a new type of color scale, with a larger perceptual length than a standard gray scale. These scales and several others were evaluated in an experiment, where the users had to find features in medical images. For this particular task, the new color scale did not yield a better performance.

Another area where optimization is often used is in graph drawing [1]. Here, upfront aesthetical and perceptual aspects of the drawings are defined, next, a lay-out is calculated such that these are met as well as possible. One example is the minimization of the number of crossing edges, but also the spring embedder or force-directed approach falls in this category. The general idea behind this approach is as follows. In principle, nodes should be located at a large distance from each other, while nodes that are endpoints of an edge should be close. This can be achieved by using repulsive forces between all nodes, and attracting springs between nodes that share an edge. Nodes are moved according to the forces, until a minimal energy in the system is achieved. Different requirements on what is considered optimal lead to different force models, and hence to different lay-outs. Figure 3 (from [5]) shows left the result of a standard electromagnetic repulsion model and linear springs, leading to a uniform distribution of the nodes. On the right the result of an alternative model, proposed by Noack [4], is shown. The aim here is to emphasize the clusters in the graph, which can be achieved if a linear repulsion force and a constant attraction force is used.

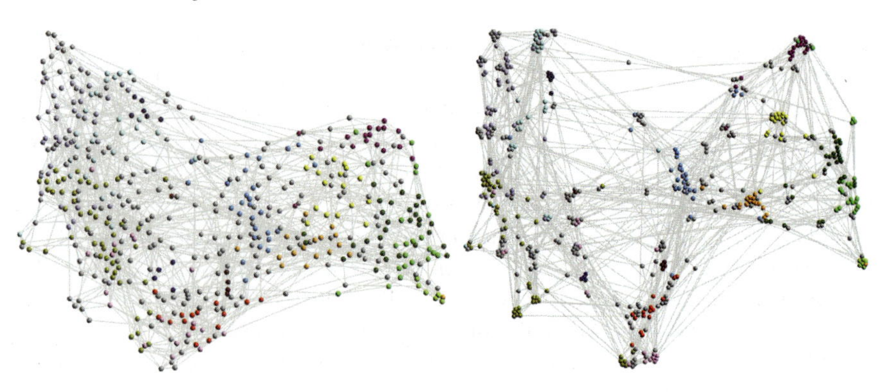

Fig. 3. Force-directed graph lay-outs, with a standard (*left*) and a modified force model (*right*) [4,5]

3 Discussion

We have proposed to develop optimal visualizations by modeling the perception, followed by an optimization of some perceptual aim, and we have given some examples. This approach has many advantages:

- It can yield good results that are widely applicable. For our zoom-pan problem we were able to define zoom-pan trajectories for arbitrary start and end-views, which were appreciated by the viewers.
- The perceptual model used or developed can be reused as a foundation for other problems as well. In our second paper on zooming and panning [8] we showed how the original ideas could be used for user-driven navigation, and extended to cope with rotation and scaling also.
- The perceptual parameters that have to be set are natural and orthogonal. At least in our case, the V parameter relates to the overall speed, while ρ denotes the trade-off between zooming and panning. Users could easily specify an personal optimal setting.

However, we cannot claim that this approach will always yield superior results. The following problems can be identified:

- Finding robust quantitative models of perceptual aspects is hard. The perception literature is vast, but hard to digest for computer scientists, few of us have the broad background and vision of Colin Ware [9]. Cooperation with experts in these fields will be vital here. Also, for many use-cases no models are available, and will have to be developed.
- Inclusion of cognitive aspects is even harder. In our zoom-pan case, users remarked that their preference would for instance depend on the complexity of the map shown, their familiarity with it, the task at hand, etc. Quantification of such aspects is harder than the average optic flow.

- Discarded aspects can spoil the results. When optimizing, one gets what one asks, but this is not always what is wanted. A simple example is 2D curve fitting. Interpolating a smooth curve through a sequence of points is trivial, and many techniques are available for this. However, such techniques usually do not enforce that the curve should not intersect itself, leading to unexpected results.
- Visualizations are complex. In the examples given, the aspects optimized were relatively simple: The path of a camera, a color scale, a lay-out of points. However, a complete visualization is a mixture of a large number of aspects. For instance, when viewing an iso-surface in 3D, decisions have to be made on the projection used, the lighting, the shading properties. These are controlled by not too many parameters, but the complexity of the resulting image is great and does not lend itself well to (analytic) computations.
- Validation remains necessary. In the case of Levkowitz and Herman [3] an optimal color scale was computed, but a user experiment did not show that it was superior.
- Optimization is expensive. In the zoom-pan case, we were lucky that we could find an analytical optimal solution. In more complex cases, one has to step back to numerical optimization. Brute-force optimization methods are computationally expensive, fast methods require expertise and tuning on the problem at hand. And, a generic hard problem is to find global optima instead of getting stuck in local ones.
- Finally, effective visualizations should be useful, accurate and clear, but to define this mathematically precise is extremely hard [6]. However, just optimizing perceptual and cognitive aspects is not enough, if the visualization is flawed then its perceptual quality is essentially no longer important.

This last list is long, and probably not exhaustive. Does this mean that the proposed approach is not effective and should be avoided? We don't think so. We have shown that it does work, at least for a simple case. More important, we think this direction could lead to an advancement of our field in general. Visualizing data nowadays is an art, rather than a science. With this approach, the level of discussion is raised. Instead of discussing on particular implementations or choices for color and geometry, the focus shifts to the aims and requirements of the viewer. Instead of discussing how to do it, the approach aims at understanding what really goes on and what is important. A challenge for our field is to hunt for the laws of visualization, and we expect that perceptual models will play an important role in this.

Given the problems mentioned when this approach is applied, for the time being its application will be limited to relatively simple problems. Nonetheless, simple problems occur often in real-world applications, so good solutions are welcome here. An example are the attributes of line graphs. The default result of a standard spreadsheet is often disappointing, and can be improved upon. How to pick the color, line width, and line style such that an optimal result is achieved? Scatterplots are another interesting case. An icon has position, color, and shape as attributes. How can these be tuned, such that the underlying data is conveyed as clearly as possible? On a more general level, can we use quantitative models of Gestalt Theory, such as given in [2], to get effective visualizations? Wattenberg and Fisher [10] have proposed to use a multi-scale approach to analyze information graphics. Can this concept be used to compute better visualizations? Many challenging problems lie ahead of us, and we hope to have given a stimulus to pursue this path further.

References

1. G. di Battista, P. Eades, R. Tamassia, and I.G. Tollis. *Graph Drawing – Algorithms for the visualization of graphs.* Prentice Hall, Upper Saddle River, NJ, 1998.
2. S. Lehar. Directional harmonic theory: A computational gestalt model to account for illusory contour and vertex formation. *Perception*, 32(4):423–448, 2003.
3. H. Levkowitz and G.T. Herman. Color scales for image data. *IEEE Computer Graphics and Applications*, 12(1):72–80, 1992.
4. A. Noack. An energy model for visual graph clustering. In G. Liotta, editor, *Proceedings of the 11th International Symposium on Graph Drawing (GD 2003), LNCS 2912*, pages 425–436. Springer, Berlin, 2003.
5. F. van Ham and J.J. van Wijk. Interactive visualization of small world graphs. In M. Ward and T. Munzner, editors, *Proceedings of the IEEE Symposium on Information Visualization 2004 (INFOVIS2004)*, pages 199–206, 2004.
6. J.J. van Wijk. The value of visualization. In C. Silva, E. Gröller, and H. Rushmeier, editors, *Proceedings IEEE Visualization 2005*, 2005.
7. J.J. van Wijk and W.A.A. Nuij. Smooth and efficient zooming and panning. In T. Munzner and S. North, editors, *Proc. of the IEEE Symposium on Information Visualization 2003 (INFOVIS2003)*, pages 15–22, 2003. Best paper award.
8. J.J. van Wijk and W.A.A. Nuij. A model for smooth viewing and navigation of large 2D information spaces. *IEEE Transactions on Visualization and Computer Graphics*, 10(4):447–458, 2004.
9. C. Ware. *Information Visualization – Perception for Design*, 2nd edition. Morgan Kaufmann, San Francisco, 2004.
10. M. Wattenberg and D. Fisher. A model of multi-scale perceptual organization in information graphics. In T. Munzner and S. North, editors, *Proceedings of the IEEE Symposium on Information Visualization 2003 (INFOVIS2003)*, pages 23–30, 2003.